ADRENOCEPTORS AND CATECHOLAMINE ACTION

NEUROTRANSMITTER RECEPTORS

Editor

GEORGE KUNOS
Department of Pharmacology and Therapeutics
McGill University
Montreal, Quebec, Canada

Volume 1

ADRENOCEPTORS AND CATECHOLAMINE ACTION

Part A

This book is due for return not later than the last date
stamped above, unless recalled sooner.

ADRENOCEPTORS AND CATECHOLAMINE ACTION

Part A

Edited by

George Kunos
McGILL UNIVERSITY

A WILEY-INTERSCIENCE PUBLICATION

JOHN WILEY & SONS

New York • Chichester • Brisbane • Toronto

Library of Congress Cataloging in Publication Data:

Main entry under title:

Adrenoceptors and catecholamine action.

   (Neurotransmitter receptors; v. 1)
   "A Wiley-Interscience publication."
   Includes index.
   1. Catecholamines—Physiological effect.
2. Adrenergic receptors.   I. Kunos, George.
II. Series.

QP801.C33A35   612'.814   81-10431
ISBN 0-471-05725-8   AACR2

Printed in the United States of America

10 9 8 7 6 5 4 3 2 1

PREFACE

Receptors are specific constituents of cells, to which a chemical substance can attach itself to initiate a biological response. This definition includes the two important features of receptors: they selectively recognize and bind ligands and, as a result of this, they trigger a change in tissue function. The receptor concept of drug and hormone action was proposed almost a century ago. Although much of the earlier knowledge of receptors was deduced from measurements of the activity of intact tissues or whole animals, it has proved to be surprisingly accurate in the light of recently developed more direct approaches. The editorial philosophy behind this series is to find a proper balance between ligand recognition and regulation of biological functions, in order to provide comprehensive information on receptors for various neurotransmitter substances.

Since their isolation from the adrenal medulla in 1895, catecholamines have been identified as neurotransmitters in both the central and peripheral nervous systems. They have been implicated in the regulation of a wide variety of biological processes. The first two volumes in this series will cover most of the important aspects of their actions and their receptors. Physiologists, pharmacologists, and biochemists have contributed to this volume. It is hoped that the quality and diversity of their expertise will not only provide a broad cross section of current knowledge, but will also stimulate new research and new ideas.

GEORGE KUNOS

Montreal, Canada
July 1981

v

CONTENTS

CHAPTER ONE

ADRENOCEPTORS IN SMOOTH MUSCLE

Ian K. M. Morton and John Halliday

Department of Pharmacology, King's College, London, England

1 INTRODUCTION AND SCOPE

Our intention is to review certain aspects of the classification and mechanisms of action of adrenoceptors in smooth muscle. Since this topic, together with similar and parallel studies of the cholinergic system, has engaged the energies of physiologists, histologists, pharmacologists, and now biochemists since the end of the nineteenth century, the most we can do is to throw some light on selected areas of current interest.

Of the many tissue types innervated by the sympathetic nervous system or responsive to epinephrine (E) and norepinephrine (NE) released from the adrenal medulla, smooth muscle, with its remarkably varied complement of adrenoceptors, has been studied most thoroughly. Although adrenergic modification of smooth muscle activity throughout the body is, undoubtedly, physiologically and clinically relevant, a principal reason for the vast extent of the work in this field must be the relative ease of measurement of the physiological correlates of drug or nerve action, both *in vivo* and *in vitro,* in the form of mechanical changes in muscle tension or length, changes in blood pressure, alterations in airway resistance, and so on. Since quantification of these changes is relatively straightforward, studies of this type have led to considerable progress in our understanding of drug–receptor interactions and in the characterization and classification of receptor classes and subclasses. More recently, electrophysiological, ion flux, and biochemical measurements have complemented the mechanical studies, with the result that smooth muscle has become a "model" system with respect to receptor-mediated events, giving useful insights that can be carried over to other tissues.

In view of the wealth of information, we have decided to limit our coverage to two main areas. First, we deal with the general principles of receptor classification from the standpoint of smooth muscle (SM), since it was with this tissue type that much of the pioneering work was done, and a good deal has been learned that can, with care, be applied to less readily accessible areas, such as the central nervous system.

Second, we give a general account of the cellular mechanisms that may underlie those actions of epinephrine and related compounds on SM that are mediated through either the α- or the β-adrenoceptor.

It is, of course, impossible in the space available to give encyclopedic coverage or reference to all the published work involved, and many of the papers cited are more by way of example than the only work on the subject. To those authors not acknowledged, we apologize. Also, certain areas must inevitably overlap other chapters in this book, and here we leave detailed citation to others. Useful review articles and similar sources are referred to under the appropriate headings throughout the chapter.

2   CLASSIFICATION OF ADRENOCEPTORS

2.1   Development of the Concept of Adrenoceptors

2.1.1   Early Work

By 1905, epinephrine had been identified in extracts of mammalian adrenal medulla, and its powerful effects in raising blood pressure and exciting and inhibiting SM had been well described (1–5). The general correspondence between the effects of epinephrine and stimulation of nerves of the sympathetic nervous system (SNS) led Elliott (4) to suggest in a preliminary communication made in 1904, that "adrenalin might . . . be the chemical stimulant liberated on each occasion when the impulse arrives at the periphery." However, the idea was not met with much enthusiasm, so he did not carry the idea to his main paper of 1905 (5). It was not until Loewi's successful demonstration in 1921 (6,7) of the release of "Vagusstoff" on the stimulation of parasympathetic nerves to the frog heart, and its later identification as acetylcholine (ACh) (8,9), that the search for the identity of the correspondingly released "Acceleransstoff" (6) from the sympathetic nerves got under way.

The very diversity of action of "sympathin," as Cannon and Rosenblueth (10) were to rename the sympathetic transmitter, proved the stumbling block. In an effort to account for both the excitatory and the inhibitory actions of sympathetic stimulation, and the noncorrespondence of the actions of epinephrine with those of sympathin, it was proposed (11) that sympathin released from nerve endings was converted into two separate mediators, sympathin E and sympathin I, which, in turn, served the two respective roles. It seems probable that this conceptual cul-de-sac, which, as Ahlquist (12) has pointed out, became "widely quoted as a 'law' of physiology," inhibited recognition of the true explanation: that the "duality" lies in the postsynaptic receptors and not in the transmitter substance.

Of course, we now know that the peripheral neural transmitter in mammals is the demethylated amine norepinephrine (13,14), but since both epinephrine and norepinephrine are present in the adrenal medulla in relative proportions that depend on species and other factors (see Ref. 15), the effects of activation of the SNS can be very varied, with a given organ potentially affected both by nerve-released transmitter (NE) and by blood-borne hormones (E and NE), each with different concentrations, time courses, and distributions.

During these difficulties in the identification of the neurotransmitter in the sympathetic nervous system, considerable progress was made in characterizing the structures on which epinephrine and its congeners act to produce their characteristic effects. Langley (16) postulated that effector cells have "receptive substances" with which parasympathomimetic, sympathomimetic, and other agents might react spe-

cifically such that "compounds are formed according to some law in which their
relative mass and chemical affinity for the substance are factors" (17), a statement
we would not quarrel with more than a century later. Even the misconception by
some workers of that period [e.g., Brodie and Dixon (18)] that epinephrine acted
by stimulating nerve terminals—a view that was promulgated well into the 1930s
(see Ref. 19)—is seen now to have an element of truth, in that presynaptic adrenoceptors have since been demonstrated (20).

The contributions of Sir Henry Dale to this area are fundamental. In his 1906
paper (21), he showed "that ergot contains a principle which has a paralytic action
on the motor elements of that . . . substance which is excited by adrenaline and
by the impulses in fibres of the true sympathetic nervous system; the inhibitor
elements of the same being relatively or absolutely unaffected."

Later, Barger and Dale (22) compared the sympathomimetic actions of a large
series of catechol and other amines and were able to show that "motor and inhibitor
sympathomimetic activity vary to some extent independently" and that "approximation to adrenine in structure is, on the whole, attended with increasing specificity
of the action." Among the catechol derivatives studied was norepinephrine, obtained
by Dale from the Höchst Dye Company as Arterrenol (23), of which compound
Barger and Dale note in Ref. 22: "The action . . . corresponds more closely with
that of sympathetic nerves than does that of adrenine" (epinephrine). Dale was later
to observe (23) that "much trouble might, perhaps, have been saved in after years
for my late friend, Walter B. Cannon," had the true meaning of the correspondence
been recognized.

Between them, the 1906 and 1910 papers (21,22) contain the two essential tenets
of receptor classification that still hold today: (*a*) comparison of the relative pharmacological potency of a series of agonists (which may initiate a response that can
be either motor or inhibitory), and (*b*) the use of more-or-less selective antagonists
to block some or all of the responses to the agonists. What is remarkable is that
Dale made so many essentially correct observations and deductions with the poor
pharmacological tools available, in a field that is by no means unraveled even today.
Such advances as were to come have largely depended on the development of new
and more specific agonists and antagonists.

2.1.2 Ahlquist and α- and β-Adrenoceptors

In 1948, Ahlquist (12), in investigating the antispasmogenic properties of certain
sympathomimetic amines, found anomalies in their activities that could not be
explained by current theory (24–26). In examining this problem, Ahlquist
(12,24–26) made the simple postulate that if, when a series of catecholamines
structurally related to epinephrine are tested on a number of different tissues to
determine their relative potencies, the order of potency is the same on all the tissues,
then the differences in activity must be entirely attributable to the differences in the
chemical structure of the compounds. However, if the order of potency differs from
tissue to tissue, these variations in relative potency must be due in part to differences
in the receptors (25,26). With the substances studied, only two orders of potency

were observed (12). For excitatory events, such as vasoconstriction and stimulation of the uterus, nictitating membrane, dilator pupillae, but excepting stimulation of the heart, the order of potency observed was $(-)$-E $> (\pm)$-E $> (\pm)$-NE $> \alpha$-methyl-NE (α-MeNE) $> \alpha$-methyl-E $> (\pm)$-isoproterenol (ISO). Of the inhibitory events studied, such as vasodilatation, bronchodilatation, and relaxation or inhibition of uterus (but excluding inhibition of the gut and including excitatory actions on the heart), the order of potency for the same compounds was found to be ISO $> (-)$-E $> \alpha$-methyl-E $> (\pm)$-E $> \alpha$-methyl-NE $>$ NE.

Ahlquist (12) concluded that the differences in orders of potency of the amines could be explained only by assuming differences in the receptors themselves, and pointed out that this hypothesis made the sympathin E and sympathin I theory untenable. There appeared to be two main receptor populations: the one that was mainly excitatory but that included inhibition of the gut was designated "α"; and the other, which was inhibitory with the exception of the heart, was designated "β." This dual-receptor concept was more-or-less ignored at the time, especially in the United States (see Refs. 26–28), and did not become generally accepted and clinically exploitable until β-blocking agents became available. The terminology has now survived several challenges (e.g., Refs. 29 and 30) and has become standard usage.

When in 1957 Powell and Slater (31,32) introduced the first β-blocking agent, dichloroisoproterenol (DCI), this gave Ahlquist and Levy (33) the opportunity to reexamine one of the "exceptions" of the earlier work—the classification of inhibition in the gut as α-type. They were able to show (33) in the canine ileum *in situ* that although the inhibitory action of the specific α-agonist phenylephrine (PE) could be blocked by the fairly specific α-blocker dibozane alone, epinephrine's action could be effectively blocked only by a combination of dibozane and DCI, and isoproterenol could be blocked by DCI alone. Thus it seemed likely that both α- and β-adrenoceptors mediated relaxation, and since epinephrine is an agonist for both, a combination of an α- and a β-blocking agent would be required to block its actions. The basic observation was soon confirmed in a number of other preparations, such as rabbit intestine (34–36), human jejunum (37), ileum (38), and taenia coli (39), and in the guinea pig taenia coli (40,41).

It is now realized that there is an additional subtype of adrenoceptor, which mediates relaxation in, for example, the guinea pig ileum, but it is located on cholinergic neurones (42,43) and inhibits by reducing ACh release from these neurones (42,43). These receptors have somewhat different properties and have been designated α_2 to distinguish them from the "classical" α_1-receptors on SM (see Ref. 20). The relative importance of the α_1- and α_2-mechanisms seems to vary with tissue, species, and conditions (see Section 2.3.1).

2.2 Principles and Practice of Receptor Classification

It is worthwhile at this stage to examine more closely the principles involved in classification. Unfortunately, a form of circular argument has become common in this context, as discussed, for example, by Moran (44). Thus an action of an amine is said to be, for instance, a β_1-effect because it is blocked by a β_1-antagonist, and a new drug may be classified as a β_1-antagonist because it blocks the β_1-effect. Such a system swiftly becomes self-supporting, but its validity may well be doubtful. The classification of adrenergic receptors involved in a given situation should preferably follow a two-stage process, as exemplified by much of Ahlquist's work. The first step is to set up a hypothesis on the basis of the order of potency of a number of agonists. This hypothesis should then be tested with the use of antagonists, preferably in a quantitative manner.

2.2.1 Classification of Adrenoceptors by Agonist Potency: Receptor Theory

The original division into α- and β-adrenoceptors, and the subsequent subdivisions into β_1- and β_2-receptors by Lands and his co-workers (45–47; see Section 2.3.2) and into α_1- and α_2-receptors, (20,48,49; see Section 2.3.1) have all relied heavily on the use of the relative activities of agonists. However, the assumption that agonists acting on similar receptors will show the same relative activities when tested on different preparations is not necessarily correct, as Schild has pointed out (50). According to classical receptor theory, the response depends on both affinity and efficacy, whereas it is presumably the affinity of the drug for the receptor that is the characteristic of prime interest for the purposes of classification. The question is central to the development of new agents and to the differentiation of receptor subtypes, so it merits further discussion.

The proportion of receptors occupied at equilibrium, on application of a given concentration of agonist to the preparation, is given by the affinity constant K_a of the drug for that receptor (51,52), although it should be noted at this stage that the relation between the applied concentration of drug and the concentration in the "biophase" may be complicated by factors, such as uptake and breakdown, that indicate the need for special precautions without which conclusions may not be valid (53–55). Isolated preparations are preferred in such studies, because maintenance of a known concentration of the drug is more readily achieved.

Receptor theory supposes that the response of the preparation depends on the initial stimulus S, a hypothetical quantity that is the product of the proportion of receptors occupied by the drug y and its efficacy e (56). The efficacy of a drug, or its "intrinsic activity" as redefined by van Rossum and Ariëns (57), or its equivalent in other theories (see Ref. 58), reflects the effectiveness of that drug–receptor complex as a stimulus, which, in turn, initiates a series of events leading to the response. In fact, the relationship between the pharmacological response R and the stimulus is unknown and undefined, except that R is some monotonous function of S and may be expected to differ according to the preparation studied or even the conditions of the experiment.

Difficulties arise because we are not readily able to separate the terms in the relationships mentioned.

$$S = ey \qquad\qquad (1)$$

$$R = f(S) \qquad\qquad (2)$$

The unknown function in Equation 2 can be satisfactorily circumvented in a given preparation only by applying the "null-method," where it is assumed that equal R is produced by equal $S;$ hence comparisons of potency should be made in terms of concentrations required to produce the same submaximal R rather than, say, differences in R produced by the same concentration of agonists (55,56,58).

The relationship in Equation 1 presents more of a problem, because only rarely can we obtain independent estimates of y and e. When $e = 0$, we have a competitive antagonist, and as detailed in Section 2.2.2, it is quite straightforward to determine the affinity constant by a variant of the null method. However, with agonists it can be seen that a given R, and therefore S, can result from any feasible combination of e and y values, which, as Mackay (58) has pointed out, can be taken to imply at the molecular level that a large number of less effective agonist–receptor complexes can produce the same S as can a smaller number of more effective complexes. It is helpful to regard S as the first step in a sequence of events that lead to the final response observed.

In view of these facts, it will be appreciated that the relative activities of a series of agonists will be the same as their relative affinities for the receptor in question only if their efficacies are equal, but we have no a priori reason to suppose that this is so. Hence, if a series of agonists is reexamined in another tissue, even if the affinities are similar the efficacies may not be, particularly so if the view is taken that e is in part determined by properties of the tissue at a stage distal to the initial recognition unit, as recently suggested by Pike et al. (59) for the β-adrenoceptor, although this idea would seem to depart somewhat from Stephenson's original theory (56).

It is therefore important to obtain estimates of K_a and e. It may be possible to gain some idea of the former directly from radioligand binding studies (see Section 2.2.3), but these are *in vitro* values obtained from broken-cell preparations and may not accurately represent the *in vivo* values, so that an independent estimate under physiological conditions is still required. Methods have been developed for determining affinity constants and relative efficacies of full agonists in isolated pharmacological preparations, but these depend on the use of irreversible antagonists, which although available for α-adrenoceptors, are not well established for β-adrenoceptors (for a discussion, see Ref. 60). In addition, the methods are laborious and have been applied in very few instances (61–66). For the present, therefore, one must be cautious in deducing the properties of receptors from orders of activity of agonists.

An analogous complication arises in considering the "receptor reserve" in various tissues. It is generally accepted that a maximum response may be produced by a strong agonist occupying only a small fraction of the available receptors (55,56,-62,67). Therefore, agonists with high, but different, efficacies cannot be distinguished simply by comparing their respective maximum responses. If, however, for a particular agonist in a particular effector system e is so low that a maximum response cannot be elicited, even when the agonist occupies all the receptors, then that agonist is termed a "partial agonist" in that system (56). In these circumstances the maxima and slope diminish as e diminishes, and dose–response lines are no longer parallel, even though this may not be apparent within experimental error or experimental design. Therefore, the activity of such an agonist on that system, relative to others differs with the level of response, and it is for this reason that the term "relative activity" has been used in the foregoing discussion instead of the term "relative potency," which implies equal slopes and maxima. Clearly, this presents difficulties in determining relative orders of activity, and in practice this problem is readily apparent with some "selective" β-adrenoceptor agonists, which are partial agonists in some systems but not in others (see Section 2.3.2).

A less obvious difficulty arises when determining the relative potencies of agonists with differing efficacies in different preparations, since the relation between S and R may vary according to the nature of the tissue. If we imagine a hypothetical curve that relates S to R, it is not unreasonable to suppose that S must exceed some threshold value before any response is apparent, or that S may continue to increase with increased receptor occupancy even after R has reached its maximum imposed by the physical constraints of the system (68,69).

Such a relationship, which is unlikely to be linear (58,70), goes a long way toward explaining such everyday observations as differing sensitivities and slopes of dose-response lines in different preparations. In relation to this, the possibility has been raised by Jenkinson (71) that if a comparison of agonist activities is made in two tissues, of which one is highly "excitable" in that a small increase in S leads to an appreciable R, and the other has an effective threshold in that S must exceed some certain value before any R is observed, then an agonist with low efficacy may generate a response as great as another agonist with high efficacy in the first tissue, but produce little or no response in the second tissue. Jenkinson (71) has extended this interesting line of argument to the suggestion that "selective" β-agonists such as salbutamol may have a lower efficacy on all β-receptors than may nonselective ones such as isoproterenol. This point is examined further in Section 2.3.2.

2.2.2   Classification of Adrenoceptors by Antagonist Affinities: Receptor Theory

We turn now to the use of antagonists in receptor classification. Here a single quantity, the affinity constant of the antagonist for the receptor, serves to define the activity. The theoretical basis has often been described (55,58,67,73) and is not reiterated here. The affinity constant of a competitive antagonist, K_b, which is the reciprocal of its dissociation constant, may be both defined and measured in terms of the parallel shift of the log concentration–response line of a full agonist in the presence of an antagonist. If a null method is used, whereby the ratio of the doses of agonist to obtain the same response in the presence and absence of a molar concentration B of antagonist is defined as the dose ratio x, then at equilibrium

$$x - 1 = K_b B \tag{3}$$

If values of the dose ratio are determined for several values of B, then the Schild plot of log $(x - 1)$ versus $-\log B$ should give a linear isobol with a slope of -1 and an intercept on the abscissa of log $K_b = \mathrm{pA_2}$ (72,73).

The index pA_2 (the negative $\log_{10}$ of molar concentration of antagonist such that the agonist concentration has to be doubled to obtain the same response), as defined by Schild (72), is an empirical measure, and it should be noted that some reported pA_2 values do not reflect an equilibrium situation, having been determined after only short times of exposure to the antagonist. Since antagonists may equilibrate only slowly with the biophase, direct translation of such a pA_2 value into a K_b value will underestimate the true K_b. In the same cautionary vein, a simple pA_2 determination is not itself indicative of competitive antagonism, since such values can also be measured in the case of functional antagonism (see Section 2.3.2). Before we can reasonably claim a competitive nature for the antagonism, we must demonstrate linearity and unity slope of the Schild plot.

However, even in seemingly competitive situations, departures from unity slope and linearity have been observed (e.g., Refs. 50 and 55). Several explanations have been advanced, including the suggestion that the reaction of drug with receptor is not a simple bimolecular reaction (74), and certainly a number of "allosteric" two-state models have now been proposed that may give relationships close to those predicted by classical occupation theory (75,76). Interestingly, in the latter models the efficacy of an agonist simply reflects its relative affinity for the two states of the receptor (75,58).

On a more technical front, atypical Schild plots may also be seen as a consequence of uptake of agonists and antagonists into tissue sites in the region of the receptor (55,77–80). Furchgott et al. (53–55,80) have made an outstanding contribution to this field by defining the necessary criteria and optimal experimental conditions for quantitative drug–receptor interaction studies. In directing the reader to Furchgott's definitive review (55), it is only necessary here to outline the points of particular importance in SM experiments:

1 In the investigation of any single receptor system, great care should be taken to use a highly specific agonist or, alternatively, to combine the use of a less specific agonist with specific antagonists to block other receptors with which the agonist may interact.

2 Indirectly acting agents should not be used (81).

3 Consideration should be given to the possibility that the antagonist under study may at some concentrations start to exert other effects on the tissue, either through another receptor or by some nonspecific means.

4 All uptake processes (particularly those for the agonist) should be blocked, as these may greatly affect the apparent applied concentrations, and to different degrees at different concentrations.

Given such conditions, the calculated values of K_b for an antagonist at a receptor should be the same with any agonist acting specifically on that receptor, thus affording a useful means of confirming which receptor the agonist is occupying. Such an approach should be especially useful with agents, such as dopamine (DA), that may act on either α-adrenoceptors or dopamine receptors (see Refs. 62 and 82). Furthermore, if the determined values of K_b for a given agonist–antagonist pair are similar in different tissues, this is strong presumptive evidence that the receptors are similar (55,62,73; but see Section 2.3.2).

In his major review of adrenoceptor classification work, Furchgott (55) points out that only a small proportion of published papers in this field represent adequately controlled experiments. Some of the conclusions to be drawn from these papers are discussed in Section 2.3.

2.2.3 Classification of Adrenoceptors in Smooth Muscle by Radioligand Binding

Since radioligand binding is dealt with in some detail in other chapters in this volume and several excellent reviews have covered the field, particularly with respect to β-adrenoceptors (83–87), we consider only the relationship between the binding method, using cell membrane fragments, and the pharmacological method, using agonist and antagonist actions in living tissue.

Only in the latter case can one be sure the receptors remain functional, although in some cell fractionation procedures an early effector step, such as activation of adenylate cyclase, may stay coupled and provide useful information. With the pharmacological method, however, uncertainties over the relationships between the steps linking receptor occupation and the end response may make the determination of affinity constants for agonist difficult, although antagonist values are probably reliable. On the other hand, although the basic binding technique cannot distinguish in any direct way among agonists, partial agonists, and antagonists, it can provide affinity values for all three types of agent on the basis of radioligand displacement. Thus, unfortunately, neither approach can give a direct measure of that quality of an agonist which proves so difficult to identify and quantify—its efficacy.

It is to be hoped that an amalgamation of the pharmacological and radioligand approaches will answer some of the outstanding questions. This expectation must depend, in part, on receptor material removed from the physiological environment retaining its important characteristics. Although early reports are encouraging, in that affinities calculated by the two methods are similar (see Refs. 67 and 87), Kleinstein and Glossmann (88) have reported that agonist (but not antagonist) affinities change when the β-receptor is solubilized.

A valuable feature of the radioligand technique is that it can provide an estimate of the tissue density of specific binding sites and also an estimate of relative proportions of β- (e.g., Refs. 89 and 90) and α- (91) receptor subtypes, thus adding a new dimension to classical pharmacological studies.

2.3 α- and β-Adrenoceptors in Smooth Muscle

2.3.1 α-Adrenoceptors in SM: General Aspects

Some progress has been made since the classical studies described earlier in defining the properties of α-receptors as recognition sites. Furchgott (53) selected three tissues from one species, the rabbit, and measured both agonist potencies and antagonist affinities in a series of carefully controlled experiments. The order of potency of the now-standard amines was E > NE > phenylephrine (PE) >> ISO in every case, and the concentrations of phentolamine required to antagonize the selective α-agonist phenylephrine nearly identical (pA$_2$ 8.08) in all three tissues. This is of particular interest because although in two of the tissues (the aorta and stomach) the response to α-receptor activation was a tonic contraction, in the third (the duodenum) the response was inhibition of the amplitude of spontaneous contractions. In other words, the operational characteristics of the recognition unit seem to be the same regardless of the polarity of response, a point important to the discussion in Section 3.

Summarizing reports in the literature up to about 1971 that satisfied the criteria outlined in Section 2.2, Furchgott (55) tabulated values of relative potencies for agonists and pA$_2$ values for some antagonists. The number of species and different tissue types was not very great, but in all cases the order of potency E > NE > PE >> ISO appeared; further, the pA$_2$ values for phentolamine were generally close to 8.0 where the antagonism had been allowed to proceed to equilibrium (which is in excess of 1 hr in some tissues). Patil et al. (92) also reported pA$_2$ values for phentolamine of 7.9–8.0 in the aorta of cat, guinea pig, rat, and rabbit.

Although a much lower value was reported by Green and Fleming (93) for phentolamine on the α-adrenoceptor of cat spleen after 5 min of incubation, in a later study of the rabbit spleen, Sheys and Green (63) reported a value of 7.95 after 2 hr of incubation. In this interesting paper the authors compare the potency of five α-antagonists in rabbit spleen and aorta, and although some of the Schild plots show departures from unity slope, the pA$_2$ values in the two tissues differ by less than 0.5 unit in every case. The dissociation constants and efficacies of several full agonists were also estimated, using the method of fractional receptor inactivation with an irreversible antagonist, by which means it was shown in the aorta that the order of agonist affinities was the same as the order of potencies. However, this was not so in the spleen, where the efficacies varied more and the order of potency reflected a combination of K_a and e. An overall order of potency E > NE >> PE > α-MeNE > DA was found, and metaraminol was a partial agonist.

In a similar series of experiments on the rabbit aorta, Besse and Furchgott (61,62) came to much the same conclusion, although some uncertainty remained as to the relative efficacies of NE and E. Their order of potency of full agonists was E > NE > PE > norphenylephrine > α-MeNE > epinine >> DA, which is in good agreement with the findings of Sheys and Green (63). The K_b value of phentolamine was the same regardless of which full agonist was used, as receptor theory predicts. That dopamine is a full α-agonist is interesting in view of its relaxant actions mediated by its own receptors in some vascular beds (see Ref. 94). Thus in the aorta Besse and Furchgott (62) found that dopamine after α-blockade produced relaxation at higher concentrations and that this effect was not blocked by propranolol.

Hence classical pharmacological techniques would seem to indicate a fair degree of uniformity among α-adrenoceptors on SM itself. It is not yet clear if this view is substantiated by the radioligand binding approach. Indeed, even general assessment of the information provided by this approach is not straightforward. If ligand binding can be equated to *in vivo* receptor binding, then, as Furchgott has pointed out (67), affinity constants derived from classical and binding methods should show good agreement.

Thus in a recent study of α-agonists and antagonists on the guinea pig vas deferens, Holck et al. (95) have shown that when *in vitro* displacement of [^{3}H]dihydroergocryptine is compared to potency in contracting the isolated organ, there is good agreement between the two apparent K_b values for the α-antagonist azapetine. In the case of full agonists, the order E > NE > PE > α-MeNE was followed both by K_a for radioligand displacement and EC$_{50}$ for contractile potency. The K_b for phentolamine ($\equiv$ pA$_2$ 8.2) by the radioligand method was higher than that derived from antagonism of epinephrine contractions ($\equiv$ pA$_2$ 6.6), but the radioligand value is, in fact, close to other estimates by Barker et al. (96) in the rat (pA$_2$ 8.22) and mouse (pA$_2$ 8.31) using the pharmacological technique.

Interestingly enough, in a series of papers on [^{3}H]dihydroazapetine binding in the rat vas deferens, Ruffolo et al. (97) have shown that imidazolines such as phentolamine inhibit binding, as might be expected, but quite paradoxically, ($-$)-phenylethylamine agonists increase binding of the ligand. This led the authors to suggest that imidazolines bind to a different part of the receptor from the catechol recognition site, with an "allosteric" interaction between the two sites. It was further suggested that the ligand binding site may be the "calcium-mobilization site" postulated by Swamy and Triggle (98). Although any direct evidence linking calcium with receptors is, of course, of the greatest interest, it should be noted that Kunos has expressed some reservations over the specificity of dihydroazapetine (60).

Similar difficulties in the interpretation of ligand binding studies have been raised by data from Snyder's group that seems to indicate two different types of α-adrenoceptor, one that binds agonists such as [^{3}H]clonidine and the other that binds [^{3}H]WB-4101, a benzodioxane-type α-antagonist. It is now suggested as coincidental that the receptors display reciprocal affinities for the agonist and antagonist, respectively (99). The method indicates only clonidine binding sites in the rabbit duodenum and only [^{3}H]WB-4101 binding sites in the rat vas deferens (100). As the authors point out, it is not yet possible to explain these findings within the classical pharmacological framework. In this context it is noteworthy that in a pharmacological study, Butler and Jenkinson (101) showed that WB-4101 was as effective in antagonizing the contractile effects of α-agonists in the rat vas deferens as their inhibitory effects in the guinea pig taenia coli (pA$_2$ 8.9 in both), but was very much less effective in antagonizing the presynaptic inhibitory effects of clonidine.

Several groups have used the radioligand technique to look at the influence of hormones on adrenoceptor populations in rabbit uterus (102,103), and such studies are discussed in detail in other chapters. The preliminary work is of great interest but it will suffice here to note that existing knowledge of the changes in adrenergic response may now be complemented by quantitative information regarding numbers and affinities of binding sites, which is a considerable advance. However, caution should be exercised in the interpretation of such findings at this stage, because "specific" ligand binding may occur to other sites, such as presynaptic α_2-receptors (see, e.g., Refs. 104 and 105).

In summary, therefore, it would seem that the radioligand binding technique is providing previously unattainable information, not all of which can be readily assimilated into the classical framework. In general, however, no compelling evidence has yet been advanced, from either pharmacological or binding studies to suggest that postsynaptic α-adrenoceptors situated on the SM membrane need be further classified into distinct subtypes, although there may well be some heterogeneity between organs and species (e.g., Refs. 55, 66, and 106).

2.3.2 β-Adrenoceptors in SM: General Aspects

Classical Approach. Much more work has been devoted to the differentiation of postsynaptic β-adrenoceptors at the sympathetic neuroeffector junction than to postsynaptic α-adrenoceptors in SM, and the literature is too extensive to attempt other than a brief summary of the important points. A number of reviews of this developing area, seen from different standpoints, have appeared in recent years (e.g., Ref. 55 for an early review; and Refs. 60, 67, 83, 107–111).

Apparent heterogeneity among β-receptors had been noted by several authors, among them B. Levy (112), Moran (113), and Furchgott (53), before Lands, Arnold, and their colleagues proposed the β_1 : β_2 subclassification. In their extensive investigations (45–47) using both *in vivo* and *in vitro* preparations, this group compared the relative potencies of a series of different β-agonists, and on this basis classified a particular tissue response as either β_1 or β_2. Although some of the experiments were carried out under less-than-optimal conditions for critical receptor studies (see Ref. 55), there can be little doubt of the overall significance of their hypothesis in terms of receptor classification and the development of useful therapeutic agents with cardiac or bronchial specificity.

Nevertheless, there was an early challenge to the proposal from Buckner and Patil (114), who demonstrated that the isomeric activity ratios of a number of β-agonists and antagonists were very similar in both guinea pig trachea (β_2) and guinea pig atria (β_1), and the interpretation was that the adrenoceptors in these tissues were essentially similar (see Ref. 115). However, it is the view of Harms et al. (116) that such an inference may not be justified, as stereoisomerism probably affects only one of several binding groups involved in the drug–receptor interaction.

Another anomaly in the original β_1 : β_2 proposals was highlighted by Carlsson and colleagues (e.g., Refs. 117 and 118), who found that the effectiveness of selective β-antagonists in antagonizing the chronotropic effects of β-agonists on cat heart preparations was dependent on the agonist used. Thus the selective β_1-antagonist practolol was found to be a good inhibitor of the effect of norepinephrine, whereas H35/25, a selective β_2-antagonist, was more effective against salbutamol; this pattern was taken to indicate the presence of both β_1 and β_2 adrenoceptors in the cat heart. Similar deductions have been made regarding the β-adrenoceptors in guinea pig trachea by Furchgott's group (e.g., Refs. 67 and 119), who made the additional observation that the relative proportions of the two adrenoceptor types in a tissue may vary considerably between individual animals of the same species. As pointed out by O'Donnell and Wanstall (120), such findings have an important bearing on the design of experiments to determine pA_2 values and the selectivity of β-antagonists, since the combination of a selective antagonist with a nonselective agonist in a tissue with a mixed β-adrenoceptor population will give rise to unsatisfactory Schild plots and misleading estimates of selectivity.

Acceptance of Carlsson's modification of the original β_1 : β_2 hypothesis, together with the idea of a mixed adrenoceptor population in a tissue with a large receptor reserve, could explain why occupation of even a small number of receptors of the minority type could evoke a large response from a preparation, and why a selective antagonist of the minority receptor type might have relatively little effect on the response evoked by a nonselective agonist. However, distinguishing such explanations from the perhaps more obvious one of a continuous spectrum of receptor types requires a more rigorous definition of the β_1 and β_2 subtypes.

G. P. Levy and colleagues have made extensive investigations in this sphere (see Refs. 109 and 110) using carefully controlled conditions (see Section 2.2.2) and suggest the use of the bronchioselective agonist salbutamol and the cardioselective antagonist practolol, together with other agents in the scheme (109) shown in Table 1. Clearly, the scheme can be extended to include other agonists and antagonists (110), particularly selective ones, although heed should be taken of the warnings of O'Donnell and Wanstall (120).

The use of agonists in β-adrenoceptor differentiation raises greater problems than in the α-adrenoceptor situation, because although it is clear that many agonists, such as salbutamol, show some selectivity, it is also clear that they may be partial agonists, so their order of potency need not be the same as their order of affinities, which raises the problems mentioned in Section 2.2.1. However, to estimate their relative affinities pharmacologically, it is necessary to use an irreversible β-antagonist, and such agents are still not fully developed, although there have been encouraging reports (see Ref. 60).

Functional Antagonism.   Recently, there have been attempts to derive values for the relative efficacy and affinity of β-agonists through use of the model of "functional antagonism" developed by van den Brink (69,121). A study by O'Donnell and Wanstall (122) of the antagonism by β-agonists of carbachol contractions in the guinea pig tracheal preparation exemplifies one approach to the estimation of relative efficacies. The log dose–response curve to carbachol is shifted to the right in an approximately parallel fashion with increasing incremental doses of β-agonist until eventually a concentration of "antagonist" is reached at which the dose–response curve of the spasmogen moves no farther. This limit to the degree of antagonism is supposed to reflect the limit of the "functional reserve" of the antagonist and presumably occurs at full β-adrenoceptor occupancy, and in that sense it can be regarded as a null method. Both O'Donnell and Wanstall (122) and van den Brink (121) found that different β-agonists gave different degrees of shift, and the extent of the shift by a given β-agonist compared to, say, isoproterenol

TABLE 1 Various Agonists and Antagonists[a]

	β₁-Adrenoceptors				β₂-Adrenoceptors			
Agonist	ISO >	NE >	E >>	SAL	ISO >	E >	SAL >	NE
EC[b]	1	5–20	10–40	500 P	1	3–15	10–35	60–400
Propranolol pA₂	8.3–8.8				8.3–9.4			
Practolol pA₂	6.5–8.9				4.6–5.1			

[a]Studies by G. P. Levy and co-workers, in a wide variety of isolated preparations, under standard conditions with blocked a-adrenoceptors and uptake, showed two patterns of activity (see Refs. 109 and 110 for full details). Tissues showing the b_1-pattern included atrial and intestinal SM preparations, and those showing the b_2-pattern included trachea, aorta, and myometrial SM. See the text for a discussion.

[b]EC is the equipotent concentration of the agonists relative to ISO z 1. SAL is (A)-salbutamol, which is a weak partial agonist (P) on b_1-adrenoceptors.

was taken as an unequivocal measure of the relative efficacy of that compound. On this basis both studies demonstrated compounds with efficacies higher than isoproterenol, which itself showed a considerable receptor or stimulus reserve in the system.

A quite different approach is that taken by Buckner and Saini (123) and Broadley and Nicholson (124), who have equated the depression of the maximum response seen with some functional agonist–antagonist pairs with the similar depression that occurs with irreversible antagonists in the receptor occlusion technique referred to in Section 2.2.1. The latter method has been used by Furchgott and others (61–66) to estimate the affinity and efficacy of adrenoceptor agonists. However, although the appearance of the dose–response lines in the presence of antagonist may be very similar in the two cases, the null method can be applied only in the case of the irreversible occlusion method, where it may reasonably be assumed that a given response before and after the antagonist reflects the same occupancy of receptors by the agonist, allowing the relevant equations to be solved for values of K_a and e (55,58,62). In the functional antagonism situation, more receptors need to be occupied by the agonist in the presence of the functional antagonist so as to increase S to such a level that R is returned to its preantagonism level. Hence it is not a null method and requires, among other things, a knowledge of the relationship between S and R that the method does not provide. It is therefore of interest to note that treatment by Broadley and Nicholson (124) of the functional antagonism data as if this were a receptor occlusion experiment gave a paradoxical finding on the heart. Salbutamol, a partial agonist giving a smaller maximal chronotropic response than isoproterenol, nevertheless appeared, using the functional antagonism method of analysis, to have an efficacy greater than that of isoproterenol. Such data seem to suggest that some alinearity, possibly in the relationship between S and R, distorts the derived values so as to make the estimates of e or K_a unreliable.

Radioligand Binding Studies. In view of the difficulties in estimating β-agonist affinities and efficacies by either the classical or functional antagonism approach, it is not surprising that the radioligand binding method has come to the forefront in recent years. After early troubles (see 125) solved by the introduction of highly active and specific ligands such as the antagonists $(-)$-[^{3}H]dihydroalprenolol and [^{125}I]iodohydroxybenzylpindolol and the agonist $(\pm)$-[^{3}H]hydroxybenzylisoproterenol (none of which are $\beta_1 : \beta_2$-selective), the technique can now give reliable estimates of the characteristics (see Ref. 126) and relative numbers of β-adrenoceptor sites in different tissues or in the same tissue after different treatment (see Ref. 127). In conjunction with studies of the coupled adenylate cyclase (e.g., Ref. 128), it can provide parallel estimates of β-agonist affinities and activities, the latter in terms of activation of adenylate cyclase; indeed, the maximal level of adenylate cyclase activation produced by a β-agonist has been taken by some (e.g., Ref. 59) as a measure of efficacy.

It is most significant that the radioligand binding technique has been able to confirm the presence of two different β binding sites in certain tissues. This development, forecast by Furchgott (67), came in 1978 with the report by Barnett et al. (129) of two binding sites showing the characteristics of β_1 and β_2 in membrane preparations of rat lung. The same group has now used the technique to examine the effect of sympathectomy on the β-adrenoceptor population in rat spleen (130), and two other groups have now reported similar techniques for the analysis of mixed binding-site populations (90,131). Thus the hypothesis established initially by inference from drug effects on biological responses has now been given direct biochemical support.

A more critical test, perhaps, is the comparison of affinity estimates provided by the two approaches. Taking the study of Bristow et al. in rabbit atria (132) as a representative pharmacological study, one obtains an apparent dissociation constant (K_D = reciprocal of affinity constant) for propranolol of 1 nM (pA_2 9.0): this figure may be compared with radioligand K_D estimates for ($-$)-propranolol of 1.7 nM in rat heart (133), 0.4 nM in guinea pig lung (87), and 0.6 nM in rat spleen (130). Unfortunately, one does not always find such good agreement. For example, in their ligand study on rat heart and lung membranes, Minneman et al. (133) detected very little selectivity with butoxamine and none with H35/25, although both are β-selective *in vivo;* and other antagonists, such as metoprolol, atenolol, and practolol, did show their expected β_1 selectivity. An even more striking disparity can be seen in the study by Hancock et al. (90), in which butoxamine appeared to be β_1-selective. The selective β_2 agonist salbutamol has also posed problems, because on simple ligand displacement analysis it does not show any selectivity between heart and lung β-sites (87,133). However, parallel adenylate cyclase studies reveal that it is a partial agonist in the heart, which could account for its high level of binding but limited biological effect (133). In such cases, then, noncorrespondence may be explained satisfactorily, and indeed may be seen to confirm the classical concept that both affinity and efficacy contribute to activity. In other cases, though, for instance with butoxamine, one might wonder whether the analysis used in one or the other approach is inadequate, or whether each approach is in fact relaying a true picture of different drug–receptor interactions in the two environments. The latter seems more likely, in which case studies of the difference could be of great assistance in the development of new drugs with even greater *in vivo* selectivity.

2.3.3 Distribution of α- and β-Adrenoceptors in Some Commonly Investigated Tissues

Table 2 shows the polarity of action and relative predominance of α- and β-adrenoceptors in a number of SM types commonly investigated or referred to extensively in the text. Most SM tissues seem to have both α- and β-adrenoceptors, although the effect of one type may well be so dominant that the effect of the other type is only "unmasked" in the presence of appropriate blocking agents (see Ref. 55) or under special conditions. In general, it is advisable to conduct quantitative experiments, including those relating to mechanisms of action, in the presence of a suitable concentration of a specific antagonist to the adrenoceptor type *not* being studied.

The predominance of receptor type shown in the table may be taken as the typical

TABLE 2   Distribution of Adrenoceptors in Some Frequently Investigated SM Preparations

Preparation	Adrenoceptor Action		Overall effect[a]	References (Representative)
	Contributary Receptors			
Alimentary Tract				
Stomach (circ. fundus): rabbit	α (+) >	β (−)	(+)	53, 54; and see 55
Ileum (long): guinea pig	β₁ (−) >	α (+)?	(−)	43, 106, 110, 134
Small intestine (long): rabbit	β₁ (−) ,	α (−)	(−)	34, 35, 36, 53; and see 55, 109, 110
Cecum (long) "taenia": guinea pig	β (−) ,	α (−)	(−)	40, 41, 134a
Respiratory Tract				
Trachea (circ.): guinea pig[b]	β₂ (−) ≫	α (+)	(−)	106, 116, 135; and see 55, 110
Bronchioles; rat[b]	β₂ (−) ≫	α (+)	(−)	125, 136, 137
Urogenital Tract				
Urinary bladder, detrusor	β (−) ≫	α (+)	(−)	96; and see 55, 138

Organ					Reference
Urinary bladder, trigone	α	(+) >	β	(−)	(+)	See 138
Vas deferens (long): most	α	(+) ⋙	β_2	(−)	(+)	96; and see 55
Uterus (long): most[c]	α	(+) >/<	β_2	(−)	(±)	12, 106, 235, 236, 237, 309; and see 55, 110

Vasculature

Portal vein; most	α	(+) >	β	(−)	(+)	227
Skeletal muscle bed: cat, dog	α	(+) >/<	β_2	(−)	(±)	12, 139a; and see 55, 110
Mesenteric bed: dog, rabbit	α	(+) >	β_2	(−)	(+)	12, 138a
Coronary artery: pig, rabbit, dog	β_1	(−)			(−)	106, 136; and see 55, 110
Aorta (circ.): rabbit	α	(+) ⋙	β_2	(−)	(+)	53, 106; and see 55, 110

Others

Anococcygeus (long): rat	α	(+)			(+)	284
Iris (sphincter): bovine	β	(−)			(−)	See 55, 139
Iris (radial): cat	α	(+) ⋙	β	(−)	(+)	12, 139a
Splenic capsule: most	α	(+) >	β	(−)	(+)	66, 93, 96; and see 55
Nictitating membrane: cat	α	(+) >	β	(−)	(+)	12; and see 55

[a] Overall effect taken as that typical for epinephrine [(+) denotes excitation and (−) inhibition].
[b] But β_1/β_2 proportion varies with species and with individual animals; see the text.
[c] In rat, $\beta \gg \alpha \rightarrow (-)$; in rabbit, $\alpha > \beta \rightarrow (+)$; in guinea pig, $\alpha \simeq \beta$ (may be biphasic).

effect of epinephrine, but this will depend on a number of variables. For instance, the experimental conditions, the age of the animal (particularly with regard to β-effects in the vasculature), the sex and hormonal state of the animal (e.g., the marked differences in reactivity of the myometrium under estrogen or progesterone domination; 235–237), and above all, the species studied have been shown to affect sensitivity and α : β balance.

In some cases the β_1 or β_2 subtype has been indicated, although a mixed population seems to occur in most tissues, as judged from pharmacological and radioligand studies (see Section 2.3.2). For instance, on the basis of the latter approach the β_1/β_2 ratio varies from 25 : 75 in the rat lung to 60 : 40 in the rabbit lung (87). Also, large differences in the ratio in the trachea have been shown between individual animals (119), so clearly the classification of tissues into β_1 and β_2 types must be treated with caution (see Ref. 110).

3 MECHANISMS OF ACTION OF CATECHOLAMINES IN SM AT THE CELLULAR LEVEL

3.1 Introduction

The patterns of innervation of the sympathetic nervous system, the cell types innervated or affected by epinephrine, and the many varieties of cellular reaction all show such diversity that it would seem unlikely that a single cellular mechanism could account for the link between activation of the recognition site on the cell and final physiological response. However, in SM taken as a whole, many lines of evidence suggest that the final coupling agent between the physiological initiation of the event and the cellular manifestation of the response is a change in the level of availability of free calcium ions within the cell.

Perhaps the first question to be asked is: Does the pharmacological differentiation into α- and β-adrenoceptor types, and indeed further subtypes, reflect corresponding differences in underlying cellular mechanisms? While realizing that such an approach has both merits and demerits, we hope to show that in SM the α- and β-mediated actions differ from one another quite markedly in their primary causal mechanism, but that within each class there is considerable similarity in the initial cellular mechanisms even when the final response may be either excitation or inhibition.

In brief, we suggest that in SM there is good evidence that α-mediated events are brought about by alterations in the permeability of the cell membrane to inorganic ions, in which process calcium is intimately involved, and that the observed tissue response stems ultimately from these changes. The final mechanical response may be either excitatory or inhibitory. Evidence for this viewpoint is discussed in Section 3.3.

In contrast, β-mediated actions result, in the first instance, from metabolic changes that are intimately concerned with the control of free intracellular calcium, the end result invariably being inhibition or relaxation. Increased adenylate cyclase activity usually seems to be involved at a very early stage, so that this may be a prerequisite for the response. The evidence for this is discussed in Section 3.4.

Before going on to examine the justification for proposing these hypotheses, it will be necessary to consider the nature of excitation–contraction coupling and the role of calcium in SM.

Clearly, it is possible to envisage a very large number of mechanisms whereby a substance may influence the contractility of SM. The situation can be simplified, conceptually and operationally, by making two working hypotheses:

1   That the α- and β-receptor classes correspond to two fundamentally different cellular mechanisms.

2   That each of these two mechanisms has a fundamental, characteristic step early in the chain of reactions following receptor activation that is essential if the succeeding chain of events and the final response is to follow normally.

The first hypothesis seems to present fewer problems since α- and β-mediated effects differ in a number of respects, not the least of which is their opposite polarity of action in most instances (see Section 2.3.3 and Table 2). The second, however, presents obvious conceptual difficulties in that α-mediated actions in SM can be either excitatory or inhibitory, whereas β-mediated actions are inhibitory in SM but excitatory in cardiac muscle.

3.2   Excitation–Contraction Coupling in SM: The Importance of Calcium

The term "excitation–contraction (E–C) coupling" takes on a broader meaning in SM than in skeletal muscle, for which it was introduced (140), and the term is used here to refer to all events interposed between chemical or electrical excitation of the membrane and contraction of the muscle. Our understanding of the process in SM is at a rather early stage, although a fairly recent symposium on the subject (141) has made it abundantly clear that calcium is important at several stages. Certainly, there is little doubt that tension in SM, as in skeletal muscle, is a function of the intracellular free-calcium level (142); and although there is debate as to which of the steps in the myofibrillar contractile process are calcium-sensitive (see Ref. 143), the issue of greater concern here is that of how excitation raises $[Ca]_i$ from subthreshold levels of 0.1 μM into the estimated contractile range (0.2–100 μM) (144–146).

In view of the great functional diversity in SM, no single scheme is adequate to describe the various mechanisms of excitation, conduction, and contraction, although several classifications have been proposed at one time or another (147–153). The following classification is based on action potential (AP) type (152,154), and since it broadly reflects SM properties, may be helpful for the purposes of further discussion.

1   Smooth muscle discharging APs

   a   Short APs (spikes of ~200 msec); AP frequency determines the size of contraction (e.g., most longitudinal SM of intestine, myometrium, and vas deferens).

   b   APs with long plateau, sometimes with initial spike, where size and duration of plateau determine the size of contraction [e.g., ureter of guinea pig, and some stomach muscle (152)].

2   Smooth muscle not normally discharging APs, although some show small or "graded" depolarization with transmitter action (e.g., rabbit aorta, pulmonary and carotid arteries, bovine and canine trachea, and rat anococcygeus).

APs are the most obvious way of synchronizing contraction in SM; in some tissues APs may arise through "myogenic" pacemaker activity, whereas in others APs are normally "neurogenic" in that they result only from the action of excitatory neurotransmitters (see Refs. 153, 155, and 156).

By analogy with skeletal muscle, it might be expected that the AP would cause release of calcium from the sarcoplasmic reticulum (SR), but it should be noted that not only is the volume of the SR (2–7% of total cell volume, 157) much less than in skeletal muscle, but SM lacks a T-tubule system to carry the wave of depolarization into the cell. Indeed, it has yet to be convincingly demonstrated in SM that depolarization per se can release calcium from internal stores, which in SM is largely in the mitochondrial compartment (158,159). However, SM cells are smaller and contract more slowly than does skeletal muscle, so it seems quite possible that their large surface area/volume ratio (160,161) and deeply invaginated plasma membrane (see Ref. 161) could allow sufficiently rapid access of extracellular calcium to the contractile elements by simple diffusion.

In the context described above, it is very significant that there is a body of experimental evidence showing that at least in some SMs the fast inward current for the upstroke of the AP is carried by calcium ions (see Refs. 155, 162, and 163), and calculations suggest that sufficient calcium may enter the cell with the spike to directly activate the contractile proteins of visceral SM (155,164,165), although the point is difficult to investigate by direct uptake experiments with ^{45}Ca since these are limited, at present, by technical difficulties (see Refs. 166–171a). Alternatively, rapid influx of calcium may release additional calcium from internal stores (calcium-released calcium), as in the myocardium (173) and at one time proposed in skeletal muscle (but see Ref. 172).

The type of coupling in which APs are involved has been termed "electromechanical coupling," and it seems likely that in many SMs the critical factor is the opening of voltage-dependent calcium channels. The role of the excitatory transmitter is to increase the frequency (in type **1a** tissues) or the duration (in type **1b** tissues) of the AP.

However, another aspect of E–C coupling is seen in those electrically inexcitable SMs that do not generate APs (type **2**). Although large "graded" depolarization in these SMs may well open voltage-dependent calcium channels, in some instances, particularly in vascular SM, α-adrenoceptor agonists and other agents seem to cause contraction with little or no change in membrane potential (see Section 3.3.4). This has led Somlyo and Somlyo (174–176) to suggest the term "pharmacomechanical coupling" (see Figure 1) for "the process through which drugs can affect smooth muscle contraction without a necessary change in membrane potential." In this respect early observations of Schild and his colleagues that K-depolarized visceral SM could still respond to a number of hormone and transmitter substances with characteristic contraction (177–182) or relaxation (40,181–183), as long as calcium was present in the medium, provided support for such a concept. Under these conditions the membrane potential is unlikely to alter much as a result of permeability changes and it is likely that here, as under more physiological conditions, contractile agents open what have been termed "receptor-operated channels" (ROCs; 155), which admit calcium from the extracellular medium and thus lead to contraction. However, particularly in vascular SM, contractions can be elicited for longer than might be expected on this basis in calcium-free edetic acid (EDTA)-containing solutions, raising the possibility that α-receptor activation may release

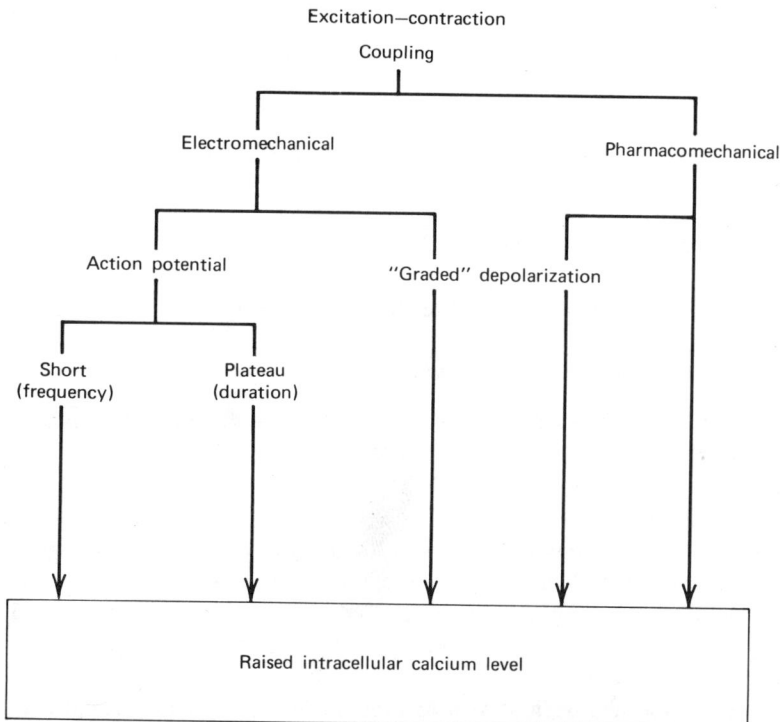

Figure 1   Scheme outlining the main mechanisms that link membrane excitation with alterations in the intracellular calcium level. It is proposed that in tissues showing graded depolarization, such as large arteries, it may be difficult to distinguish the relative contributions of the two coupling mechanisms to the final mechanical response. See the text for a discussion.

calcium bound within the cell into the cytosol or reduce calcium sequestration within the cell (see Ref. 155 and Section 3.3.5).

In summary, then, possible sources of the calcium that is initiating contraction (see Ref. 171a) may be listed:

1 Extracellular calcium admitted to the myoplasm via:
   a Voltage- and possibly time-dependent ion channels opened during the AP or with graded depolarization. (These channels may also admit sodium.)
   b Channels opened by receptor occupation (ROCs). In some tissues these may be voltage- and time-dependent.
   c Leakage channels, particularly when membrane function has been impaired.

2 Intracellular, or membrane, calcium released from stores by:
   a A rapid influx of extracellular calcium (calcium-released calcium).
   b Depolarization (e.g., following the AP; not yet proven in SM, but exemplified by skeletal muscle).
   c Direct result of drug action in occupying membrane receptors.
   d Intracellular drug action (e.g., caffeine; see 155).

Epinephrine and norepinephrine can, at least in principle, utilize a number of these mechanisms to raise $[Ca]_i$ in SM. Certain of the steps may be in sequence and others in parallel. Since the cell maintains very low $[Ca]_i$ against a high electrochemical gradient, it must also have highly developed systems for sequestering calcium and pumping it out of the cell, and stimulation of these processes may lead to relaxation. This may be important in β-adrenoceptor-mediated relaxation and is dealt with in more detail in Section 3.4.

3.3 α-Adrenoceptor-Mediated Mechanisms in SM

3.3.1 Introduction: Some Possibilities

The suggestion has been made (Section 3.1) that α-adrenoceptor-mediated effects in SM may be explained by the initial step of drug–receptor interaction, leading to changes in the ion permeability of the membrane. This idea is attractive because so many α-adrenoceptor effects are now known to be intimately associated with the electrical events at the membrane, particularly in tissues that generate APs. The simplest scheme would involve the receptors being linked to ion channels in the plasma membrane (ROCs) in such a way that passive changes in membrane permeability to specific ions would lead to excitation or inhibition according to the ion species involved.

Such a scheme has been shown to account for the essential features of neurotransmission, both motor and inhibitory, in a wide range of effector sites in both vertebrates and invertebrates, with neurotransmitters as varied as acetylcholine, serotonin, dopamine, γ-aminobutyric acid, glycine, and glutamic acid. For an excellent general account of this field, the reader is referred to the reviews of Ginsborg (184,185).

However, electrical events at the membrane may also be brought about by more indirect means, particularly through the influence of ion pumps, and some of these possibilities will be discussed more fully in Section 3.4 in relation to β-adrenoceptor-mediated actions. In fact, many mechanisms have been suggested as our knowledge of SM has developed, and Table 3 summarizes some hypotheses that have been selected on the basis that at some stage they have been proposed in the literature. It should be stressed that these mechanisms are not necessarily exclusive; more than one might operate in series or in parallel.

TABLE 3   Some Possible Mechanisms of Adrenoceptor Modification of Electrical Excitability and Contractility of Smooth Muscle

	Mechanism		ΔG_m
	Inhibition, with hyperpolarization		
a	↑ P_K	(where $E_K > E_m$)	↑
b	↑ P_K + P_{Cl}	(where ε > E_m)	↑
c	↓ P_{Na}	(so E moves toward E_K)	↓
d	↑ electrogenic Na pump	[ouabain-sensitive (Na, K)ATPase]	o
e	↑ electrogenic Ca pump	(Ca-ATPase)	o
	Inhibition, with little change in E_m		
f	↑ P_K	(where $E_K \simeq E_m$ or > E_t)	↑
g	↑ P_K + ↑ P_{Cl}	(where ε $\simeq E_m$ or > E_t)	↑
h	↑ nonelectrogenic Ca pump		o
i	↑ Na/Ca exchange	(?, electrogenic)	o
j	↑ intracellular Ca binding		o
	Excitation, with depolarization		
k	↑ P_{Na}	(since $E_{Na} < E_m$ and E_t)	↑
l	↑ P_{Cl}	(where $E_{Cl} < E_m$ or E_t)	↑
m	↑ P_{Cl} + P_K	(where ε < E_m or E_t)	↑
n	↑ P_{Na} + ↑ P_{Cl}	(where ε < E_m or E_t)	↑
o	↑ P_{Na} + ↑ P_{Cl} + ↑ P_K	(where ε < E_m or E_t)	↑
p	↑ P_{Ca} + ↑ P_{Na} + ↑ P_{Cl} + ↑ P_K	(where ε < E_m or E_t)	↑
q	↓ P_K	(so $E_m < E_t$)	↓
	Excitation, with little or no depolarization		
r	↑ P_{Ca}	(if small ΔG_m)	(↑)
s	↑ P_{Ca} + ↑ P_K	(where ε $\simeq E_m$)	↑
t	↑ P_{Ca} + ↑ P_K + ↑ P_{Cl} + ↑ P_{Na}	(where ε $\simeq E_m$)	↑
u	intracellular Ca release		o

[a] E_K, E_{Na}, E_{Ca}, and E_{Cl} are equilibrium potentials for K⁺, Na⁺, Ca²⁺, and Cl⁻; E_m is the resting membrane potential; E_t is the threshold membrane potential for initiation of APs; ΔG_m is the change in membrane conductance (reciprocal of membrane resistance); ε is the transmitter equilibrium potential.

It may be seen that changes in electrical excitability or mechanical response may be achieved in four basic ways:

1   Alteration of the passive permeability of the resting membrane to ions.
2   Alteration of voltage- or time-dependent ion permeabilities.
3   Alteration of ion transport, either electrogenic or neutral.
4   Alteration of ion binding at the cell membranes or elsewhere.

3.3.2   α-Adrenoceptor-Mediated Inhibition of SM

As shown in Section 2, it is only in the gastrointestinal tract that α-adrenoceptor activation leads to inhibition or relaxation. It might seem illogical to begin an account of α-mediated actions with an exception, but the modern biophysical investigation of SM began with investigations into the membrane properties and biochemistry of the guinea pig taenia coli, and it was with this preparation that a permeability mechanism was first established for catecholamines acting on any tissue type.

Intestinal muscle is, in any case, a good object lesson in the application of pharmacological analysis to biophysical and biochemical techniques, since a "mixed" agonist such as epinephrine may produce inhibition through both α- and β-adrenoceptors simultaneously, and in consequence its relaxant action may not be blocked by α- or β-adrenoceptor antagonists acting alone. Consequently, those who have studied epinephrine's biophysical and metabolic actions in intestinal muscle without the use of receptor antagonists have observed *causa sine qua non* the results of two quite different initial events, with subsequent confusion over the causal relationships. The use of more selective agonists has greatly helped to resolve this difficulty.

The guinea pig taenia coli (more precisely, taenia caeci, since it is found as three bands of longitudinal muscle on the cecum) shows a high degree of tone and spontaneous electrical and mechanical activity, which makes it very suitable for studying the inhibitory actions of catecholamines. Largely through the work of Bülbring, who established the preparation in 1954 (186), and her many co-workers since, the membrane properties, cellular and subcellular morphology, innervation, and biochemical characteristics are particularly well established (187–197).

The spontaneous electrical activity, which is not seen in some of the tissues to be discussed later, takes the form of APs superimposed on pacemaker potentials (193–196); and the higher the spike frequency, the higher the tension developed in the preparation (164,187,192,195).

Since the pacemaker potentials are myogenic in origin and seem to be associated with small rhythmic fluctuations in the membrane potential—the slow waves— investigation of the mechanisms of action of substances that influence excitability must take account of the possibility of alterations in the slow wave itself or in the events linking it to the firing of APs. In fact, very little is known about the genesis of slow waves other than that they may be sodium-dependent and are sensitive to temperature and metabolic poisons and so may reflect some instability of metabolic processes of the membrane (155,156,163,192,194,198).

Synchrony of myogenic or neurogenic contractions is achieved by a high degree of electrotonic coupling between cells in bundles that anastomose and interconnect with adjacent bundles, forming an electrical syncytium (149,150,153,164). Hence the "unit" of visceral muscle must consist of millions of cells, and this unity will be disrupted by influences that inhibit conduction and propagation of APs.

Application of epinephrine to the taenia leads to inhibition or cessation of spike activity, preceded or followed by hyperpolarization of the membrane and relaxation of the tissue (186,188,190,199). Inspection of Table 3 will show that, at least in principle, a number of mechanisms listed (and there are many more in addition) could lead to some or all of these effects. The most straightforward explanation, bearing greatest resemblance to, for example, peripheral cholinergic systems, can be couched in terms of membrane permeability changes, but perhaps because the metabolic actions of adrenaline had been recognized since the beginning of the century, early proposals invoked a metabolic component, even though the outcome was a change in electrical parameters. Hence initial suggestions favored the idea of epinephrine contributing metabolic energy to an active (outward) electrogenic sodium pump (mechanism d, Table 3) and so hyperpolarizing the membrane (e.g., Refs. 189, 190, and 200). Another suggestion postulated a metabolic component in fixation of calcium to the membrane reducing sodium permeability (P_{Na}) (mechanism c, Table 3) (199,201). As we shall see in Section 3.4, a rationale still exists for a "metabolic" stage in the action of epinephrine through β-receptors, but this can probably be divorced from passive permeability changes.

An argument against a simple permeability change to potassium (mechanism a, Table 3) being involved in epinephrine's hyperpolarizating action was the demonstration in tracer studies by Born and Bülbring (202) of an increased uptake of ^{42}K in response to epinephrine but little consistent change in efflux. However, a likely explanation of the lack of effect on efflux is that in the spontaneously active preparation, epinephrine inhibits APs, thus reducing loss of ^{42}K and masking the increase due to the underlying permeability change (ΔP_K). Moreover, the increase in membrane potential ($\uparrow E_m$) would tend to oppose outward movement of potassium, so that the overall effect of epinephrine on efflux of ^{42}K would be small and variable. However, were the same experiment performed with relaxed quiescent preparations (17°C under little tension), a greater observable increase in efflux of ^{42}K would be expected, and this was in fact demonstrated by Jenkinson and Morton (40).

Changes in spike frequency and E_m complicate all attempts to measure drug-induced alterations in membrane permeability, but if the membrane potential is abolished by the somewhat drastic expedient of bathing the preparation in potassium-rich medium, it becomes possible to demonstrate permeability changes, albeit at the cost of a departure from physiological conditions. By this means Jenkinson and Morton (40,203,204) were able to show that application of norepinephrine (0.9 μM) to the depolarized taenia caused an equal increase in both uptake and efflux of ^{42}K, so establishing a permeability change. Similar experiments with ^{36}Cl showed a small but statistically insignificant increase in efflux and uptake, but ^{24}Na uptake was not affected by norepinephrine under conditions where carbachol produced a highly significant increase. Taken overall, these findings indicate that norepinephrine increases P_K without a change in P_{Na}, which under physiological conditions would lead to stabilization of the membrane and to hyperpolarization of the membrane, since in this tissue, as in most SM, the potassium equilibrium potential is more negative than is E_m. Further, Jenkinson and Morton (183) were able to show that isoproterenol was without effect on P_K and that phentolamine (0.3 μM) blocked the action of norepinephrine, whereas the β-adrenoceptor blocker pronethalol was without effect, thus establishing that α-adrenoceptors mediated the permeability change (see Figure 2).

The studies cited above, confirmed by Bülbring et al. (205), demonstrated unequivocably that norepinephrine could change membrane permeability, an important principle, but it remained to be shown that membrane conductance fell during the action of the amine under more physiological conditions. The parallel approaches of direct determination of permeabilities to ions by means of tracers, and the measurement of the resultant changes in membrane conductance by electrophysiological means, have not always shown good correspondence (184,185). However, quite independently, Bogach and Klevets (206), using the double-sucrose-gap tech-

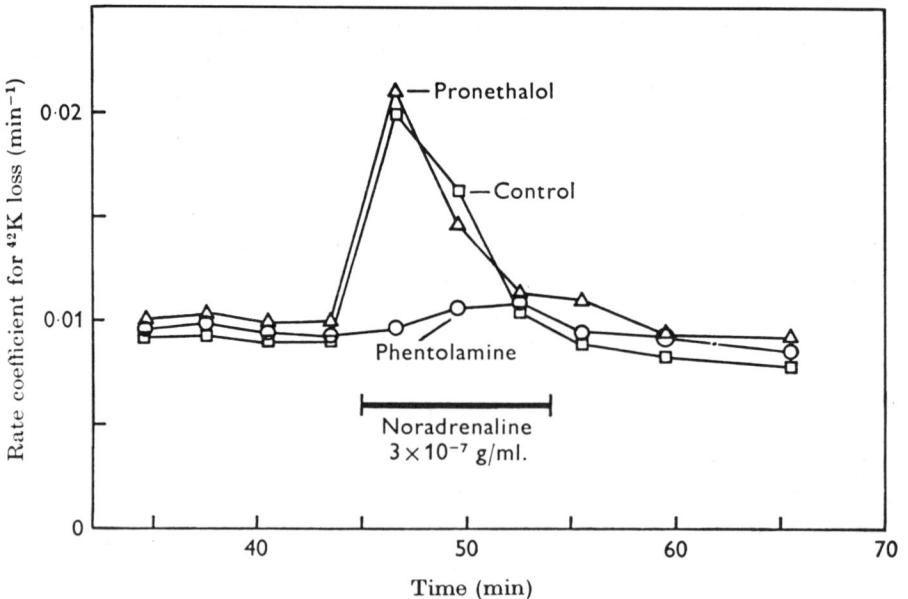

Figure 2 Comparison of the antagonism by phentolamine (0.32 μM) and pronethalol (0.38 μM) of the action of norepinephrine (0.89 μM) in increasing the rate of loss of ^{42}K from the depolarized taenia coli of the guinea pig. The antagonists were included in the potassium-rich bathing fluid from 5 min after removal of the tissues from the load solution (zero time). It may be concluded that the action of norepinephrine in increasing P_K is mediated through α-adrenoceptors. From Ref. 183, with permission of the *Journal of Physiology.*

nique, were able to build on the early work of Shuba (see Ref. 207 for references) with epinephrine on frog stomach, and demonstrate in the guinea pig taenia that norepinephrine caused a rise in G_K associated with an increase in E_m.

These findings of the Kiev group were elegantly confirmed by Bülbring and Tomita (208), who also used the double-sucrose-gap technique to show a fall in membrane resistance with epinephrine (0.1–0.3 μM). The size of the fall in resistance was affected by altering $[Cl]_o$ as well as $[K]_o$, suggesting that increased G_{Cl} also contributes to the increase in membrane conductance (G_m). In addition, Bülbring and Tomita (209) were able to confirm the observations of Jenkinson and Morton (183) that only the α-receptors mediate the conductance change; this was through the use of isoproterenol, which was without effect on conductance, and phentolamine, used to block the α-action of norepinephrine. However, isoproterenol did have effects, mediated through β-adrenoceptors, on spontaneous spike generation: these are discussed in Section 3.4. The nature of these observations are shown in Figure 3.

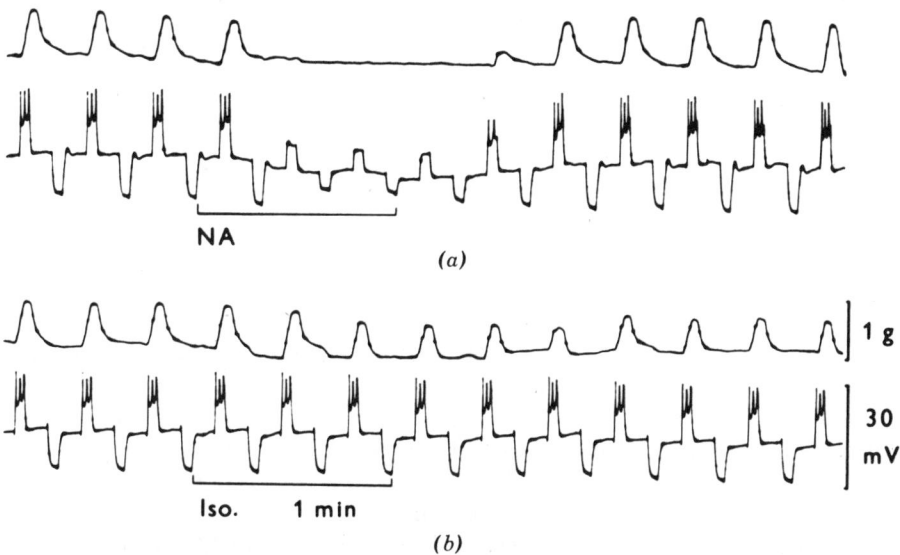

NA

(a)

1 g

30 mV

Iso.   1 min

(b)

Figure 3 Tension (upper) and electrical (lower) records recorded with the double-sucrose-gap technique in the guinea pig taenia coli. Constant current pulses, with alternating polarities, of 3-sec duration applied every 10 sec. Responses to norepinephrine (NA; 0.8 μM) and isoproterenol (Iso; 1 μM) applied for 1 min, as indicated by horizontal bar. (a) Note that norepinephrine hyperpolarizes the membrane and increases its conductance (as shown by the decreased amplitude of the voltage deflections produced by constant current pulses). Depolarizing current no longer generates APs, and consequently produces no mechanical response. (b) In contrast, isoproterenol diminishes the mechanical response without affecting the electrical properties of the membrane. These findings suggest that α-adrenoceptor-mediated events involve an increase in membrane permeability to an ion with an equilibrium potential more negative than the membrane potential, whereas the β-mediated action involves some facet of excitation–contraction coupling. Reproduced, with permission of the author and Macmillan, London and Basingstoke, from E. Bülbring, in H. P. Rang, Ed., *Drug Receptors*, 1973.

With respect to the change in chloride conductance, Ohashi (210) used a microelectrode to measure changes in the electrotonic potential produced by extracellularly applied current pulses, and estimated that the membrane resistance fell to half with epinephrine, owing to increases of G_K and G_{Cl} in the relative proportion 1 : 0.36. The equilibrium potential for epinephrine was estimated as -75 mV, which is at least 8 mV more negative than E_m. From the analyses of Casteels and Kuriyama (211), the value of E_K is -92 mV, and E_{Cl} is between -15 and -31 mV, according to the assumptions made about chloride binding.

Voltage-Clamp Studies.   Evidence discussed so far from tracer flux observations of α-adrenoceptor ROCs and electrical measurements in terms of the corresponding changes in membrane resistance have been in reasonable agreement that mainly potassium channels are involved. However, the most certain way to analyze membrane conductance changes is by clamping the membrane potential at selected values and measuring the time course of the resultant ion current under various conditions. In particular, this method allows the resolution of time- and voltage-dependent conductance changes. The method has allowed the specific conductances associated with the AP, and the effects on membrane conductance of excitatory and inhibitory transmitters, to be resolved in a number of excitable tissues (184,185). In the case of a syncytial tissue, special problems are posed because of the low-resistance pathways between cells and the consequent difficulty of uniformly clamping the membrane voltage.

Application of this technique to SM by Anderson in 1969 (212) and others (see Ref. 213) has been attended with lively controversy. Kao and his co-workers (e.g., Refs. 214–216) report that catecholamines have no effect on membrane conductance of the taenia coli and the rat uterus, and suggest instead that the hyperpolarization results from a change in the potassium equilibrium potential (E_K) by some mechanism involving either a change in $[K]_i$ or the "ion selectivity of the membrane." There is some evidence for changes in $[K]_i$ in the rat uterus, but since this involves a β-adrenoceptor-mediated mechanism, it will be discussed under that heading (Section 3.4). It is hard to see how the membrane can become more selective without a change in resistance, and these workers have not detected any. This negative finding has been challenged by two other groups, Tomita et al. (217,218) and Shuba et al. (207), who have also used the voltage-clamped taenia and find that the hyperpolarization with catecholamines is associated with a marked fall in membrane resistance. Shuba et al. (207) observed no current flow at a holding potential of -75 to -80 mV, which they presumed corresponded to the transmitter equilibrium potential. It should be noted that Ohashi (210) estimated a similar value using a different technique. Tomita's group (217,218) used a clamped node of >700 μm, as compared to that of Kao's group of about 50 μm (214–216), and have explained the negative effect of the latter workers as an artifact arising due to the gap width being less than the space constant for the tissue, with the result that leakage current through the tissue–sucrose interface obscures the resistance change due to epinephrine, a situation they were able to mimic in their apparatus on narrowing the gap to 100 μm. These difficulties also have a considerable bearing on the interpretation of drug action as assessed by this important technique in other tissues (see Ref. 155).

Role of Calcium.   Since changing the external concentration of an ion may alter the permeability of the membrane to other ions or change the activity of a linked ion pump, and is also very likely to change internal ion concentrations, interpretation of the results of the maneuver can prove very difficult and uncertain; for a critical review, see Bolton (155).

Bülbring and Tomita (219,220) have made the important suggestion that calcium may link α-adrenoceptor activation to the potassium conductance change. Increasing external calcium potentiates the effect on conductance, and reducing $[Ca]_o$ to 1/10 of normal diminishes it (219). Shuba's group found that the conductance effect became smaller in zero $[Ca]_o$ (207), but when high $[Mg]_o$ was used to prevent the depolarization and reduction in membrane resistance caused by removal of calcium itself, Bülbring and Tomita (220) were able, on total removal of calcium in ethyleneglycol-bis(β-aminoethyl ether)N,N′-tetraacetic acid (EGTA)-treated tissues, to abolish the effect of epinephrine on membrane conductance, although with a transient reintroduction (15–30 sec) of calcium the response was renewed. Further, Mn^{2+} blocked the action of epinephrine, and Ba^{2+} was not able to substitute for calcium in this respect (219). Bülbring and Tomita (219,220) concluded that calcium is essential for the α-mediated action, and might be acting at the "gate mechanism" for the α-adrenoceptor-mediated potassium permeability increase, or may be necessary for the reaction between epinephrine and the receptor. It is of interest in this respect that a similar "linking" role for calcium has been suggested for a variety of other tissues (for references, see 221–224).

To summarize, catecholamines acting through α-receptors are able to selectively increase the membrane permeability of the taenia coli, opening ROCs to certain ions; the resultant rise in G_m is due principally to increased G_K with only a small contribution from G_{Cl}, so the membrane potential moves toward an equilibrium potential lying between E_K and E_{Cl} (mechanism b, Table 3). The opening of additional channels will tend to decrease membrane excitability as long as the transmitter equilibrium potential is more negative than the threshold for excitation (184,185). In the taenia the transmitter equilibrium potential is about -75 mV, and E_m is normally more positive than this, so that opening of the ROCs results in hyperpolarization, but Bülbring and Kuriyama (199) noted that the hyperpolarization may be very small in unstretched preparations or at lowered temperature. The stabilization or hyperpolarization of the membrane in the normally spontaneously active taenia results in a reduction or cessation of spike activity, although slow-wave activity may continue.

It would be of interest to know whether similar ROCs are involved in α-adrenoceptor-mediated inhibition in other regions of the intestine, but detailed studies seem not to have been made. Although a good deal is known about muscarinic cholinergic ROCs in guinea pig ileum (see Ref. 155), α_1-adrenoceptors in this tissue seem either not to be coupled to ROCs or to be aberrant in behavior (see Section 2.3.3). An increase in ^{42}K efflux with norepinephrine has been observed in a preliminary study of potassium-depolarized circular muscle from the cecum of the rabbit and longitudinal taenia muscle from the colon of the same species (225), so increased P_K may well be generally involved in relaxation of SM.

3.3.3 α-Adrenoceptor-Mediated Excitation in Muscles Generating APs

If we accept the argument of the preceding section that α-adrenoceptors may inhibit SM by means of a selective increase of ion permeabilities through ROCs, then a natural extension would be to consider whether there are instances where the ion permeability extends to ions whose equilibrium potentials are less negative than the membrane potential, so that depolarization ensues. If the SM type is able to generate APs, then excitation will occur if the depolarization is sufficient to take E_m past its threshold, and a contraction will result, the size of which will depend on spike frequency. This situation is seen in the guinea pig vas deferens, which is not normally spontaneously active. In preparations, such as the myometrium or portal vein, that are spontaneously active, excitation takes the form of an increase in the number of spikes in a burst and hence an increase in tension develops. In the ureter it is the duration rather than the frequency of APs that seems to determine the increase in tension. These various examples will be discussed separately. Tissues not developing APs are dealt with in Section 3.3.4.

Rat Portal Vein.   This tissue is representative of spontaneously active vascular SM and has been studied extensively, together with the superior mesenteric vein, and has similar electrophysiological properties in several species (see Ref. 226). Spontaneous mechanical activity in the longitudinal layer originates from fast my-ogenic spike activity, often in bursts on top of slow waves, termed "multispike complexes" (MSC) by Holman et al. (227) and Weston (228). Low concentrations of norepinephrine depolarize the membrane a little, and the increase in tension is related to an increase in the frequency and duration of the MSC discharge, but at concentrations higher than about 1 μM, the membrane is depolarized by 20–30 mV and the MSC become continuous, although greatly reduced in magnitude at peak depolarization; later this gives way to repolarization and burst of MSCs associated with large-amplitude tension changes (228). In the presence of "calcium antagonists" such as verapamil and D600, this phasic activity is converted to tonic contractions with no APs and a somewhat diminished peak depolarization and contraction (228–230). These findings suggest that calcium spikes are generated in this tissue when E_m falls below threshold as a result of slow-wave or ROC activity, and also that when the membrane is greatly depolarized, spike electrogenesis is suppressed, as indeed is to be expected. In these respects the longitudinal muscle of the portal vein behaves in a way entirely analogous to the longitudinal layer in intestinal muscle with respect to acetylcholine, and the underlying mechanisms may be use-fully compared (see Ref. 155).

Wahlström (231,232) examined the effect of norepinephrine on ^{36}Cl, $^{?}K$, and ^{24}Na efflux from the rat portal vein and concluded that P_{Cl} was markedly increased with a maximal effect at 6 μM norepinephrine, at which concentration a decrease in $[Cl]_i$ but an increase in $[Ca]_i$ was observed after 30 min. Although increased efflux of ^{42}K was observed with concentrations of the amine as low as 6 nM, the effect was not dose-dependent and was explained as a consequence of depolarization. No effect of norepinephrine was observed on the efflux of ^{24}Na, but this ion has a complex washout curve, and there are difficulties in identifying and measuring flux from the intracellular compartment. Wahlström (231) estimated values of E_K $= -93$ mV, $E_{Cl} = -20$ mV, and $E_{Na} = +63$ mV under the conditions of his experiments. On this basis E_m was calculated as -42 mV and measured as -45 mV; hence an increase in P_{Cl} alone due to norepinephrine would give a depolarization to a calculated equilibrium level of -28 mV, which was close to the measured level. Anion substitution experiments (233) also lent support to the idea that chloride was important for the effect of the amine (mechanism l, Table 3). A recent preliminary report by Tsay and Jones (234), comparing rat aorta and portal vein, confirms the finding that ^{36}Cl efflux is increased by the amine, but these authors also found some effect on ^{42}K and ^{24}Na fluxes (mechanism o, Table 3).

Shuba et al. (207), using the double-sucrose-gap technique with the portal vein, report that α-adrenoceptor activation caused a fall in E_m and a rise in G_m and increased spontaneous activity. However, ion-substitution experiments with decreased $[Na]_o$ and $[Ca]_o$ led them to conclude that P_{Na} and P_{Ca} increased as well as P_{Cl}.

Taken overall, this work suggests that the opening of chloride ROCs is essential to the α-mediated depolarization in this tissue, although other ions may also play a part. Increased entry of calcium would be expected either through ROCs or normal voltage-dependent channels.

Myometrium. The mechanical reactivity, electrophysiology, and influence of catecholamines vary greatly between species and according to hormonal state in this organ, making generalizations difficult (235–237). However, as shown in Section 2, some differences in response to the amines are explicable in terms of changes in relative effect of the α- and β-mediated actions, which are in opposition. It is noteworthy that the absolute numbers of receptors have been claimed to change under hormonal influence (102,103) and that there are certainly marked changes in morphology and electrophysiology (236,237).

In the cat myometrium, where, as Dale pointed out (21), the effect of epinephrine changes from inhibition to excitation during pregnancy, it was demonstrated by Bülbring et al. (238) that pregnancy caused $[Cl]_i$ to double; E_{Cl} consequently fell from -25 to -11 mV, while E_m rose from 48 to 64 mV. This raised the interesting possibility that if epinephrine increases P_{Cl}, the resulting depolarization in the pregnant uterus should be greater than in the nonpregnant state, although the effect of the amine on membrane conductance was not determined. In the guinea pig myometrium under comparable conditions (238), no changes in electrolyte levels were detected.

In a later study in the estrogen-dominated guinea pig myometrium, Bülbring and Szurszewski (239,240) used the double-sucrose-gap technique to compare the actions of epinephrine and acetylcholine (ACh). The value of E_m was -55 mV, with evidence of a contribution from an electrogenic sodium pump, and E_K was thought to be as high as -90 mV and E_{Cl} as low as -20 mV (238,239). When tested in a medium containing propranolol (4 μM), norepinephrine was found, at concentrations that produced reversible effects, to be less effective than ACh. Unlike those with ACh, these contractions reached their peak about 1 min after application of the amine and lasted 5 min or more. In the majority of preparations, norepinephrine depolarized the membrane to about half the extent of ACh (30 mV), with a much smaller decrease in the electrotonic potential. Both depolarization and reduction in electrotonic potential with the amine depended directly on $[Cl]_o$ but not $[Na]_o$.

These and other observations led the authors to conclude that norepinephrine, acting through α-adrenoceptors, increases P_{Cl} and to a lesser extent P_K (but not P_{Na}, in contrast to ACh), which in view of the low value of E_{Cl} leads to depolarization and increased AP frequency, and contraction of the preparation.

In view of the long latency and duration and the temperature dependency of the effect, Bülbring and Szurszewski (239,240) did not rule out the possibility that the changes in P_{Cl} were secondary to some intermediate process such as mobilization of sequestered calcium. However, another possibility worth considering in situations such as this is that release of local hormones could contribute to the effects. In particular, catecholamines have been shown to release prostaglandins (PGs) in a number of tissues (e.g., Refs. 241 and 242), including the myometrium (243), and slow onset, prolonged duration, and potentiation of other spasmogens are characteristic of the action of these agents on the myometrium (e.g., Refs. 244–246; see 155 for references on the electrophysiology of PGs).

Guinea Pig Vas Deferens. Although this muscle conducts APs well, it is normally quiescent. However, stimulation of the hypogastric nerve elicits excitatory junction potentials (EJPs) of short latency (6–10 msec on field stimulation) and fast rise time (~40 msec in guinea pig and 10–20 msec in rat and mouse). These EJPs may summate and exceed threshold, leading to the generation of propagated APs giving a fast phasic contraction, in a way rather analogous to other neuromuscular transmission systems involving permeability changes in the postsynaptic membrane (164,185,193).

Magaribuchi et al. (247) have investigated the ionic basis of this depolarization using the double-sucrose-gap technique and find that norepinephrine causes a fall in membrane resistance due to increased P_{Na} and P_K. This action of the amine is enhanced if chloride is replaced with a less permeant anion, suggesting that the high chloride conductance of this membrane tends to shunt the effect of opening these ROCs. The effect of norepinephrine was mediated through α-adrenoceptors since it was antagonized by phentolamine, and isoproterenol did not affect membrane resistance. It is difficult to establish whether P_{Ca} increases with this technique, but calcium would be expected to enter the cell to cause a contraction, although whether through ROCs or voltage-dependent cation channels cannot readily be determined.

Guinea Pig Ureter. It is of interest that the history of smooth muscle physiology effectively began with two papers by Engelmann in 1869 and 1870 (see Ref. 248), who observed that electrical stimulation of the ureter caused a myogenic peristaltic wave to pass down, such that "under all conditions the ureter responds to stimulation exactly as if it were a colossal hollow muscle fibre." The preparation has continued to intrigue electrophysiologists and pharmacologists since, and has proved unusual in many ways.

As has been pointed out, the AP is unusual in that a spike is followed by a long plateau (e.g., Ref. 249). It appears that in the guinea pig the spike component of the AP is carried by calcium and the plateau current by sodium (249,250), since if sodium is removed, the duration of the AP is shortened by suppression of the plateau, whereas Mn^{2+} decreases the phasic part of the contraction, although it increases the amplitude and duration of the remaining part of the AP. Removal of calcium decreased the spikes of the early phasic response (251,252).

Allen and Bridges (251) used the double-sucrose-gap technique to show that in the guinea pig ureter, norepinephrine increased the duration of the plateau and increased the amplitude and duration of the tonic contraction even in the presence of Mn^{2+}. These findings were confirmed by Shuba (253), who also found that catecholamines and histamine were quite alike with respect to their effects on the AP, and further, that both slightly depolarized the resting membrane potential, although norepinephrine increased the membrane resistance whereas histamine reduced it. Shuba (253) concluded that "these drugs affect the passive ionic permeability of the membrane in a manner that results in depolarization" but that "they specifically activate the potential-dependent conductance of the slow Na channels, thereby increasing the plateau duration. The increased amplitude and duration of the contraction is the result of their primary effect on the plateau of the AP."

These observations are unusual in several respects. First, the weak depolarizing effect of norepinephrine on the resting membrane potential is associated with a *decrease* in membrane conductance. Second, the effects on the AP seem to be a prolongation of a voltage- or time-dependent G_{Na} increase. This does not seem to be the result of a delay in the late increase in outward current since the effect was still seen in the presence of tetraethylammonium ions (TEA), which apparently block this channel in some excitable tissues. Bury and Shuba (252) showed that this slow Na current was not blocked by tetrodotoxin (TTX), although it was somewhat affected by prolonged withdrawal of calcium, suggesting that it is controlled by calcium in the membrane. Since, as already pointed out, calculations suggest that enough calcium can enter during the fast phasic component of the AP to trigger contraction, the effect of norepinephrine in prolonging the slow sodium current may be to increase $[Ca]_i$ due to release of stored calcium within the cell following depolarization, or, alternatively, it could be that calcium is also able to enter through the slow sodium channel. Further study is clearly warranted.

3.3.4 α-Adrenoceptor-Mediated Excitation in Muscles Not Generating APs

Although some small arteries and arterioles (175,254–259), and the portal and superior mesenteric veins (256), have been shown to generate APs and are sometimes spontaneously active, with α-adrenoceptor activation causing "phasic" contraction by initiating or increasing AP frequency, this does not seem to be true for the large arteries.

Large arteries that have been studied, usually as circular muscle preparations, are mostly electrically inexcitable, mechanically quiescent, and respond to vasoconstrictor agents with only small depolarizations and do not normally initiate APs. They do, however, contract if depolarized by external influences such as a high-potassium solution or a depolarizing current passed across the membrane. These points are particularly well established for rabbit aorta (260–262), pulmonary artery (174,175,207,263–265), carotid artery (266,267), and ear artery (268–270).

However, some arteries may generate APs [e.g., sheep carotid artery (259, 271–274) and rat aorta (275)], and most develop oscillatory behavior, not unlike slow waves, with large doses of norepinephrine and other vasoconstrictor agents (174,259,262,267,276). The phenomenon is similar to that seen in the rat anococcygeus with norepinephrine (277) or in visceral muscle (e.g., the guinea pig ileum) with large doses of carbachol (155).

The lack of APs is now known not to be related to a lack of conduction between cells, as was implied by Bozler (147,148) in his original classification of these arteries as "multi-unit" muscles, since they have now been shown to have cable properties (153,264,266), so should probably be regarded as "unitary" muscles (264). In fact, when treated with TEA or similar agents, some tissues of this type become more excitable, are able to generate APs, and even become spontaneously active. Examples include the rabbit ear artery (269,270), guinea pig superior mesenteric artery (278,279), as well as nonvascular inexcitable tissues such as bovine trachea (280–282) and rat anococcygeus (277). TEA is thought to reduce the resting P_K of the membrane that depolarizes the tissues, but more important, it suppresses the voltage-dependent rise in P_K that normally underlies the repolarization phase of the AP. Presumably, the latter mechanism is highly developed in these inexcitable tissues and normally prevents spiking. In some instances "calcium antagonists" such as D600 and manganese have been shown to block TEA-induced spikes, whereas TTX does not, so presumably calcium carries the inward current (269, 270,282).

With regard to the contractile action of norepinephrine on large arteries mediated via α-adrenoceptors, it is now well recognized that with low concentrations this occurs with little or no depolarization, as originally shown in rabbit pulmonary artery by Su et al. (263) and confirmed by several groups (174,175), including Casteels et al. (264,265), who showed that low concentrations of norepinephrine (60 nM) cause contractions with little depolarization, and that even high concentrations (3 μM) depolarize by 9 mV at most. Similar findings with respect to the effect of low and high concentrations are reported for the rabbit carotid artery (267), but the rabbit ear artery was shown by Casteels and colleagues (269,270) not to depolarize with "maximal" concentrations of norepinephrine, although histamine did depolarize it somewhat. Extensive work by Somlyo et al. (174–176) has demonstrated similar behavior with other vasoconstrictor agents, such as angiotensin and serotonin, which has led to this process being termed "pharmacomechanical coupling" (174–176), as discussed in Section 3.2.

If these arteries contract with norepinephrine with little change in E_m, it is of the greatest interest to determine the mechanism involved. An increase in membrane conductance has been demonstrated in these tissues, associated with a rise in P_K, P_{Cl} or P_{Na} (225,234,263,267,269,270,283). Presumably, the particular combinations of changes in ion permeabilities accounts for the small or absent change in E_m (267,269) (mechanisms m, n, o, Table 3).

However, this does not explain why the tissues contract, although this must certainly involve a rise in the intracellular level of calcium, presumably either through an increase in P_{Ca}, allowing more calcium to enter the cell, and/or by the release of calcium from intracellular storage sites. There is a body of evidence for both views, particularly in experiments with the aorta, and these uncertainties exemplify the difficulty of studying calcium fluxes and distribution within the SM cell, as mentioned previously (166–171a). Norepinephrine contraction in these tissues shows both phasic and tonic components, which seem to utilize different calcium "pools," although the methods of study are complex and involve the use of potassium-depolarized preparations, sometimes in conjunction with La^{3+} to block transmembrane calcium flux (see Refs. 155 and 169). Evidence relating to this question is summarized in Section 3.3.5.

Rat Anococcygeus. This muscle is a paired structure associated with the alimentary tract. It has been studied in several species by Gillespie and colleagues and shown to differ in several respects from typical visceral muscle (277,284,285). As mentioned previously, in the rat this muscle is quiescent, electrically inexcitable, and does not generate APs except when treated with TEA. Creed et al. (277,284) have shown that stimulation of intramural sympathetic nerves produces a graded depolarization of the membrane with a decrease in membrane resistance. Application of a supramaximal concentration of norepinephrine (9 μM) also produced a graded depolarization of similar magnitude, sometimes accompanied by membrane oscillations, and the effect was blocked by phentolamine. This concentration of the amine markedly increased membrane conductance and took E_m close to the estimated transmitter equilibrium potential of -21 mV (285). Preliminary values of the ion equilibrium potentials have been reported as $E_{Na} = 13$ mV, $E_K = -81$ mV, and $E_{Cl} = -13$ mV (285), so presumably increases in P_{Na} or P_{Cl} would depolarize E_m from its resting value of -62 mV; comparison with α-adrenoceptor-mediated events in other SM suggest that either or both is possible.

3.3.5 *General Comments on α-Adrenoceptor Mechanisms*

From the evidence discussed in the preceding parts of Section 3 and summarized in Table 4, it can be seen that there is considerable evidence demonstrating ROC-induced changes (almost invariably increases) in membrane permeability to inorganic ions following α-adrenoceptor activation in SM. In the case of muscles capable of generating APs, the species of ion involved in the permeability increase corresponds well with the observed end response. Thus inhibition or relaxation of intestinal muscle is related to an increase, predominantly in P_K, which would be expected to hyperpolarize or stabilize the membrane and reduce spike frequency. Conversely, in a number of tissues exemplified by the portal vein, myometrium,

TABLE 4 Suggested ROCs Involved in α-Adrenoceptor-Mediated Inhibition or Excitation in SM[a]

Tissues	ΔMembrane Properties ΔE_m	ΔG_m	ΔP_K	ΔP_{Cl}	ΔP_{Na}	Method	Notes	References
Inhibition								
Guinea pig taenia coli			↑↑	0	0	T	K-depolarized preparation	Jenkinson and Morton (40,203,204)
	↑	↑	↑↑	0	0	T	Various $[K]_o$	Bülbring et al. (205)
			↑↑	0		E	Double sucrose gap	Shuba et al. (see 207); Bogach and Klevets (206)
	↑↑	↑	↑↑	↑	0	E	Ca essential	Bülbring and Tomita (208,209,220)
	↑↑	↑	↑↑	↑		E	$G_K/G_{Cl} = 0.36$; $\varepsilon = -75$ mV	Ohashi (210)
	↑	↑	0			E	Voltage clamp	Tomita et al. (218,219)
	↑	0	0			E	Voltage clamp, ↑E_K suggested	Kao et al. (214–216)
Excitation								
Rat portal vein	↓		↑?	↑↑	0	T	↑^{42}K efflux suggested due ↓E_m	Wahlström et al. (231–233)
Guinea pig myometrium	↓	↑	0	↑	↑	E	↑G_{Ca} suggested	Taranenko and Shuba (see 207)
	↓	↑	(↑)	↑↑	0	E	Estrogen-dominated	Bülbring and Szurszewski (239,240)

Guinea pig vas deferens	↓	↓ ↑				E	↑G_{Ca} suggested	Maragaribuchi et al. (247)
Guinea pig ureter	↓	↓↑	0	(↑)*		E	↓G_K resting E_m; (↑)G_{Na}* of AP plateau	Shuba et al. (207,250,253)
	↓			(↑)*		E	Prolonged AP plateau*	Allen and Bridges (251)
Rabbit pulmonary artery	↓	↓	0	↑		E		Takanenko and Shuba (see 207)
	(↓)	↑		↑		E	↑P_{Ca}? No ↓E_m with low conc. NE	Su et al. (263); Casteels et al. (264,265)
Rabbit ear artery	0	↑		↑		E, T	↑P_{Ca}? No ↓E_m with low conc. NE	Droogmans et al. (269,270)
Rabbit carotid artery	(↓)	(↑)	(?)	↑		E	No ↓E_m low conc. NE, but contracts	Mekata and Niu (267)
Rat anococcygeus	↓		↑?	↑?		E	ε = −21 mV suggests ↑G_{Na} or ↑G_{Cl}	Creed et al. (277,284,285)

aThe changes (Δ) shown are in permeability (ΔP) or conductance (ΔG) according to whether tracer (T) or electrical (E) methods of estimation were used. Changes are in resting values [except for the ureter, where a change (G_{Na}*) is indicated in the voltage- or time-dependent conductance of the plateau of the AP], and a depolarization is shown as a decrease in the membrane potential (E_m). The symbol ε denotes the transmitter equilibrium potential. P_{Ca} and G_{Ca} are not tabulated separately because the methodology does not normally allow calcium movement through normal voltage-sensitive channels, as used by the AP, to be distinguished from movement through ROC channels [although it may be presumed that P_{Ca} increases for either reason where a tissue is contracted or depolarized by norepinephrine (NE)].

and vas deferens, the increase in membrane permeability extends to P_{Na} and P_{Cl}, which tends to depolarize the membrane and favor spike electrogenesis and hence contraction.

The ureter seems to be a special case where the increase in P_{Na} is time- or voltage-dependent, so tending to increase the duration of the AP and thus contraction, but a small depolarization of the resting E_m was reported to be associated with a decrease in G_m. Other examples of SM showing a plateau-type AP are known; for instance, the human and canine stomach antrum (286,287) and the actions of norepinephrine on the myocardium are known to involve time- and voltage-dependent ion conductance changes (see, e.g., Ref. 288), albeit through β-receptor mediation, and more examples of this mechanism may well be discovered in SM (see Ref. 152).

In the case of muscles not discharging APs, the situation is more complex. Although graded depolarization is seen, for instance, in some of the larger arteries and the rat anococcygeus, and seems attributable, at least in part, to an increase in P_{Na} and/or P_{Cl}, it seems that depolarization is not a necessary prerequisite for contraction in all such tissues, since in the rabbit pulmonary, carotid, and ear arteries, contraction was observed in the absence of depolarization. The most likely explanation is that P_{Ca} also increases through ROC action in these, and perhaps all, tissues contracted by the α-adrenoceptor mechanism, although another possibility is that the α-adrenoceptors cause calcium to be released from intracellular stores.

However, methods of measuring P_{Ca} are not yet sufficiently advanced to allow these mechanisms to be distinguished. Electrophysiological methods show that the presence of some calcium is normally necessary for the α-adrenoceptor-mediated changes in G_m or E_m to be manifested, but interpretation of such experiments is complicated by the fact that calcium may play a part in controlling membrane permeability to other ions (155,222–224,289,290) and may also have what has been termed a "permissive" role with regard to the action of norepinephrine on the SM membrane (154,155,219,220).

Tracer methods are complicated by certain technical difficulties, as already pointed out, particularly in identifying flux from different compartments and by the large amount of extracellular binding, although the "lanthanum method" may be used to abolish the latter, and to block transmembrane influx and to some extent efflux, which may simplify analysis. Studies with potassium-depolarized preparations may also yield useful information, since under these conditions the opening of ROCs that admit calcium will cause increased contracture and may be taken as evidence of increased P_{Ca}, particularly if the effect depends on external calcium.

Studies using the foregoing methods suggest that norepinephrine and other contractile agents are able to utilize calcium from different sources according to conditions. The initial response in SM not generating APs seems to involve, in part, a store of calcium that is difficult to deplete even in calcium-free solutions containing chelating agents, although the store becomes exhausted on repeated application of an agonist, unless external calcium is present (e.g., Refs. 265, 269, and 291–293). The maintenance of contraction requires external calcium; it seems to be associated with passage of calcium into the cell, and may be related to an increase in P_{Ca}.

A number of possible sources of intracellular calcium that may be released by procedures that lead to contraction were discussed earlier, but there is much to recommend the view that at least some "trigger" calcium may be released from the inside of the plasma membrane as a more-or-less direct consequence of receptor occupation, and this same calcium source may be utilized by agents acting through different receptors. Thus Deth and van Breemen (294) suggest that norepinephrine, angiotensin, and histamine utilize the same store as a coupling signal in rabbit aorta to link receptor occupation and intracellular events. It is not clear if such a mechanism would involve a change in G_m, but it is of interest that recent observations in porcine coronary artery by Ito et al. (295) show that ACh may contract the muscle in low concentrations without changing E_m or G_m. In these experiments with ACh, as with norepinephrine in sheep carotid artery (296,297), very prolonged washing in calcium-free solution containing EGTA or EDTA was required before the contractile response was lost, but this quickly returned when calcium was readmitted.

In experiments with arterial SM loaded with ^{45}Ca, application of maximal concentrations of norepinephrine caused a transient, although marked, rise in calcium efflux (265,269,294), which was taken as evidence of a large intracellular release of stored calcium. However, not all workers have shown increased calcium efflux, although there are a number of reports of increased influx (e.g., Refs. 232, 265, and 298; see Ref. 155) as would be expected if α-adrenoceptor activation increases P_{Ca}.

Is the Permeability Change the First Step after Receptor Occupation?   It may be concluded that there is reasonable support for the hypothesis advanced in Section 3.1: that α-adrenoceptor activation leads to permeability or conductance changes, which result in alterations of membrane potential. This leaves open the question of whether opening of the ion channels is the primary event following receptor occupation.

One possibility is that some metabolic step intervenes, such as a change in cyclic nucleotide levels. However, the evidence for this particular mechanism in α-mediated responses is not nearly as strong as it is for β-mediated responses (see Section 3.4.5); further, it should be noted that the α-response is faster than the β-response under equivalent conditions, such as in intestinal muscle (e.g., Refs. 35, 36, 41, and 225). Indeed, in tissues with close synaptic relations, stimulation of sympathetic nerves may produce fast-rising α-mediated EJPs with latencies of <10 msec, as has been pointed out already. Technical difficulties alone are likely to make it difficult to demonstrate metabolic changes with such a rapid time course.

Phospholipid metabolism is affected by autonomic neurotransmitter action in several tissue types, including exocrine glands, hepatocytes, and SM, and the suggestion has been made by Michell and co-workers that increased phosphatidylinositol turnover (the "PI effect") actually precedes and "gates" ion permeability changes, possibly through increased $[Ca]_i$ (for references, see Refs. 224, 299, and 300). Jafferji, Michell, and others have demonstrated the "PI effect" in SM with both muscarinic and α-adrenoceptor agonists (see Refs. 224 and 300), but the time course seems much slower than that of contraction, so the metabolic changes could be a consequence rather than the cause of the motor effect. However, a recent report by Salmon and Honeyman (300*a*) suggests that in disaggregated SM cells, phosphatidate levels rise very rapidly (<10 sec) in response to carbachol, and precede contraction. Low concentrations of exogenous phosphatidate contract these SM cells, and Putney et al. (300*b*) have demonstrated calcium-dependent α-adrenoceptor-type actions with this metabolite in the parotid gland.

Thus, although it seems feasible that short-term alterations in membrane phospholipids may modulate membrane calcium gates, it would, at this stage, be premature to suggest this as the primary causal event in α-adrenoceptor-mediated SM contraction.

It seems equally possible that the ROCs are influenced or "gated" in some more direct way by membrane calcium, and, as already noted, this mechanism has been suggested to operate in SM and other tissues following α-adrenoceptor activation (see Refs. 155, 223, 224, and 289).

Ion Selectivity of α-Adrenoceptor ROCs.   Given that ROCs are opened by some process initiated, directly or otherwise, by α-adrenoceptor activation, the important questions become: (*a*) Does α-adrenoceptor occupation open similar ROCs in all SM types, and (*b*) are the ROCs opened by norepinephrine the very same ones as those opened by other neurotransmitters or hormones? Unfortunately, these questions cannot yet be answered (for a critical review, see Ref. 155). Inspection of Table 4, which summarizes much of the available information, shows a diversity of ion channels opened by α-adrenoceptors, although the data are not in full agreement. The possibility cannot be ruled out that in those situations where excitation occurs there is a nonspecific increase in ion permeability that extends to P_K, P_{Na}, and P_{Cl}.

It certainly seems that P_K increases, whether inhibition or excitation follows, although the increase in ^{42}K efflux is frequently illsustained (e.g., Refs. 204, 205, 232, 264, 269, and 283). The role of chloride is of special interest because in many SM tissues, E_{Cl} is a good deal more positive than E_m since this ion is not in Donnan equilibrium. Hence a rise in P_{Cl} alone would depolarize and contract the muscle, as has been suggested for guinea pig myometrium (239,240) and rat portal vein (231–233). An increase in P_{Na} would depolarize in all instances and has been implicated in the majority of instances where SM is contracted by norepinephrine.

The ion of greatest interest, calcium, is also the most problematic to study. There seems little doubt that it can enter the cell through voltage-dependent channels during electrically or chemically induced depolarization, but the relative contribution of these channels to calcium entry in different SM types as opposed to calcium entry via ROCs cannot readily be assessed. That calcium entry through ROCs is important can be gauged from the fact that in some vascular muscles, and in potassium-depolarized preparations of many different SMs, α-adrenoceptor activation can lead to tonic contractions with no change in membrane potential. Such contractions cannot be sustained in the absence of external calcium, suggesting that at least some of the calcium must be entering through ROCs.

Overall then, potassium, chloride, sodium, and calcium have all been implicated in various studies of α-mediated contraction of SM, and it does seem possible that, in those tissues contracted by norepinephrine, there could be an essentially non-selective increase in permeability to both anions and cations.

3.4 β-Adrenoceptor-Mediated Mechanisms in SM

3.4.1 Introduction

Although activation of α-adrenoceptors in SM invariably causes inhibition, two different types of inhibitory effect may be distinguished in an operational sense. First, and most obviously, there is depression of membrane electrical excitability. This is best exemplified by β-mediated reduction in spontaneous AP frequency in a myogenically active preparation such as the rat myometrium. Second, in circumstances where the free intracellular calcium level in SM has been raised above contraction threshold by any means (including AP, depolarizing current, high-potassium solutions, or spasmogen action), activation of β-adrenoceptors diminishes the force of the contractile response. Of course, both effects may be involved simultaneously in any particular instance, but for the purposes of analysis it will be useful to differentiate between the two. It is, as yet, not clear whether the two effects are indicative of two distinct β-mechanisms or simply reflect different facets of the same fundamental β-mechanism.

3.4.2 Electrophysiological Observations and β-Adrenoceptor-Mediated Inhibition of Spontaneous Activity in SM

β-Adrenoceptor activation in spontaneously active tissues generally causes a gradual lessening of spike activity and associated mechanical activity. In the rabbit portal vein, Holman et al. (227), using a sucrose-gap technique, showed that isoproterenol caused spontaneous bursts of spike activity to become less frequent, with fewer spikes in each MSC. In a similar study on rat portal vein, Johansson et al. (301) observed a decrease in the number of spikes per MSC and eventually complete absence of spike activity. In pregnant rat myometrium, Diamond and Marshall (302), using intracellular recording, showed that epinephrine abolished spontaneous activity and hyperpolarized the membrane, these effects being blocked by a β-antagonist. However, as already mentioned, Bülbring and Tomita (209) found that β-adrenoceptor-mediated inhibition of spontaneous electrical activity in the taenia coli was not associated with any significant alteration in G_m (see Figure 3). The authors suggested that the critical β-action was inhibition of a "generator" or "pacemaker" potential, which initiated APs in such tissues.

As already pointed out in Section 3.3.1 with respect to α-mediated mechanisms, many different ionic mechanisms could account for inhibition of membrane electrical activity (see Table 3), among them actual hyperpolarization of the membrane, or, less obviously, stabilization of the membrane potential through increased conductance to an ion such as potassium whose equilibrium potential is more negative than the threshold for AP generation. Both these mechanisms have been advanced in explanation of β-mediated inhibition of spontaneous activity in SM.

Some degree of hyperpolarization has been reported with β-mediated inhibition in the guinea pig taenia coli (303,304), the guinea pig vas deferens, (305), the rat (302,306–309) and mouse (310) myometrium, and porcine coronary artery (295). Among the most detailed studies of this phenomenon have been those carried out by Marshall and her colleagues on the myometrium of the pregnant rat (302,306–308), and two important points emerge from their investigations.

First, whereas hyperpolarization is prominent with high levels of β-stimulation (11 mV with 600 nM epinephrine), low concentrations of epinephrine (e.g., 20 nM) reduce spontaneous activity without causing any membrane hyperpolarization (302,306). Second, even when concentrations of β-agonist high enough to completely inhibit spontaneous activity are used, hyperpolarization follows rather than precedes the inhibitory effect (308). The question then arises whether hyperpolarization is an essential factor in β-adrenoceptor-mediated inhibition of spike activity; it may be significant that several studies, in guinea pig myometrium (311), guinea pig vas deferens (247), guinea pig taenia coli (209,311), guinea pig ureter (253), and rat portal vein (301), have detected little or no change in the level of membrane polarization during β-mediated inhibition.

In view of the foregoing findings, it seems unlikely that hyperpolarization per se is a necessary prerequisite for β-adrenoceptor-mediated inhibition of spike activity in spontaneously active SM tissues.

Interestingly, Bülbring (197) has suggested that the small hyperpolarization that is often observed with β-mediated inhibition may be a consequence of the inhibition rather than the cause (i.e., the cessation of APs and other membrane activity may contribute directly to the hyperpolarization). Certainly, it is easy to imagine that reduced action potential frequency, by reducing potassium loss from the cell, may raise $[K]_i$ and hence increase E_K. However, if the absence of APs were the critical factor, one might expect other drugs that inhibited spike activity to cause a similar hyperpolarization. This is not the case, as Diamond and Marshall clearly showed in the rat myometrium, where papaverine completely abolished spike activity without any subsequent hyperpolarization, although the tissue hyperpolarized when epinephrine was administered (302). Thus the loss of APs did not seem to contribute directly to the hyperpolarization, at least in this instance.

As discussed in Section 3.3, increased P_K would inhibit membrane activity and would also be expected to cause membrane hyperpolarization (mechanism a, Table 3); and while this is known to be the α-adrenoceptor-mediated inhibitory mechanism in guinea pig taenia coli, it has also been suggested as a β-adrenoceptor-mediated inhibitory mechanism.

In investigations on pregnant rat myometrium using intracellular electrodes, Kroeger and Marshall (307) and Kawarabayashi and Osa (309,312) have detected a β-adrenoceptor-mediated increase in G_m and an associated hyperpolarization, suggesting an increase in P_K, and a similar finding has been reported in the porcine coronary artery by Ito et al. (295). Further, Marshall and Kroeger showed that the hyperpolarization decreased as $[K]_o$ was increased, suggesting that the hyperpolarization was due to an increase in P_K (307,313). However, with regard to this particular finding it should be noted that Somlyo and Somlyo, in experiments on pulmonary artery strips (314), observed the same phenomenon (i.e., decreasing β-mediated hyperpolarization as $[K]_o$ was increased, but no change in G_m). Their interpretation was that β-adrenoceptor activation stimulated an electrogenic sodium pump (see Section 3.4.4) and that the coupling ratio of this (i.e., ratio of Na^+ transported out to K^+ transported in) varied with $[K]_o$, to the extent that in 10 mM $[K]_o$ the direction of the current carried by the pump reversed, resulting in depolarization, which was in fact observed.

Nevertheless, Marshall's findings certainly suggest that an increase in P_K plays some part in β-adrenoceptor-mediated inhibition of the pregnant rat myometrium (313), although Bülbring and her colleagues have not observed a decrease in membrane resistance in the guinea pig taenia coli (209,303,304,311) nor have Magaribuchi et al. in the guinea pig vas deferens (247). Also, Jenkinson and Morton (40,168) were unable to demonstrate any β-mediated increase in P_K in tracer flux studies on the guinea pig taenia coli. Similarly, Daniel et al. (315) found no increase in potassium efflux during β-adrenoceptor activation in the pregnant rat myometrium.

Reference was made in Section 3.3.1 to work by Kao and co-workers (e.g., Refs. 214–216) in voltage-clamped preparations of guinea pig taenia coli and rat myometrium in which catecholamines, apparently acting through β-adrenoceptors, caused membrane hyperpolarization without any detectable change in G_m. Their interpretation of these electrophysiological data, and parallel intracellular ion concentration studies (214), was that β-adrenoceptor activation raised $[K]_i$ by some undetermined mechanism [not (Na^+, K^+) adenosine triphosphatase (ATPase)], thus increasing E_K and hyperpolarizing the cell. However, as Bolton has pointed out (155), such a mechanism would need to involve a very great increase in $[K]_i$ to change E_K to an extent that would account for the observed degree of hyperpolarization. Such an effect should disappear in K-free solution, but Marshall did not find this to be the case (313).

Summarizing the electrophysiological investigations on β-adrenoceptor-mediated inhibition, it seems that in some tissues under certain conditions, both hyperpolarization and an increase in P_K may be involved, but in general terms the evidence is not compelling enough to consider them essential to the β-adrenoceptor inhibitory effect. This has served to concentrate attention on other possible mechanisms, such as increased calcium sequestration within the cell, increased activity of an electrogenic sodium pump, and increased calcium efflux from the cell. These are discussed in Section 3.4.4.

3.4.3 β-Adrenoceptor-Mediated Action on Development of Tension in SM

In addition to the inhibitory effect on membrane electrical activity dealt with in the preceding section, β-adrenoceptor activation in SM exerts an attenuating effect on the force of contraction generated by contractile stimuli. Situations where this effect can be recognized include β-adrenoceptor-mediated (*a*) inhibition of contractile responses to an electrical stimulus, (b) inhibition of calcium contractures in potassium-depolarized tissues, and (c) inhibition of contractile responses to spasmogens.

β-Mediated reduction in the force of contraction generated by an electrical stimulus has been demonstrated in pregnant rat myometrium by Feinstein et al. (316) and Diamond and Marshall (306), in guinea pig myometrium by Bülbring and Hardman (311), in mouse myometrium by Magaribuchi and Osa (310), in guinea pig vas deferens by Magaribuchi et al. (247), and in guinea pig taenia coli (see Figure 3) by Bülbring and her colleagues (209,303,304,311). In the pregnant rat myometrium, higher levels of β-stimulation were required to achieve this effect than were required to inhibit spontaneous activity (302,306) and it was also found that the β-adrenoceptor-mediated inhibitory effect was not associated with depression of the rate of rise of the AP, as seen with some other types of SM relaxants (306). Johansson et al. observed in the portal vein that the effect on tension was more prolonged than was the inhibitory effect on spontaneous activity (301). In the taenia coli, the reduction in the force of evoked contractions was not due directly to β-adrenoceptor-mediated hyperpolarization, because when Bülbring and den Hertog used current injection to counteract the slight hyperpolarizing action of the β-agonist, the inhibitory effect on contraction strength was undiminished (303,304). All the tissues investigated in these studies generate APs, and it may be assumed that depolarization opens voltage-dependent channels in the membrane, allowing entry of calcium ions into the cell (see Section 3.2). The β-action must either decrease the rate, amplitude, or duration of the rise in free intracellular calcium, or alter the response of the contractile elements to this calcium.

An analogous, and perhaps more readily investigated situation is provided by potassium-depolarized preparations. These contract when calcium is present in the external medium and the amplitude of the contractures is directly related to $[Ca]_o$ (167). Schild and his co-workers (177,178,181,317,318) and many other investigators (e.g., Refs. 183, 302, 306, 315, and 319–324) have shown that such contractures are depressed by β-adrenoceptor agonists (as well as certain other smooth muscle relaxant agents). An interesting finding is that although isoproterenol can inhibit a calcium contracture, it cannot inhibit one caused by barium, and Schild (318) suggested that this might reflect a much lower affinity of the SR for barium than for calcium, as is the case in skeletal muscle. Schild's general conclusion was that increased calcium sequestration within the cell was a likely mechanism for the β-adrenoceptor-mediated relaxant effect on calcium contractures in depolarized preparations.

It is interesting, therefore, that after comparing the size of calcium contractures in depolarized rat myometrial preparations in the presence and the absence of epinephrine (17 nM), Feinstein (319) concluded that the "antagonism" between calcium and epinephrine was noncompetitive, in the sense that they did not appear to compete at the same critical site, as calcium and local anesthetics appeared to. Indeed, in experiments on polarized rat myometrial preparations, Diamond and Marshall (306) found that increased $[Ca]_o$ had little effect on the inhibition by epinephrine of contractions caused by field stimulation, although calcium did antagonize epinephrine's inhibitory action on AP frequency.

Sodium may be involved in some way in the relaxant effect on depolarized tissues since both Magaribuchi et al. (247), in studies on guinea pig vas deferens, and Magaribuchi and Kuriyama (320), in studies on guinea pig taenia coli, found that calcium contractures in fully depolarized tissues were not relaxed by isoproterenol when sodium was totally absent from the external medium.

These and many similar studies provide ample evidence that β-adrenoceptor-mediated inhibitory effects on SM are not confined to actions on membrane electrical activity.

The third type of experimental situation to consider, and undoubtedly the most complicated, is that in which β-adrenoceptor activation antagonizes the contractile response of SM to a spasmogen. Two problems are evident. First, if the spasmogen acts by increasing electrical activity in the tissue [e.g., acetylcholine in the guinea pig taenia coli (188)], it will be difficult to distinguish between an effect on spike frequency and a separate effect on mechanical tension development. Second, spasmogens with different mechanisms of action, that is, with different means of raising the internal calcium activity, may well be antagonized to quite different degrees. However, such differences may give some useful insight into the point, or points, of intervention of β-mechanisms in the chain of events leading to contraction. A recent study by Collis and Shepherd (325) on dog saphenous vein, in which they showed that isoproterenol was more effective in antagonizing acetylcholine contractures than either norepinephrine (α) or calcium (in high [K]$_o$) contractures, provides an interesting example of this approach.

The inhibition of the action of a spasmogen by a β-adrenoceptor mechanism is, in fact, an example of "functional" antagonism, the analysis of which is discussed in Section 2.3.2. Although this can be investigated in any type of SM preparation, it is most straightforward and valuable in tissues that do not develop APs; respiratory-tract SM is particularly suitable because it neither develops spontaneous APs nor is it readily excited by electrical stimuli (153,280). However, the most noteworthy feature of this tissue is that it develops tone "spontaneously" (although in practice a muscarinic or histaminic agonist is often used to produce a constant level of tone). Where an *in vitro* preparation such as the guinea pig tracheal muscle preparation of Castillo and De Beer (326) or the guinea pig lung strip of Siegl et al. (327) is allowed to develop tone spontaneously, it is likely that this tone is in fact induced by prostaglandin (328,329) release from the tissue. Thus whether the tone is allowed to develop "spontaneously" or is carbachol-induced, β-relaxation can be seen in both instances as functional antagonism against, respectively, prostaglandins or a muscarinic agonist.

3.4.4 β-Adrenoceptor-Mediated Effects on Cation Transport and Binding in SM

Whereas the investigational problems of α-adrenoceptor mechanisms in SM are those of membrane permeability and calcium release, the main problems of β-adrenoceptor mechanisms are those of ion transport and calcium binding. Here the small size and internal complexity of the SM cells present immediate difficulties and a striking contrast to the standard "model" systems for investigating such phenomena: the red blood cell and the squid axon. It is hardly surprising that there is considerable "borrowing" of mechanisms first investigated in these simpler systems. Despite the difficulties, progress is being made and there have been several excellent reviews covering various aspects of electrolyte control in SM from Weiss (330), van Breemen et al. (331), Bolton (155), and Brading (170).

Electrogenic Sodium Pumps.   An electrogenic ion pump or exchange mechanism is one that by its action generates a current across a membrane; it may be uncoupled or incompletely coupled, and the potential difference resulting from its action will depend upon the size of the current and the resistance of the membrane. As Rang and Ritchie showed in nerve fibers (332), the removal of a permeant anion (chloride) from the bathing solution may accentuate the potential generated by such a pump by producing an effective increase in membrane resistance.

Electrogenic sodium pumps have been demonstrated in many SM tissues (see Ref. 170), and the enzyme responsible would seem to be a ouabain-sensitive (Na^+, K^+)ATPase, working at a coupling ratio estimated as 3 Na : 2 K (333,334). Such pumps may contribute to the resting membrane potential in SM (335,336).

A modification of this basic scheme has come from Somlyo and co-workers (314,337), who have proposed that β-agonists may activate an electrogenic sodium pump whose coupling ratio varies with $[K]_o$, the K/Na exchange ratio increasing as $[K]_o$ increases. Thus β-adrenoceptor activation at low $[K]_o$ might cause hyperpolarization, but at high $[K]_o$ the coupling ratio would swing toward electroneutrality, or even go beyond it, giving rise to depolarization. This is an interesting mechanism, but in SM it has not been rigorously tested in its biochemical aspects.

The idea that an increase in electrogenic sodium pumping might be responsible for some adrenoceptor-mediated inhibitory effects, notably hyperpolarization (Table 3, mechanism d), was first suggested by Burnstock (338) with respect to epinephrine's action on the guinea pig taenia coli. Although it soon became clear that hyperpolarization in this tissue was probably mainly α-adrenoceptor-mediated, the possibility that β-adrenoceptor effects might operate through such a mechanism has continued to attract attention (e.g., Refs. 314, 337, 339, and 340). Perhaps the strongest evidence in support of the mechanism has come from Scheid et al. (340), who demonstrated in tracer flux studies that isoproterenol increased both potassium influx and sodium efflux, in addition to inhibiting carbachol contractures, in an isolated cell preparation of toad stomach (341), these effects being inhibited by ouabain. Considering the inherent simplicity of the preparation for flux studies, their findings seem incontrovertible, although it is perhaps surprising that the observed cation flux changes were so transient. One might have expected that a maintained increase in pump activity would be necessary to maintain the increased cation gradients across the membrane. The calculated increase in E_K alone may not be sufficient to give maintained hyperpolarization.

In view of these apparently clear-cut findings, it is perhaps surprising that Mueller and van Breemen (342) and Bülbring and den Hertog (303,304), both working with guinea pig taenia coli, and Bose and Innes (343), working with cat carotid artery, found no evidence, as judged from the effect of ouabain, that isoproterenol was acting through the (Na^+, K^+)ATPase in either of these tissues. Marshall and Kroeger, by the use of potassium- and chloride-free solution, came to the same conclusion regarding rat myometrium (308). The part played by electrogenic sodium pumping in β-adrenoceptor-mediated SM effects must therefore be considered uncertain, although it seems to be important in the toad stomach cell preparation of Scheid et al. (340). The case will be examined further in the following section on Na/Ca exchange mechanisms. As a final point it should be noted that potassium transport in turkey erythrocytes is activated by β-adrenoceptor stimulation, through the mediation of adenosine $3'$, $5'$-cyclic monophosphate (cyclic AMP) (see Ref. 344).

Na/Ca Exchange in SM. A Na/Ca exchange mechanism has been demonstrated in nerve cell membranes (345; also see Refs. 346 and 347), where the large sodium gradient fuels sodium influx and calcium efflux at a coupling ratio estimated as $4Na_{in}$: 1 Ca_{out} (348). The mechanism is therefore electrogenic and voltage-dependent in this tissue. There is evidence for a similar mechanism in SM (349–351) (Table 3, mechanism i), although its importance in normal situations has been questioned (352).

Van Breemen et al. (331), on consideration of likely Na/Ca exchange ratios and intracellular calcium levels in SM, postulate that the exchange could be an effective calcium efflux mechanism only if it operated on a level of free calcium higher than the cytoplasmic level. On this basis they suggest that it may operate in series with a Ca-ATPase mechanism, which would accumulate calcium into the SR, the Na/Ca exchange then being responsible for calcium efflux from the SR rather than directly from the cytoplasm. However, such a system would require considerable (although far from unlikely) sophistication in compartmentalization of internal calcium and as yet there is insufficient evidence to judge the issue. In a sense such a hypothesis seems unnecessary, since the squid axon, in which sodium-dependent calcium efflux has been most intensively investigated, has an equally low $[Ca]_i$ and yet has no endoplasmic reticulum. It may simply be that the exchange mechanism is a reserve one that normally plays a subordinate role to adenosine triphosphate (ATP)-dependent, uncoupled calcium efflux (353). Whichever view is taken, the interest with respect to β-effects is that the exchange mechanism provides a possible explanation of why sodium-pump activity might cause inhibition: pump activity would increase the transmembrane sodium gradient, giving a greater driving force to the Na/Ca exchange mechanism and thus lowering $[Ca]_i$. This is the chain of events postulated by Scheid et al. (340) to explain the sodium-pump activation seen with isoproterenol inhibition of toad stomach cells.

Sodium-free solutions, which would not be expected to support Na/Ca exchange and which in themselves may have significant actions on membranes (e.g., Refs. 304 and 311), have been reported to reduce the effects of isoproterenol in the guinea pig taenia coli (339,354) and guinea pig vas deferens (320), although other studies in the same tissue (304,342) and in rat myometrium (313) have detected little change in low $[Na]_o$ solutions. Clearly, the pressing need is for a selective inhibitor of this exchange mechanism, and until one is available there may be little progress in deciding the role of this mechanism in β-actions in SM.

  Calcium Sequestration and Efflux Mechanisms.   Free calcium in SM cells may be rapidly "inactivated" by calcium-binding proteins, binding to the inner surface of the cell membrane, uptake into SR, uptake into mitochondria, or extrusion from the cell (mechanisms e, h, j, Table 3).

  Little is known of the various cytoplasmic proteins that may bind calcium; presumably the contractile proteins are the dominant type (see Ref. 155) although even in noncontractile cells such as nerve axons, there is very effective buffering of free calcium by non-energy-dependent mechanisms (355). Neurotransmitters are not known to influence such binding.

  Mitochondrial calcium uptake (see Ref. 356) may be fueled directly by substrate oxidation or ATP hydrolysis and is obviously a priority for the mitochondrion, as it will take up calcium in "preference" to forming ATP. The uptake is balanced by a countercurrent of protons. Although the system has considerable capacity, there are doubts whether it has a high-enough affinity ($K_m \simeq$ 5–50 μM) to be responsible for the fine control of intracellular calcium activity in contractile tissues (357; see Ref. 159). Batra has investigated calcium uptake in mitochondrial preparations from human myometrium and considers it sufficiently active to play a part in relaxation in this tissue (158). There is no evidence that adrenoceptor mechanisms can affect mitochondrial uptake directly; indeed, very little is known about the controls on this system (356).

  Uptake by "microsomes" has received considerably more attention; however, such preparations normally contain a mixture of membranes and membrane-bound vesicles (possibly some "inside out") derived from both SR and the plasma membrane; therefore, interpretation of findings is not straightforward (see Ref. 358). SR membranes are readily permeated by anions, and microsomal calcium uptake is usually potentiated in the presence of a permeant anion such as oxalate, which can cross the membrane together with calcium to maintain electroneutrality inside membrane-bound vesicles. It would seem that the enzyme involved in the uptake is a membrane-bound Ca-ATPase similar but not identical to that found in red blood cell membranes (e.g., Refs. 359 and 360), cardiac SR (e.g., Ref. 361), and squid axolemma (see Ref. 347). Recently, Wuytack has demonstrated excellent correlation between calcium uptake and Ca-ATPase activity in a microsomal preparation from porcine coronary artery (362).

  A significant advance has been the combination of "marker" enzyme studies (363; and see Ref. 171a) with sucrose gradient fractionation techniques to enable preparation of microsomes derived largely from the plasma membrane (e.g., Ref. 364 and 365). The fact that these actively transport calcium in the presence of ATP indicates the likelihood of an ATP-dependent calcium efflux mechanism in the plasma membrane. This would be in addition to other calcium binding sites on the inner surface of the plasma membrane (366) associated perhaps with a receptor-operated store of intracellular bound calcium, which can be released (even in calcium-free solutions) on activation of the receptor by a spasmogen (e.g., Refs. 367, 368, and 171a; see also Section 3.3).

β-Adrenoceptor agonists have been reported to stimulate SM microsomal calcium uptake (358,364,369–372), although in the study by Batra and Daniel (369), the delay in response to epinephrine was over 5 min. Whether the increased "uptake" reflects simply increased calcium binding, or increased calcium storage within some membrane-bound pool, or increased activity of some calcium extrusion mechanism, it is not possible to say.

Increased calcium extrusion is suggested by the findings of Marshall and her colleagues (308,313,373). For instance, Marshall and Kroeger (308) showed that β-adrenoceptor-mediated hyperpolarization in rat myometrium was blocked in calcium-free or lanthanum-containing solutions, and lanthanum is known to be a potent inhibitor of ATP-dependent, uncoupled calcium efflux in the red blood cell (374) and the squid axon (see Ref. 347). Also, isoproterenol, papaverine, and cyclic AMP all caused a reduction in the excess calcium content of rat myometrial tissue exposed to a calcium-containing, depolarizing high-potassium solution (308). Finally, Kroeger et al. (373) demonstrated a β-adrenoceptor-induced increase in ^{45}Ca efflux in the same tissue. Although these findings certainly point to the involvement of calcium in the β-mediated response, it seems unlikely that electrogenic calcium efflux could make much direct contribution to the sizable β-mediated hyperpolarization observed in this tissue, particularly since chloride-free solutions, which effectively decrease G_m, have little influence on the hyperpolarization (308,313). However, evidence for β-adrenoceptor activation of an electrogenic calcium pump (mechanism e, Table 3) has also come from the studies of Bülbring and den Hertog in guinea pig taenia coli (303,304), in which several likely inhibitory mechanisms were considered and eliminated before it was shown that isoproterenol (1.4 μM) caused a 20% increase in ^{45}Ca efflux from the tissue with no change in ^{45}Ca influx.

Mueller and van Breemen (342) favor the idea of β-activation of increased calcium sequestration (mechanism g, Table 3) within the cell rather than activation of calcium efflux. In contrast to the findings of Marshall and Kroeger in the rat myometrium (308), they have reported that isoproterenol and phosphodiesterase inhibitors antagonize calcium contractures (in high $[K]_o$) in guinea pig taenia coli without causing actual calcium efflux (342). Casteels and Raeymaekers (368) have reported that β-activation in the same tissue increases calcium binding to an intracellular calcium store from which calcium can subsequently be released by spasmogen action. Their data do not, however, exclude the possibility of increased calcium efflux in addition to increased sequestration.

In summary, then, it would seem that β-adrenoceptor activation of SM may result in both increased calcium sequestration within the cell and increased ATP-dependent calcium efflux from the cell. Whether cyclic AMP is involved in these actions will be examined in Section 3.4.5.

3.4.5 Cyclic Nucleotides and β-Adrenoceptor-Mediated Actions in SM

The involvement of cyclic AMP in β-adrenoceptor-mediated inhibition of SM has been a major area of controversy in the last decade and, despite improvements in methodology, some fundamental questions still remain unanswered. The early history has been excellently reviewed by Andersson (375) and Bär (376), and the proceedings of the first symposium on "The Biochemistry of Smooth Muscle" held in Winnipeg, August 1975, contains many relevant contributions: glycolysis and glycogenolysis in SM (377); cyclic nucleotides and mechanical activity in SM (358); cyclic AMP in bronchial tissue (378); cyclic AMP in cell-free vascular systems (379); evidence for dissociation between cyclicnucleotides and tension in SM (380); hormones, relaxants, cyclic AMP, and calcium in SM (381,382).

As indicated in Section 3.1, it is our intention to examine the hypothesis that the full spectrum of β-adrenoceptor-mediated effects in SM is dependent on one initial step: the activation of the membrane-bound adenylate cyclase which forms a complex with the β-adrenoceptor (e.g., Refs. 59, 383, and 384). There seems little disagreement that β-adrenoceptor activation per se does, in fact, increase activity of the adenylate cyclase (but see Ref. 385), and the critical question is whether this action alone is sufficient to explain all observed β-mediated effects, or whether there is some other, parallel, and as yet unrecognized step responsible for some of the β-mediated effects.

Before reviewing the evidence in detail, it will be helpful to consider some common experimental problems encountered in this field. For instance, few studies have been carried out in preparations that are homogeneous with respect to cell type, leaving the possibility that cyclic AMP changes in SM cells themselves could either be diluted or swamped by those in other cell types. There are also difficulties in working with cell-free, membranous preparations, in that in some tissues (e.g., vascular SM) the isolated adenylate cyclase is very labile (see Ref. 379). In addition, there are difficulties in establishing the appropriate environment in terms of ionic concentrations, cofactor requirements, substrate supply, and enzymic complement (see Ref. 358). For example, Fitzpatrick and Szentivanyi (386) found that the presence of protein kinase was necessary before they could demonstrate cyclic AMP stimulation of calcium uptake into aortic microsomes.

Notwithstanding these reservations, numerous studies in normal, polarized SM preparations have demonstrated some degree of correlation between β-mediated inhibition and increases in tissue cyclic AMP content (340,387–393; and see Refs. 358, 375, 376, and 378). Especially interesting are those which have demonstrated that the correlation still holds when the conditions of the experiment are altered, as in the study by Marshall and Kroeger (308), where decreasing the temperature to 10°C slowed down the relaxation of the rat myometrium preparation and the rise in tissue cyclic AMP content, both to the same degree.

Of course, for cyclic AMP to have a causal role, one would expect to see the rise in cyclic AMP before, or at least at the same time as the relaxation. The most convincing evidence for this has come from experiments on disaggregated toad stomach cells, in which Honeyman et al. showed that changes in cyclic AMP reached a peak at 5 sec, before the inhibitory effect was fully established (340,341). Such a finding highlights the usefulness of this particular preparation, since investigations in whole tissues have failed to detect such rapid changes in cyclic AMP levels. Similarly, it is a common finding that cyclic AMP tissue levels continue to rise after the peak inhibitory effect has been seen (e.g., Ref. 394), but here again this need not be taken as evidence against the hypothesis because there may well be alinearity in coupling, and cyclic AMP could continue to be produced and not be hydrolyzed, even after the necessary next step in the chain of events has been set in motion. This is particularly true of situations where the inhibition has been maximal, when consideration of the properties of full agonists clearly indicates the possibility that cyclic AMP levels could rise well above that which would give a maximal inhibitory effect (see Ref. 378). Indeed, it has been frequently observed that the β-agonist log concentration–response curves, in terms of cyclic AMP changes, are considerably displaced to the right compared with those for inhibition of biological response (e.g., Refs. 389, 395, and 396), implying that higher levels of β-stimulation are required to activate adenylate cyclase than to inhibit contractions. This strikes at the heart of the problem, because one is attempting to follow a complex "flow" of events resulting from receptor activation by measurements of total tissue cyclic AMP. The changes in cyclic AMP responsible for initiating the end-biological response could be compartimentalized (397), and the later, more readily detectable changes in total cyclic AMP might be less relevant, perhaps linked to some other aspect of the β-response, such as metabolic changes. Indeed, in most situations, it seems that total tissue cyclic AMP levels are not directly related to the state of activation of the contractile elements in the tissue (see Ref. 380). This particular argument can, of course, work both ways and might equally be taken to indicate the likelihood that cyclic AMP does *not* bring about β-adrenoceptor-mediated relaxation. Returning to a more positive vein, we should consider whether the circumstantial evidence linking cyclic AMP with β-mediated relaxation can be supported by more direct means.

The obvious approach would be to demonstrate that cyclic AMP itself, on direct application to SM, caused inhibition of the same type as that seen with β-agonists. Unfortunately, however, cyclic AMP does not readily cross cell membranes, and a more lipid soluble derivative such as dibutyryl cyclic AMP or 8-bromo cyclic AMP is normally used (see Ref. 398), but even then these compounds produce little effective inhibition until concentrations in the millimolar range are reached [e.g., guinea pig taenia coli: dibutyryl cyclic AMP, 1 mM (311); 8-bromo cyclic AMP, 0.5 mM (342)]. This compares unfavorably with estimated intracellular cyclic AMP concentrations in the micromolar range [e.g., basal cyclic AMP $\simeq$ 4 μM; in the presence of β-agonist $\simeq$ 12 μM (340)]. External application of high concentrations of cyclic nucleotides may inhibit SM by an action on adenosine receptors (311, see Ref. 358) [which may influence adenylate cyclase (399)] or by other nonspecific means (311). Despite these reservations, it is interesting that dibutyryl cyclic AMP has been reported to mimic β-mediated hyperpolarization in vascular SM (338,400) and guinea pig taenia coli (401). However, in other studies on taenia coli and myometrium, Bülbring and Hardman (311) concluded that the inhibitory action of dibutyryl cyclic AMP did not show good qualitative correspondence with that of isoproterenol. The situation is therefore confused.

Perhaps more promising are the reports that cyclic AMP (or its dibutyryl derivative) potentiates microsomal sodium uptake (358,364,370–372,386,402). Even here, though, the effect is not great. For instance, Andersson and Nilsson (358) report that uptake in the fraction derived largely from the plasma membrane was increased 35% by 10 μM cyclic AMP, and in the predominantly SR fraction, uptake was increased only 17%. 5′-AMP in the same concentration did not stimulate binding. Such findings indicate the possibility that cyclic AMP could mediate both increased calcium sequestration and increased efflux. Certainly, in the myocardium there are many reports that cyclic AMP can influence both sarcolemmal and SR (e.g., Refs. 403, 404) Ca-ATPases.

Turning now to a completely different mechanism, it is interesting that Winegrad and McClellan (405), in investigations on cardiac muscle, have raised the possibility that cyclic AMP may be capable of modulating the sensitivity of the contractile elements to calcium. Although this would be an economical way of modulating activity, there is no evidence yet that it is an important physiological mechanism.

Finally, Scheid et al. (340) report that in isolated SM cells, 100 μM cyclic AMP caused an increase in ^{42}K influx identical to that produced by isoproterenol. They interpret this finding as a cyclic AMP-induced increase in activity of (Na^+, K^+) ATPase, which could lead in turn to an increase in calcium efflux, through the Na/Ca exchange mechanism described in Section 3.4.4. They have strengthened this argument by showing that (Na^+, K^+)ATPase activity in these cells is potentiated, in a dose-related manner, by a protein kinase, this effect being inhibited by protein kinase-inhibitory protein. Thus they have established a chain of events that could explain SM inhibition in their preparation.

The involvement of protein kinases in cyclic AMP effects is, of course, well established (e.g., Refs. 406–410), and we need only emphasize that by this mechanism, β-stimulation, leading in turn to elevated cyclic AMP, can exert powerful modulating effects on glucose metabolism (e.g., Refs. 411 and 412) and perhaps also on membrane function (e.g., Refs. 409 and 413). In SM, protein kinase has been shown to be involved in the phosphorylation of membrane proteins and in microsomal calcium uptake (386; see Ref. 370).

However, although the various strands of evidence can be brought together to produce a reasonable case for the involvement of cyclic AMP in β-mediated relaxation of SM, one must enter a note of caution as there have been several observations that cannot be readily fitted into the scheme outlined. For instance, there is the well-documented lack of correlation between total tissue cyclic AMP levels and SM tension (see Ref. 380). Daniel and his colleagues (414–416) have demonstrated circumstances where the relaxant effects of phosphodiesterase inhibitors, and even isoproterenol, do not correlate well with cyclic AMP changes. Using a different approach, Honda et al. (417) exposed guinea pig taenia coli to 2°C for periods of up to 14 days and then found dissociation between isoproterenol effects on cyclic AMP and tension in these preparations. Even more problematical are the findings of Harbon and co-workers with rat myometrium (418–421). They have shown that both epinephrine and E series prostaglandins increase cyclic AMP through an action on the same pool of adenylate cyclase. In both cases the cyclic AMP produced appears to bind to the same intracellular binding sites. The dilemma is that epinephrine inhibits the preparation, whereas the E series prostaglandins stimulate it.

The other major line of evidence against the cyclic AMP hypothesis comes from studies in fully and partially potassium-depolarized tissues (see Ref. 380). Verma and McNeill (322), Diamond and Holmes (422), Meisheri and McNeill (323), and Meisheri et al. (423) have all found that although the relaxant effects of β-agonists persist in depolarized tissues, there is poor or nonexistent correlation with changes in cyclic AMP. Specifically, in depolarized rat uterus with a stable cyclic AMP level, addition of isoproterenol relaxes the preparation but does not increase either cyclic AMP or phosphorylase activity (322). Overweg and Schiff (424), after comparing isoproterenol's action with those of other relaxants in depolarized rat ileal preparations, conclude that the initial rapid relaxation produced by isoproterenol may be independent of cyclic AMP. It is possible, of course, that the abnormal ionic environment both inside and outside the membrane in potassium-depolarized tissues may have complicating effects on drug–receptor interactions, β-adrenoceptor-adenylate cyclase coupling, or other undetermined functions. In view of such evidence running counter to the cyclic AMP hypothesis of β-mediated inhibition, it would seem wise at present to reserve judgment.

Before leaving this section on cyclic nucleotide involvement we should make mention of guanosine 3',5'-cyclic monophosphate (cyclic GMP) lest it be allowed to cloud the issue. This cyclic nucleotide entered into the argument with the suggestion by Lee et al. (425) that the ratio of intracellular cyclic GMP to cyclic AMP might be an important determinant of cell function: the yin–yang hypothesis (426). The suggestion was based on observations that bethanechol (a muscarinic agonist) increased the cyclic GMP level in guinea pig ileum and isoproterenol inhibited this rise (425). However, it is now clear that the rise in intracellular cyclic GMP is a phenomenon merely *associated* with the contraction (380,427–429). For instance, it has been shown that cyclic GMP does not increase during α-adrenoceptor activation in calcium-deficient solutions (428), and it seems much more likely that the increased free calcium levels occurring during contraction are the cause rather than the consequence of guanylate cyclase activation, especially when one considers that this enzyme is cytoplasmic and not likely to be readily influenced by extracellular neurotransmitters. Further, the lack of involvement of cyclic GMP in SM relaxation has recently been reported by Janis and Diamond (430).

3.4.6 Summary

The quest for the mechanism of β-adrenoceptor-mediated inhibition has carried researchers into all the nooks and crannies of SM function, yet the final answer still proves elusive. Much the most efficient and economical mechanism of β-action seems to be inhibition of AP generation in spontaneously active tissues, and it is likely that this is achieved by some alteration, as yet undetermined, in the electrical properties of the membrane. In the biochemical sphere there are grounds for thinking that β-adrenoceptor activation can increase both internal calcium sequestration and calcium efflux, but there is uncertainty over how the receptor could transmit the signal to the enzymes involved, unless it be by some mediator such as cyclic AMP. This is the obvious candidate and yet, often, its levels seem poorly correlated to the muscle's state of activity. The only alternative on the horizon is some mechanism such as that hinted at by Axelrod's group (431), where drug–receptor combination disinhibits methyltransferases in the membrane before the link with adenylate cyclase takes place.

3.5 Conclusions

The approach adopted in Section 3 has been to examine the possibility that the α- and β-adrenoceptors have fundamentally different modes of action on the SM cell. It was proposed that the α-mediated events are brought about primarily by an ionic permeability change in the plasma membrane, whereas the β-mediated response results, in the first instance, from a metabolic change intimately related to the calcium economy of the cell, with an increase in cyclic AMP playing an early part.

We believe that the data presented justify this approach, and that the two receptor types do, in fact, utilize quite separate mechanisms, in SM at any rate. However, several qualifications must be made.

1   Although the α-mediated mechanism does generally involve an increase in the permeability of the membrane through the opening of ROCs, it is clear that in some SMs the effects of α-agonists on the mechanical response *may* occur with little or no change in membrane potential. Such findings, and also experiments that show that catecholamines may still act when the membrane potential has been abolished in potassium-rich solutions, serve to remind us that it is the change in intracellular calcium that is important rather than the effect on the membrane permeability per se. Indeed, it is possible that at least some of the effects on the membrane are a consequence of the increase in cytosolic calcium, and it may be that the primary function of the α-adrenoceptor is to mobilize calcium, which then serves as a coupling agent to open ROCs or influence other cellular processes.

2   In contrast, the evidence for β-mediated changes in ionic permeability is poor, at least as a primary event. It seems more likely that alterations in membrane potential, spike electrogenesis and ion transport are secondary to certain metabolic changes, and the same may be true for increased calcium sequestration within the cell. The latter factor may be the main cause of the relaxation seen in SM types that do not generate APs, whereas in those that do, inhibition of spike electrogenesis may be at least as important.

3   If the receptor types are, in fact, associated with distinct cellular mechanisms, then it is of great interest to compare the mechanisms in different tissue types to see if this generalization holds true. Although exocrine glands and liver hepatocytes obviously differ from SM in many respects, it seems significant that here, too, α-adrenoceptors appear to mobilize calcium and increase P_K, suggesting that these aspects of α-activation are linked even though the events that follow differ (223,224,437). Also, although this chapter has concerned itself only with what might be termed α_1-mechanisms, it will be of interest to compare these with α_2-

mechanisms on neurones and at other sites. For example, it is already clear that the α_2-mediated hyperpolarization in the rat superior cervical ganglion (82) shows some similarities to the α_1-mediated hyperpolarization in the guinea pig taenia coli.

With regard to β_1- and β_2-mediated mechanisms, it is not yet clear if they differ in any important respects; in fact, it is conceivable that the pharmacological distinctions between these receptors could reflect differences in the membrane matrix surrounding the same macromolecule. Related to this point, it is not clear to what extent other hormone recognition units embedded in the same membrane can independently influence adenylate cyclase and cyclic AMP levels. If compartmentalization within the cell is substantial, this might explain some of the lack of correlation between β-mediated relaxation and whole-cell cyclic AMP levels.

4 Finally, it should be pointed out that several potentially important aspects of the physiological organization of the adrenoceptors have not been dealt with in this chapter. For instance, most of the studies discussed have been of the actions of applied norepinephrine and other catecholamines, but it is clearly of interest to know how well these observations relate to the effect of nerve-released transmitter. For instance, the reactivity of the inner and outer layers of the tunica media of the large blood vessels suggest that the α-adrenoceptors of the former are adapted to be responsive to low levels of circulating epinephrine, whereas the latter are activated by quite high levels of nerve-released norepinephrine (see Refs. 432 and 433). Further, in smaller vessels the effect of nerve-released and applied norepinephrine does not correspond well, and this may reflect differences between "junctional" and "extrajunctional" receptor populations (434,434a).

Similarly, it has been suggested that β_1-adrenoceptors are suited to activation by nerve-released norepinephrine, whereas the β_2-adrenoceptors may best be stimulated by blood-borne epinephrine (110,435,436).

Clearly, much remains to be learned before the physiological roles of the mechanisms discussed in Section 3 can be fully related to the properties of the adrenoceptor types dealt with in Section 2.

ACKNOWLEDGMENTS

We should like to express our appreciation to Professor D. H. Jenkinson and Dr. J. M. Littleton for their careful reading of the manuscript and helpful comments.

REFERENCES

1 G. Oliver and E. A. Schäfer, *J. Physiol. (Lond.)*, **18**, 230 (1895).

2 M. Lewandowsky, *Arch. Anat. Physiol., Physiol. Abt.*, 360 (1899).

3 J. N. Langley, *J. Physiol. (Lond.)*, **27**, 237 (1901).

4 T. R. Elliott, *J. Physiol. (Lond.)*, **31**, 20P (1904).

5 T. R. Elliott, *J. Physiol. (Lond.)*, **32**, 401 (1905).

6 O. Loewi, *Pflügers Arch. Ges. Physiol.*, **189**, 239 (1921).

7   O. Loewi, *Pflügers Arch. Ges. Physiol.*, **193**, 201 (1921).

8   O. Loewi and E. Navratil, *Pflügers Arch. Ges. Physiol.*, **214**, 678 (1926).

9   O. Loewi and E. Navratil, *Pflügers Arch. Ges. Physiol.*, **214**, 689 (1926).

10 W. B. Cannon and A. Rosenblueth, *Autonomic Neuroeffector Systems*, Macmillan, New York, 1937.

11 W. B. Cannon and A. Rosenblueth, *Am. J. Physiol.*, **104**, 557 (1933).

12 R. P. Ahlquist, *Am. J. Physiol.*, **153**, 586 (1948).

13 U. S. von Euler, *Acta Physiol. Scand.*, **12**, 73 (1946).

14 U. S. von Euler, *Noradrenaline*, Charles C Thomas, Springfield, Ill., 1956.

15 M. Holzbauer and D. F. Sharman, in H. Blaschko and E. Muscholl, Eds., *Catecholamines*, Handbook of Experimental Pharmacology, Vol. 33, Springer-Verlag, Berlin, 1972, pp. 110–185.

16 J. N. Langley, *J. Physiol. (Lond.)*, **33**, 374 (1905).

17 J. N. Langley, *J. Physiol. (Lond.)*, **1**, 339 (1878).

18 T. G. Brodie and W. E. Dixon, *J. Physiol. (Lond.)*, **30**, 476 (1904).

19 G. Campbell, in E. Bülbring, A. F. Brading, A. W. Jones, and T. Tomita, Eds., *Smooth Muscle*, Edward Arnold, London, 1970, pp. 451–495.

20 S. Z. Langer, *Br. J. Pharmacol.*, **60**, 481 (1977).

21 H. H. Dale, *J. Physiol. (Lond.)*, **34**, 163 (1906).

22 G. Barger and H. H. Dale, *J. Physiol. (Lond.)*, **41**, 19 (1910).

23 H. H. Dale, *Adventures in Physiology*, Pergammon Press, Oxford, 1953, p. 98.

24 R. P. Ahlquist, *Pharmacol. Rev.*, **11**, 441 (1959).

25 R. P. Ahlquist, *Ann. N.Y. Acad. Sci.*, **139**, 549 (1967).

26 R. P. Ahlquist, *Am. Heart J.*, **92**, 661 (1976).

27 N. C. Moran, *Ann. N.Y. Acad. Sci.*, **139**, 649 (1967).

28 R. P. Ahlquist, *Trends Pharmacol. Sci.* **1**, 16 (1979).

29 A. M. Lands, *Am. J. Physiol.*, **169**, 11 (1952).

30 R. F. Furchgott, *Pharmacol. Rev.*, **11**, 429 (1959).

31 I. H. Slater and C. E. Powell, *Fed. Proc.*, **16**, 336 (1957).

32 C. E. Powell and I. H. Slater, *J. Pharmacol. Exp. Ther.*, **122**, 480 (1958).

33 R. P. Ahlquist and B. Levy, *J. Pharmacol. Exp. Ther.*, **127**, 146 (1959).

34 R. F. Furchgott, in J. R. Vane, G. E. W. Wolstenholme, and M. O'Connor, Eds., *Adrenergic Mechanisms*, Ciba Foundation Symposium, Churchill, London, 1960, pp. 246–252.

35 J. M. van Rossum and M. Mujić, *Arch. Int. Pharmacodyn.*, **155**, 418 (1965).

36 W. C. Bowman and M. T. Hall, *Br. J. Pharmacol.*, **38**, 399 (1970).

37 B. Whitney, *J. Pharm. Pharmacol.*, **17**, 465 (1965).

38 A. Bennett, *Nature (Lond.)*, **208**, 1289 (1965).

39 A. Bucknell and B. Whitney, *Br. J. Pharmacol. Chemother.*, **23**, 164 (1964).

40 D. H. Jenkinson and I. K. M. Morton, *Ann. N.Y. Acad. Sci.*, **139**, 762 (1967).

41 T. M. Brody and J. Diamond, *Ann. N.Y. Acad. Sci.*, **139**, 772 (1967).

42 W. D. M. Paton and E. S. Vizi, *Br. J. Pharmacol.*, **35**, 10 (1969).

43 H. W. Kosterlitz, R. J. Lydon, and A. J. Watt, *Br. J. Pharmacol.*, **39**, 398 (1970).

44 N. C. Moran, *Ann. N.Y. Acad. Sci.*, **139**, 545 (1967).

45 A. M. Lands, A. Arnold, J. P. McAuliff, F. P. Luduena, and T. G. Brown, *Nature (Lond.)*, **214**, 597 (1967).

46 A. M. Lands, F. P. Luduena, and H. J. Buzzo, *Life Sci.*, **6**, 2241 (1967).

47 A. M. Lands, G. E. Grollewski, and T. G. Brown, *Arch. Int. Pharmacodyn.*, **161**, 68 (1966).

48 S. Z. Langer, *Biochem. Pharmacol.*, **23**, 1793 (1974).

49 S. Berthelsen and W. A. Pettinger, *Life Sci.*, **21**, 595 (1977).

50 H. O. Schild, in H. P. Rang, Ed., *Drug Receptors*, Macmillan, London, 1973, pp. 29–36.

51 A. J. Clark, *J. Physiol. (Lond.)*, **61**, 530 (1926).

52 A. J. Clark, *General Pharmacology*, Handbook of Experimental Pharmacology, Vol. 4, Springer-Verlag, Berlin, 1937, pp. 1–223.

53 R. F. Furchgott, *Ann. N.Y. Acad. Sci.*, **139**, 553 (1967).

54 R. F. Furchgott, *Fed. Proc.*, **29**, 1352 (1970).

55 R. F. Furchgott, in Ref. 15, pp. 283–335.

56 R. P. Stephenson, *Br. J. Pharmacol. Chemother.*, **11**, 379 (1956).

57 J. M. van Rossum and E. J. Ariëns, *Arch. Int. Pharmacodyn.*, **136**, 385 (1962).

58 D. Mackay, in J. M. van Rossum, Ed., *Kinetics of Drug Action*, Handbook of Experimental Pharmacology, Vol. 47, Springer-Verlag, Berlin, 1977, pp. 255–321.

59 L. J. Pike, L. E. Limbird, and R. J. Lefkowitz, *Nature (Lond.)*, **280**, 502 (1979).

60 G. Kunos, *Annu. Rev. Pharmacol. Toxicol.*, **18**, 291 (1978).

61 J. C. Besse and R. F. Furchgott, *Fed. Proc.*, **26**, 401 (1967).

62 J. C. Besse and R. F. Furchgott, *J. Pharmacol. Exp. Ther.*, **197**, 66 (1976).

63 E. M. Sheys and R. D. Green. *J. Pharmacol. Exp. Ther.*, **180**, 317 (1972).

64 B. Johansson, S. R. Johansson, F. Ljung, and L. Stage, *J. Pharmacol. Exp. Ther.*, **180**, 636 (1972).

65 L. Edvinsson and C. Owman, *Circ. Res.*, **35**, 835 (1974).

66 B. Harper, I. E. Hughes, and J. Stott, *J. Pharm. Pharmacol.*, **31**, 105 (1979).

67 R. F. Furchgott, *Fed. Proc.*, **37**, 115 (1978).

68 E. J. Ariëns, A. M. Simonis, and J. M. van Rossum, in E. J. Ariëns, Ed., *Molecular Pharmacology*, Vol. 1, Academic Press, New York, 1964, pp. 394–466.

69 F. G. van den Brink, in Ref. 58, pp. 169–254.

70 R. P. Stephenson, in M. Worcel and G. Vassort, Eds., *Smooth Muscle Pharmacology and Physiology, Colloq. INSERM*, **50**, 15 (1975).

71 D. H. Jenkinson, *Br. Med. Bull.*, **29**, 142 (1972).

72 H. O. Schild, *Br. J. Pharmacol. Chemother.*, **2**, 189 (1947).

73 O. Arunlakshana and H. O. Schild, *Br. J. Pharmacol. Chemother.*, **14**, 48 (1959).

74 M. Wenka, D. Lincová, J. Čepelík, M. Černokovský, and S. Hynie, *Ann. N.Y. Acad. Sci.*, **139**, 860 (1967).

75 D. Colquhoun, in Ref. 50, pp. 149–182.

76 C. D. Thron, *Mol. Pharmacol.*, **9**, 1 (1973).

77 S. Z. Langer and U. Trendelenburg, *J. Pharmacol. Exp. Ther.*, **167**, 117 (1969).

78 U. Trendelenburg, in Ref. 15, pp. 726–761.

79 D. R. Waud, *J. Pharmacol. Exp. Ther.*, **167**, 140 (1979).

80 R. F. Furchgott, A. Jurkiewicz, and N. F. Jurkiewicz, in E. Usdin and S. Snyder, Eds., *Frontiers in Catecholamine Research*, Pergamon Press, New York, 1973, pp. 295–299.

81 U. Trendelenburg, in Ref. 15, pp. 336–360.

82 D. A. Brown and M. P. Caulfield, Chap. 3, this volume.

83 A. Levitzki, *Rev. Physiol. Biochem. Pharmacol.*, **82**, 1 (1978).

84 R. J. Lefkowitz, *Fed. Proc.*, **37**, 123 (1978).

85 B. B. Wolfe, T. K. Harden, and P. B. Molinoff, *Annu. Rev. Pharmacol. Toxicol.*, **17**, 575 (1977).

86　M. G. Caron, L. T. Williams, and R. J. Lefkowitz, in E. Szabadi, C. M. Bradshaw, and P. Bevan, Eds., *Recent Advances in the Pharmacology of Adrenoceptors,* Elsevier/North-Holland, 1978, pp. 133–144.

87　S. R. Nahorski, in Ref. 86, pp. 163–164.

88　J. Kleinstein and H. Glossmann, *Naunyn-Schmiedebergs Arch. Pharmacol.,* **305,** 191 (1978).

89　E. L. Rugg, D. B. Barnett, and S. R. Nahorski, *Mol. Pharmacol.,* **14,** 996 (1978).

90　A. A. Hancock, A. K. DeLean, and R. J. Lefkowitz, *Mol. Pharmacol.,* **16,** 1 (1979).

91　B. B. Hoffman, A. DeLean, C. L. Wood, D. D. Schocken, and R. J. Lefkowitz, *Life Sci.,* **24,** 1739 (1979).

92　P. N. Patil, K. Fudge, and D. Jacobowitz, *Eur. J. Pharmacol.,* **19,** 79 (1972).

93　R. D. Green and W. W. Fleming, *J. Pharmacol. Exp. Ther.,* **162,** 254 (1968).

94　L. I. Goldberg, *Biochem. Pharmacol.,* **24,** 651 (1975).

95　M. I. Holck, B. H. Marks, and C. A. Wilberding, *Mol. Pharmacol.,* **16,** 77 (1979).

96　K. A. Barker, B. Harper, and I. E. Hughes, *J. Pharm. Pharmacol.,* **29,** 129 (1977).

97　R. R. Ruffolo, Jr., B. S. Turowski, and P. N. Patil, *J. Pharm. Pharmacol.,* **30,** 498 (1978).

98　V. C. Swamy and D. J. Triggle, *Eur. J. Pharmacol.,* **19,** 67 (1972).

99　S. H. Snyder, *Trends Neurosci.,* **1,** 123 (1978).

100　H. Kapur, B. Rouot, and S. H. Snyder, *Eur. J. Pharmacol.,* **57,** 317 (1979).

101　M. Butler and D. H. Jenkinson, *Eur. J. Pharmacol.,* **52,** 303 (1978).

102　L. T. Williams and R. J. Lefkowitz, *J. Clin. Invest.,* **60,** 815 (1977).

103　J. M. Roberts, P. A. Insel, R. D. Goldfien, and A. Goldfien, *Nature (Lond.),* **270,** 624 (1977).

104　B. B. Hoffman, A. L. DeLean, C. L. Wood, D. D. Schocken, and R. J. Lefkowitz, *Life Sci.,* **34,** 1739 (1979).

105　G. Kunos, B. Hoffman, Y. N. Kwok, W. H. Kan, and L. Mucci, *Nature (Lond.),* **278,** 254 (1979).

106　B. Harper, I. E. Hughes, and F. H. Noormohamed, *J. Pharm. Pharmacol.,* **30,** 167 (1978).

107　R. J. Lefkowitz, *Biochem. Pharmacol.,* **24,** 583 (1975).

108　J. Zaagsma, to appear in the second volume of this series.

109　G. P. Levy and G. H. Apperley, in Ref. 86, pp. 201–208.

110　M. J. Daly and G. P. Levy, in S. Kalsner, Ed., *Trends in Autonomic Pharmacology,* Vol. 1, Urban & Schwarzenberg, Baltimore, Md., 1979, pp. 347–385.

111　E. Carlsson, *Acta Pharmacol. Toxicol.,* **44,** Suppl. 11, 17 (1979).

112　B. Levy, *J. Pharmacol. Exp. Ther.,* **151,** 413 (1966).

113　N. C. Moran, *Pharmacol. Rev.,* **18,** 503 (1966).

114　C. K. Buckner and P. N. Patil, *J. Pharmacol. Exp. Ther.,* **176,** 634 (1971).

115　P. N. Patil, D. D. Miller, and U. Trendelenburg, *Pharmacol. Rev.* **26,** 323 (1974).

116　H. H. Harms, J. Zaagsma, and J. de Vente, *Life Sci.,* **21,** 123 (1977).

117　E. Carlsson, B. Åblad, A. Brändstrom, and B. Carlsson, *Life Sci.,* **11,** 953 (1972).

118　E. Carlsson and B. Åblad, in P. R. Saxena and R. P. Forsyth, Eds., *Beta-Adrenoceptor Blocking Agents,* North-Holland, Amsterdam, 1976, pp. 305–309.

119　R. F. Furchgott, T. D. Wakade, C. E. Sorace, and J. S. Stollak, *Fed. Proc.,* **34,** 794 (1975).

120　S. R. O'Donnell and J. C. Wanstall, *Naunyn-Schmiedebergs Arch. Pharmacol.* **308,** 183 (1979).

121　F. G. van den Brink, *Eur. J. Pharmacol.,* **22,** 270, 279 (1973).

122　S. R. O'Donnell and J. C. Wanstall, *Br. J. Pharmacol.,* **60,** 255 (1977).

123　C. K. Buckner and R. K. Saini, *J. Pharmacol. Exp. Ther.,* **194,** 565 (1975).

124　K. J. Broadley and C. D. Nicholson, *Br. J. Pharmacol.,* **66,** 397 (1979).

125 E. Haber and S. Wrenn, *Physiol. Rev.*, **56**, 317 (1976).

126 E. Haber and C. Homcy, Chapter 8, this volume.

127 G. Kunos, Chapter 10, this volume.

128 R. J. Lefkowitz, L. E. Limbird, C. Mukherjee, and M. G. Caron, *Biochim. Biophys. Acta*, **457**, 1 (1976).

129 D. B. Barnett, E. L. Rugg, and S. R. Nahorski, *Nature (Lond.)*, **273**, 166 (1978).

130 S. R. Nahorski, D. B. Barnett, D. R. Howlett, and E. L. Rugg, *Naunyn-Schmiedebergs Arch. Pharmacol.*, **307**, 227 (1979).

131 K. P. Minneman, L. R. Hegstrand, and P. B. Molinoff, *Mol. Pharmacol.*, **16**, 34 (1979).

132 M. Bristow, T. R. Sherrod, and R. D. Green, *J. Pharmacol. Exp. Ther.*, **171**, 52 (1970).

133 K. P. Minneman, L. R. Hegstrand, and P. B. Molinoff, *Mol. Pharmacol.*, **16**, 21 (1979).

134 D. G. Reynolds, G. D. Demaree, and M. H. Heiffer, *Proc. Soc. Exp. Biol. Med.*, **125**, 73 (1967).

134*a* N. W. Weisbrodt, C. C. Hug, and P. Bass, *J. Pharmacol. Exp. Ther.*, **170**, 272 (1969).

135 N. Toda, S. Hayashi, Y. Hatano, H. Okunishi, and M. Mihazaki, *J. Pharmacol. Exp. Ther.*, **207**, 311 (1978).

136 D. B. Barnett, E. L. Rugg, and S. R. Nahorski, in Ref. 86, pp. 337–338.

137 P. K. S. Siegl, G. V. Rossi, and R. F. Orzechowski, *Eur. J. Pharmacol.*, **54**, 1 (1979).

138 N. Taira, *Annu. Rev. Pharmacol.*, **12**, 197 (1972).

138*a* N. Taira, Y. Yabuuhi, and S. Yamashita, *Br. J. Pharmacol.*, **59**, 577 (1977).

139 P. N. Patil, *J. Pharmacol. Exp. Ther.*, **166**, 299 (1969).

139*a* J. L. Matheny and R. P. Ahlquist, *Arch. Int. Pharmacodyn. Ther.*, **209**, 197 (1974).

140 A. Sandow, *Pharmacol. Rev.*, **17**, 265 (1965).

141 R. Casteels, T. Godfraind, and J. C. Rüegg, Eds., *Excitation–Contraction Coupling in Smooth Muscle* (Proceedings of the International Symposium on Excitation–Contraction Coupling in Smooth Muscle held in Louvain and Heidelberg, July 1977), Elsevier/North-Holland, Amsterdam, 1977.

142 J. C. Rüegg, *Physiol. Rev.*, **51**, 201 (1971).

143 S. V. Perry and R. J. A. Grand, *Br. Med. Bull.*, **35**, 219 (1979).

144 R. S. Filo, D. F. Bohr, and J. C. Rüegg, *Science*, **147**, 1581 (1965).

145 M. Endo, T. Kitazawa, S. Yagi, M. Iino, and Y. Kakuta, in Ref. 130, pp. 199–209.

146 A. R. Gordon, *Proc. Natl. Acad. Sci. U.S.A.*, **75**, 3527 (1978).

147 E. Bozler, *Biol. Symp.*, **3**, 95 (1941).

148 E. Bozler, *Experientia*, **4**, 213 (1948).

149 G. Burnstock, in Ref. 19, pp. 1–69.

150 G. Burnstock, *Br. Med. Bull.*, **35**, 255 (1979).

151 K. Golenhofen, in E. Bülbring and M. F. Shuba, Eds., *Physiology of Smooth Muscle*, Raven Press, New York, 1976, pp. 197–202.

152 J. H. Szurszewski, *Fed. Proc.*, **36**, 2456 (1977).

153 K. Creed, *Br. Med. Bull.*, **35**, 243 (1979).

154 T. B. Bolton and E. Bülbring, in Ref. 86, pp. 7–13.

155 T. B. Bolton, *Physiol. Rev.*, **59**, 606 (1979).

156 T. B. Bolton, *Br. Med. Bull.*, **35**, 275 (1979).

157 C. E. Devine, A. V. Somlyo, and A. P. Somlyo, *J. Cell Biol.*, **52**, 690 (1972).

158 S. Batra, *Biochim. Biophys. Acta*, **305**, 428 (1973).

159 E. Carafoli and M. Crompton, *Ann. N.Y. Acad. Sci.*, **307**, 269 (1978).

160 P. J. Goodford and G. S. Wooton, in Ref. 70, pp. 405–441.

161 G. Gabella, *Br. Med. Bull.*, **35**, 213 (1979).

162 T. Tomita, *Prog. Biophys. Mol. Biol.*, **30**, 185 (1979).

163 M. E. Holman and T. O. Neild, *Br. Med. Bull.*, **35**, 235 (1979).

164 M. R. Bennett, *Autonomic Neuromuscular Transmission,* Monographs of the Physiological Society, No. 30, Cambridge University Press, Cambridge, 1972.

165 V. A. Bury and M. F. Shuba, in Ref. 151, pp. 65–76.

166 P. J. Goodford, *J. Physiol. (Lond.),* **176**, 180 (1965).

167 P. M. Hudgins and G. B. Weiss, *Am. J. Physiol.*, **217**, 1310 (1969).

168 H. Lüllman, in Ref. 19, pp. 151–165.

169 C. van Breemen, O. Hwang, and B. Siegel. in Ref. 130, pp. 243–252.

170 A. F. Brading, *Br. Med. Bull.*, **35**, 227 (1979).

171 E. E. Daniel and D. M. Paton, Eds., *Smooth Muscle,* Methods in Pharmacology, Vol. 3, Plenum Press, New York, 1975, Chap. 10

171a E. E. Daniel, D. J. Crankshaw, and C-Y. Kwan, in Ref. 110, pp. 443–484.

172 M. Endo, *Physiol. Rev.*, **57**, 71 (1977).

173 A. Fabiato and F. Fabiato, *Nature (Lond.),* **281**, 146 (1979).

174 A. P. Somlyo and A. V. Somlyo, *J. Pharmacol. Exp. Ther.*, **159**, 129 (1968).

175 A. V. Somlyo and A. P. Somlyo, *Pharmacol. Rev.*, **20**, 197 (1968).

176 A. P. Somlyo, in T. Narahashi, *Cellular Pharmacology of Excitable Tissues,* Charles C Thomas, Springfield, Ill., 1975, pp. 360–407.

177 D. H. L. Evans, H. O. Schild, and S. Thesleff, *J. Physiol. (Lond.),* **143**, 474 (1958).

178 H. O. Schild, in Ref. 34, pp. 288–292.

179 R. P. Durbin and D. H. Jenkinson, *J. Physiol. (Lond.),* **157**, 90 (1961).

180 K. A. P. Edman and H. O. Schild, *J. Physiol. (Lond.),* **161**, 424 (1962).

181 K. A. P. Edman and H. O. Schild, *J. Physiol. (Lond.),* **169**, 404 (1963).

182 H. O. Schild, in E. Bülbring, Ed., *Pharmacology of Smooth Muscle,* Pergamon Press, Oxford, 1964, pp. 95–104.

183 D. H. Jenkinson and I. K. M. Morton, *J. Physiol. (Lond.),* **188**, 387 (1967).

184 B. L. Ginsborg, *Pharmacol. Rev.*, **19**, 289 (1967).

185 B. L. Ginsborg, *Biochim. Biophys. Acta,* **300**, 289 (1973).

186 E. Bülbring, *J. Physiol. (Lond.),* **125**, 302 (1954).

187 E. Bülbring, *J. Physiol. (Lond.),* **128**, 200 (1955).

188 E. Bülbring, *J. Physiol. (Lond.),* **135**, 412 (1957).

189 E. Bülbring, in Ref. 34, pp. 275–287.

190 G. Burnstock, *J. Physiol. (Lond.),* **143**, 183 (1958).

191 R. Casteels, in Ref. 19, pp. 70–99.

192 T. Tomita, in Ref. 19, pp. 197–243.

193 M. E. Holman, in Ref. 19, pp. 244–288.

194 H. Kuriyama, in Ref. 19, pp. 366–395.

195 J. Axelsson, in Ref. 19, pp. 289–315.

196 E. Bülbring, in Ref. 50, pp. 1–13.

197 E. Bülbring, *Br. Med. Bull.*, **35**, 285 (1979).

198 C. L. Prosser, *Annu. Rev. Physiol.*, **36**, 503 (1974).

199 E. Bülbring and H. Kuriyama, *J. Physiol. (Lond.),* **166**, 59 (1963).

200 J. Axelsson, E. Bueding, and E. Bülbring, *J. Physiol. (Lond.),* **156**, 357 (1961).

201 H. Kuriyama, *J. Physiol. (Lond.)*, **166,** 15 (1963).

202 G. V. R. Born and E. Bülbring, *J. Physiol. (Lond.)*, **131,** 690 (1956).

203 D. H. Jenkinson and I. K. M. Morton, *Nature (Lond.)*, **205,** 505 (1965).

204 D. H. Jenkinson and I. K. M. Morton, *J. Physiol. (Lond.)*, **188,** 373 (1967).

205 E. Bülbring, P. J. Goodford, and J. Setekliev, *Br. J. Pharmacol. Chemother.*, **28,** 296 (1966).

206 P. G. Bogach and M. Y. Klevets, *Biofizika*, **12,** 997 (1967).

207 M. F. Shuba, A. V. Gurkovskaya, M. J. Klevetz, N. G. Kochemasova, and V. M. Taranenko, in Ref. 151, pp. 347–355.

208 E. Bülbring and T. Tomita, *Proc. R. Soc. Lond. Ser. B*, **172,** 89 (1969).

209 E. Bülbring and T. Tomita, *Proc. R. Soc. Lond. Ser. B*, **172,** 103 (1969).

210 H. Ohashi, *J. Physiol. (Lond.)*, **212,** 561 (1971).

211 R. Casteels and H. Kuriyama, *J. Physiol. (Lond.)*, **184,** 120 (1966).

212 N. C. Anderson, *J. Gen. Physiol.*, **54,** 145 (1969).

213 N. C. Anderson, in Ref. 130, pp. 81–89.

214 C. Y. Kao, H. Inomata, J. R. McCullough, and J. C. Yuan, in Ref. 70, pp. 165–176.

215 H. Inomata and C. Y. Kao, *J. Physiol. (Lond.)*, **226,** 53P (1972).

216 C. Y. Kao, J. R. McCullough, and H. L. Davidson, *Fed. Proc.*, **30,** 384 (1971).

217 T. Tomita, Y. Sakamoto, and M. Ohba. *Nature (Lond.)*, **250,** 432 (1974).

218 T. Tomita, H. Tokuno, and S. Usune, *Proc. R. Soc. Lond. Ser. B*, **198,** 473 (1977).

219 E. Bülbring and T. Tomita, *Proc. R. Soc. Lond. Ser. B*, **172,** 121 (1969).

220 E. Bülbring and T. Tomita, *Proc. R. Soc. Lond. Ser. B*, **197,** 271 (1977).

221 J. W. Putney, in G. B. Weiss, Ed., *Calcium in Drug Action*, Plenum Press, New York, 1978, pp. 180–187.

222 R. W. Meech, *Annu. Rev. Biophys. Bioeng.*, **7,** 1 (1978).

223 D. H. Jenkinson, D. G. Haylett, and K. Koller, in R. W. Straub and L. Bolis, Eds., *Cell Membrane Receptors for Drugs and Hormones: A Multidisciplinary Approach*, Raven Press, New York, pp. 89–105 (1978).

224 J. W. Putney, *Pharmacol. Rev.*, **30,** 209 (1978).

225 I. K. M. Morton, Ph.D. thesis, University of London, 1969.

226 R. N. Speden, in Ref. 19, pp. 558–588.

227 M. E. Holman, C. B. Kasby, M. B. Suthers, and J. A. F. Wilson, *J. Physiol. (Lond.)*, **196,** 111 (1968).

228 A. H. Weston, in Ref. 86, pp. 15–22.

229 K. Golenhofen, N. Hermstein, and E. Lammel, *Microvasc. Res.*, **5,** 73 (1973).

230 K. Golenhofen and N. Hermstein, *Blood Vessels*, **12,** 21 (1975).

231 B. A. Wahlström, *Acta Physiol. Scand.*, **89,** 436 (1973).

232 B. A. Wahlström, *Acta Physiol. Scand.*, **89,** 522 (1973).

233 B. A. Wahlström and B. Svennerholm, *Acta Physiol. Scand.*, **92,** 404 (1974).

234 S. H. L. Tsay and A. W. Jones, *Fed. Proc.*, **38,** 603 (1979).

235 J. W. Miller, *Ann. N.Y. Acad. Sci.*, **139,** 788 (1967).

236 Y. Abe, in Ref. 19, pp. 296–417.

237 J. Marshall, *Ergeb. Physiol.*, **62,** 6 (1970).

238 E. Bülbring, R. Casteels, and H. Kuriyama, *Br. J. Pharmacol.*, **34,** 388 (1968).

239 E. Bülbring and J. H. Szurszewski, *Proc. R. Soc. Lond. Ser. B*, **185,** 225 (1974).

240 J. H. Szurszewski and E. Bülbring, *Philos. Trans. R. Soc. Lond. Ser. B*, **265,** 149 (1973).

241 S. H. Ferreira. S. Moncada, and J. R. Vane, *Br. J. Pharmacol.*, **47**, 48 (1973).

242 J. H. Botting. *J. Pharm. Pharmacol.*, **29**, 708 (1977).

243 A. Tothill, L. Rathbone, and E. Willman, *Nature (Lond.)*, **233**, 56 (1971).

244 T. Osa and K. Morita, in Ref. 70, pp. 443–458.

245 P. C. Clegg, in V. R. Pickles and R. J. Fitzpatrick, Eds., *Endogenous Substances Affecting the Myometrium, Mem. Soc. Endrocrinol.* **14**, 89 (1966).

246 E. M. Eagling, H. G. Lovell, and V. R. Pickles, *Br. J. Pharmacol.*, **44**, 510 (1972).

247 T. Magaribuchi, Y. Ito, and H. Kuriyama, *Jap. J. Physiol.*, **21**, 691 (1971).

248 E. Bozler, *Philos. Trans. R. Soc. Lond. Ser. B*, **265**, 3 (1973).

249 H. Kuriyama and T. Tomita, *J. Gen. Physiol.*, **55**, 147 (1970).

250 M. F. Shuba, *J. Physiol. (Lond.)*, **264**, 837 (1977).

251 J. M. Allen and J. B. Bridges, *J. Physiol. (Lond.)*, **263**, 245P (1976).

252 V. A. Bury and M. F. Shuba, in Ref. 151, pp. 65–75.

253 M. F. Shuba, *J. Physiol. (Lond.)*, **264**, 853 (1977).

254 S. Funaki, *Proc. Jap. Acad.*, **34**, 534 (1958).

255 M. E. Holman, *Ergeb. Physiol.*, **61**, 137 (1969).

256 W. M. Steedman, *J. Physiol. (Lond.)*, **186**, 382 (1966).

257 G. D. S. Hirst, *J. Physiol. (Lond.)*, **273**, 263 (1977).

258 D. von Loh and D. F. Bohr, *Proc. Soc. Exp. Biol. Med.*, **144**, 513 (1973).

259 W. R. Keatinge, *Br. Med. Bull.*, **35**, 249 (1979).

260 F. Mekata, *J. Physiol. (Lond.)*, **242**, 143 (1974).

261 F. Mekata, *J. Physiol. (Lond.)*, **258**, 269 (1976).

262 F. Mekata, *J. Physiol. (Lond.)*, **293**, 11 (1979).

263 C. Su, J. A. Bevan, and R. C. Ursillo, *Circ. Res.*, **15**, 20 (1964).

264 R. Casteels, K. Kitamura, H. Kuriyama, and H. Suzuki, *J. Physiol. (Lond.)*, **271**, 41 (1977).

265 R. Casteels, K. Kitamura, H. Kuriyama, and H. Suzuki, *J. Physiol. (Lond.)*, **271**, 63 (1977).

266 F. Mekata, *J. Gen. Physiol.*, **57**, 738 (1971).

267 F. Mekata and H. Niu, *J. Gen. Physiol.*, **59**, 92 (1972).

268 R. N. Speden, *Nature (Lond.)*, **216**, 289 (1967).

269 G. Droogmans, L. Raeymaekers, and R. Casteels, *J. Gen. Physiol.*, **70**, 129 (1977).

270 G. Droogmans and R. Casteels, in Ref. 130, pp. 71–78.

271 W. R. Keatinge, *J. Physiol. (Lond.)*, **174**, 184 (1964).

272 W. R. Keatinge, *Circ. Res.*, **18**, 641 (1966).

273 W. R. Keatinge, *J. Physiol. (Lond.)*, **185**, 701 (1966).

274 W. R. Keatinge, *J. Physiol. (Lond.)*, **194**, 169 (1968).

275 G. Biamino and P. Kruckenberg, *Am J. Physiol.*, **217**, 376 (1969).

276 W. R. Keatinge, *J. Physiol. (Lond.)*, **279**, 275 (1978).

277 K. E. Creed, J. S. Gillespie, and T. C. Muir, *J. Physiol. (Lond.)*, **245**, 33 (1975).

278 D. R. Harder and N. Sperelakis, *Pflügers Arch.*, **378**, 111 (1978).

279 D. R. Harder and N. Sperelakis, *Am. J. Physiol.*, **237**, C75 (1979).

280 C. T. Kirkpatrick, *J. Physiol. (Lond.)*, **244**, 263 (1975).

281 N. L. Stephens, E. A. Kroeger, and U. Kromer, *Am. J. Physiol.*, **228**, 628 (1975).

282 E. A. Kroeger and N. L. Stephens, *Am. J. Physiol.*, **228**, 633 (1975).

283 A. W. Jones, *Circ. Res.*, **33**, 563 (1973).

284 K. E. Creed and J. S. Gillespie, in Ref. 151, pp. 295–301.

285 K. E. Creed, *J. Physiol. (Lond.)*, **245**, 49 (1975).

286 J. H. Szurszewski, *J. Physiol. (Lond.)*, **252**, 335 (1975).

287 J. H. Szurszewski and T. El-Sharkawy, *Fed. Proc.*, **35**, 303 (1976).

288 M. Brown, D. Noble, and S. Noble, in C. J. Dickinson and J. Marks, Eds., *Developments in Cardiovascular Medicine*, MTP Press (Butterworth), London, 1978, pp. 31–51.

289 E. Bülbring, *Br. Med. Bull.*, **35**, 285 (1979).

290 T. Tomita and H. Watanabe, *Philos. Trans. R. Soc. Lond., Ser. B*, **265**, 73 (1973).

291 M. Vonderlage, *Eur. J. Pharmacol.*, **36**, 61 (1976).

292 C. van Breemen, B. R. Farinas, P. Gerba, and E. D. McNaughton, *Circ. Res.*, **30**, 44 (1972).

293 R. Deth and C. van Breemen, *Pflügers Arch.*, **348**, 13 (1974).

294 R. Deth and C. van Breemen, *J. Membr. Biol.*, **30**, 363 (1977).

295 Y. Ito, K. Kitamura, and H. Kuriyama, *J. Physiol. (Lond.)*, **294**, 595 (1979).

296 W. R. Keatinge, *J. Physiol. (Lond.)*, **224**, 21 (1972).

297 W. R. Keatinge, *J. Physiol. (Lond.)*, **224**, 35 (1972).

298 T. Godfraind, *J. Physiol. (Lond.)*, **260**, 21 (1976).

299 J. N. Fain and M. J. Berridge, *Biochem. J.*, **178**, 45 (1979).

300 R. H. Michell, *Biochim. Biophys. Acta*, **415**, 81 (1975).

300a D. M. Salmon and T. W. Honeyman, *Nature (Lond.)*, **284**, 344 (1980).

300b J. W. Putney, S. J. Weiss, C. M. Van De Walle, and R. A. Haddas, *Nature (Lond.)*, **284**, 345 (1980).

301 B. Johansson, O. Jonsson, J. Axelsson, and B. Wahlström, *Circ. Res.*, **21**, 619 (1967).

302 J. Diamond and J. M. Marshall, *J. Pharmacol. Exp. Ther.*, **168**, 13 (1969).

303 E. Bülbring and A. den Hertog, *J. Physiol. (Lond.)*, **268**, 29P (1977).

304 E. Bülbring and A. den Hertog, *J. Physiol. (Lond.)*, **304**, 277 (1980).

305 N. O. Sjöstrand, *Acta Physiol. Scand.*, **89**, 10 (1973).

306 J. Diamond and J. M. Marshall, *J. Pharmacol. Exp. Ther.*, **168**, 21 (1969).

307 E. A. Kroeger and J. M. Marshall, *Am. J. Physiol.*, **225**, 1339 (1973).

308 J. M. Marshall and E. A. Kroeger, *Philos. Trans. R. Soc. Lond. Ser. B*, **265**, 135 (1973).

309 T. Kawarabayashi and T. Osa, *Jap. J. Physiol.*, **26**, 403 (1976).

310 T. Magaribuchi and T. Osa, *Jap. J. Physiol.*, **21**, 627 (1971).

311 E. Bülbring and J. H. Hardman, in Ref. 70, pp. 125–133.

312 T. Osa and T. Kawarabayashi, *Jap. J. Physiol.*, **27**, 111 (1977).

313 J. Marshall, *Fed. Proc.*, **36**, 2450 (1977).

314 A. P. Somlyo and A. V. Somlyo, in J. A. Bevan, R. F. Furchgott, R. O. Maxwell, and A. P. Somlyo, Eds., *Physiology and Pharmacology of Vascular Neuroeffector Systems*, S. Karger, Basel, 1971, pp. 216–228.

315 E. E. Daniel, D. M. Paton, G. S. Taylor, and B. J. Hodgson, *Fed. Proc.*, **29**, 1410 (1970).

316 M. B. Feinstein, M. Paimre, and M. Lee, *Trans. N.Y. Acad. Sci.*, **30**, 1073 (1968).

317 H. O. Schild, *Pharmacol. Rev.*, **18**, 495 (1966).

318 H. O. Schild. *Br. J. Pharmacol. Chemother.*, **31**, 578 (1967).

319 M. B. Feinstein, *J. Pharmacol. Exp. Ther.*, **152**, 516 (1966).

320 T. Magaribuchi and H. Kuriyama, *Jap. J. Physiol.*, **22**, 253 (1972).

321 N. I. A. Overweg and J. D. Schiff, *Eur. J. Pharmacol.*, **47**, 231 (1978).

322 S. C. Verma and J. H. McNeill, *J. Pharmacol. Exp. Ther.*, **198**, 539 (1976).

323 K. D. Meisheri and J. H. McNeill, *Can. J. Physiol. Pharmacol.*, **57**, 1177 (1979).

324   K. D. Meisheri, J. H. McNeill, and J. M. Marshall, *Eur. J. Pharmacol.*, **60**, 1 (1979).

325   M. G. Collis and J. T. Shepherd, *J. Pharmacol. Exp. Ther.*, **209**, 359 (1979).

326   J. C. Castillo and E. J. De Beer, *J. Pharmacol. Exp. Ther.*, **90**, 104 (1947).

327   P. K. S. Siegl, G. V. Rossi, and R. F. Orzechowski, *Eur. J. Pharmacol.*, **54**, 1 (1979).

328   J. Orehek, J. S. Douglas, A. J. Lewis, and A. Bouhuys, *Nature New Biol. (Lond.)*, **245**, 84 (1973).

329   R. F. Coburn, *Fed. Proc.*, **36**, 2692 (1977).

330   G. B. Weiss, *Adv. Gen. Cell. Pharmacol.*, **2**, 71 (1977).

331   C. van Breemen, P. Aaronson, and R. Loutzenhiser, *Pharmacol. Rev.*, **30**, 167 (1978).

332   H. P. Rang and J. M. Ritchie, *J. Physiol. (Lond.)*, **196**, 183 (1968).

333   R. Casteels, G. Droogmans, and H. Hendrickx, *Philos. Trans. R. Soc. Lond. Ser. B*, **265**, 47 (1973).

334   J. H. Widdicombe, *J. Physiol. (Lond.)*, **266**, 235 (1977).

335   A. F. Brading and J. H. Widdicombe, *J. Physiol. (Lond.)*, **238**, 235 (1974).

336   G. Droogmans and R. Casteels in Ref. 151, pp. 11–18.

337   A. P. Somlyo, G. Haeusler, and A. P. Somlyo, *Fed. Proc.*, **29**, 2062 (1970).

338   G. Burnstock, *J. Physiol. (Lond.)*, **143**, 183 (1958).

339   H. Watanabe, *Jap. J. Pharmacol.*, **26**, 217 (1976).

340   C. R. Scheid, T. W. Honeyman, and F. S. Fay, *Nature (Lond.)*, **277**, 32 (1979).

341   T. Honeyman, P. Merriam, and F. S. Fay, *Mol. Pharmacol.*, **14**, 86 (1978).

342   E. Mueller and C. van Breemen, *Nature (Lond.)*, **281**, 682 (1979).

343   D. Bose and I. R. Innes, *Can. J. Physiol. Pharmacol.*, **50**, 378 (1972).

344   J. D. Gardner, D. R. Kiino, N. Jow, and G. D. Aurbach, *J. Biol. Chem.*, **250**, 1164 (1975).

345   M. P. Blaustein and A. L. Hodgkin, *J. Physiol. (Lond.)*, **200**, 497 (1969).

346   M. P. Blaustein, *Rev. Physiol. Biochem. Pharmacol.*, **70**, 33 (1974).

347   P. F. Baker, *Ann. N.Y. Acad. Sci.*, **307**, 250 (1978).

348   L. J. Mullins, *J. Gen. Physiol.*, **70**, 681 (1977).

349   T. Katase and T. Tomita, *J. Physiol. (Lond.)*, **224**, 489 (1972).

350   H. Reuter, M. P. Blaustein, and G. Haeusler, *Philos. Trans. R. Soc. Lond. Ser. B*, **265**, 87 (1973).

351   M. P. Blaustein, *Am. J. Physiol.*, **232**, C165–C173 (1977).

352   R. Casteels and C. van Breemen, *Pflügers Arch.*, **359**, 197 (1975).

353   P. F. Baker and P. A. McNaughton, *J. Physiol. (Lond.)*, **276**, 127 (1978).

354   T. S. Ma and D. Bose, *Am. J. Physiol.*, **232**, C59 (1977).

355   P. F. Baker and W. W. Schlaepfer, *J. Physiol. (Lond.)*, **276**, 103 (1978).

356   A. L. Lehninger, B. Reynafarje, A. Vercesi, and W. P. Tew, *Ann. N.Y. Acad. Sci.*, **307**, 160 (1978).

357   A. V. Somlyo and A. P. Somlyo, in Ref. 70, pp. 381–397.

358   R. G. G. Andersson and K. B. Nilsson, in N. L. Stephens, Ed., *The Biochemistry of Smooth Muscle*, International Medical Publishers, Baltimore, Md., 1977, pp. 263–291.

359   H. J. Schatzmann and H. Bürgin, *Ann. N.Y. Acad. Sci.*, **307**, 125 (1978).

360   F. F. Vincenzi, *Ann. N.Y. Acad. Sci.*, **307**, 229 (1978).

361   L. R. Jones, H. R. Besch, and A. M. Watanabe, *J. Biol. Chem.*, **253**, 1643 (1978).

362   F. Wuytack, *J. Physiol. (Lond.)*, **295**, 23P (1979).

363   L. Hurwitz, D. F. Fitzpatrick, G. Debbas, and E. J. Landon, *Science*, **179**, 384 (1973).

364 K. B. Nilsson, R. G. G. Andersson, S. Börjesson, L. Mackerlova, E. Mohme-Lundholm, and L. Lundholm, in Ref. 130, pp. 189–197.

365 M. A. Matlib, J. Crankshaw, R. E. Garfield, D. J. Crankshaw, C. Y. Kwan, L. A. Branda, and E. E. Daniel, *J. Biol. Chem.*, **254**, 1834 (1979).

366 H. Sugi and T. Daimon, *Nature (Lond.)*, **269**, 436 (1977).

367 N. Shibata, H. Ohashi, T. Takewaki, and T. Okada, *Jap. J. Pharmacol.*, **28**, 561 (1978).

368 R. Casteels and L. Raeymaekers, *J. Physiol. (Lond.)*, **294**, 51 (1979).

369 S. C. Batra and E. E. Daniel, *Comp. Biochem. Pharmacol.*, **38A**, 285 (1971).

370 R. Andersson and K. Nilsson, *Nature New Biol. (Lond.)*, **238**, 119 (1972).

371 R. Andersson, L. Lundholm, E. Mohme-Lundholm, and K. Nilsson, *Adv. Cyclic Nucleotide Res.*, **1**, 439 (1972).

372 M. Baudouin-Legros and P. Meyer, *Br. J. Pharmacol.*, **47**, 377 (1973).

373 E. A. Kroeger, J. M. Marshall, and C. P. Bianchi, *J. Pharmacol. Exp. Ther.*, **193**, 309 (1975).

374 B. Sarkadi, I. Szász, A. Gerloczy, and G. Gardos, *Biochim. Biophys. Acta*, **464**, 93 (1977).

375 R. G. G. Andersson, *Acta Physiol. Scand.*, **87**, Suppl. 382, 1 (1972).

376 H. P. Bär, *Adv. Cyclic Nucleotide Res.*, **4**, 195 (1974).

377 L. Lundholm, R. G. G. Andersson, H. J. Arnquist, and E. Mohme-Lundholm, in Ref. 358, pp. 159–208.

378 Y. Vulliemoz, M. Verosky, and L. Triner, in Ref. 358, pp. 293–314.

379 J. G. Hardman, J. N. Wells, and P. Hamet, in Ref. 358, pp. 329–342.

380 J. Diamond, in Ref. 358, pp. 343–360.

381 M. E. Carsten, in Ref. 358, pp. 617–639.

382 E. A. Kroeger, T. S. Teo, H. Ho, and J. H. Wang, in Ref. 358, pp. 641–652.

383 G. A. Robison, R. W. Butcher, and E. W. Sutherland, *Ann. N.Y. Acad. Sci.*, **139**, 703 (1967).

384 E. M. Ross, A. C. Howlett, K. M. Ferguson, and A. G. Gilman, *J. Biol. Chem.*, **253**, 6401 (1978).

385 M. L. Steer and A. Levitzki, *Arch. Biochem. Biophys.*, **167**, 371 (1975).

386 D. F. Fitzpatrick and A. Szentivanyi, *Naunyn-Schmiedebergs Arch. Pharmacol.*, **298**, 255 (1977).

387 G. A. Collins and M. C. Sutter, *Can. J. Physiol. Pharmacol.*, **53**, 989 (1975).

388 H. Ohkubo, I. Takayanagi, and K. Takagi, *Jap. J. Pharmacol.*, **26**, 65 (1976).

389 L. Triner, Y. Vulliemoz, and M. Verosky, *Eur. J. Pharmacol.*, **41**, 37 (1977).

390 S. Katsuki and F. Murad, *Mol. Pharmacol.*, **13**, 330 (1977).

391 R. G. G. Andersson, G. Kovesi, and E. Ericsson, *Acta Pharmacol. Toxicol.*, **43**, 323 (1978).

392 M. A. Kumar, *J. Pharmacol. Exp. Ther.*, **206**, 528 (1978).

393 D. E. Niewoehner, H. Campe, S. Duane, T. McGowan, and M. R. Montgomery, *J. Appl. Physiol.*, **47**, 330 (1979).

394 R. Andersson, *Acta Physiol. Scand.*, **85**, 312 (1972).

395 H. Yamumura, M. Rodbell, and J. N. Fain, *Mol. Pharmacol.*, **12**, 693 (1976).

396 J. E. Birnbaum, P. W. Abel, G. L. Amidon, and C. K. Buckner, *J. Pharmacol. Exp. Ther.*, **194**, 396 (1975).

397 H. S. Earp and A. L. Steiner, *Annu. Rev. Pharmacol. Toxicol.*, **18**, 431 (1978).

398 T. Posternak, E. W. Sutherland, and W. F. Henion, *Biochim. Biophys. Acta*, **65**, 558 (1962).

399 C. Londos and J. Wolff, *Proc. Natl. Acad. Sci. U.S.A.*, **74**, 5482 (1977).

400 A. P. Somlyo, A. V. Somlyo, and V. Smiesko, *Adv. Cyclic Nucleotide Res.*, **1**, 175 (1972).

401 K. Takagi, I. Takayanagi, and A. Tomiyama, *Jap. J. Pharmacol.*, **21**, 477 (1971).

402 R. W. Alexander, J. A. Fontana, and W. Lovenberg, *Fed. Proc.*, **32**, 711 (1973).

403   A. Ziegelhoffer, M. B. Anand-Srivastava, R. L. Khandelwal, and N. S. Dhalla, *Biochem. Biophys. Res. Commun.*, **89**, 1073 (1979).

404   A. M. Katz, *Circ. Res.*, **44**, 384 (1979).

405   S. Winegrad and G. B. McClellan, *Ann. N.Y. Acad. Sci.*, **307**, 477 (1979).

406   D. A. Walsh, P. Perkins, and E. G. Krebs, *J. Biol. Chem.*, **243**, 3763 (1968).

407   J. F. Kuo and P. Greengard, *Proc. Natl. Acad. Sci. U.S.A.*, **64**, 1349 (1969).

408   D. A. Walsh, *Biochem. Pharmacol.*, **27**, 1801 (1978).

409   P. Greengard, *Fed. Proc.*, **38**, 2208 (1979).

410   E. G. Krebs and J. A. Beavo, *Annu. Rev. Biochem.*, **48**, 923 (1979).

411   G. A. Robison, R. W. Butcher, and E. W. Sutherland, *Cyclic AMP*, Academic Press, New York, 1971.

412   F. G. Krebs, *Curr. Top. Cell. Regul.*, **5**, 99 (1972).

413   M. M. Hosey and M. Tao, *Curr. Top. Membr. Transp.*, **9**, 233 (1977).

414   I. Polacek and E. E. Daniel, *Can. J. Physiol. Pharmacol.*, **49**, 988 (1971).

415   I. Polacek, J. Bolan, and E. E. Daniel, *Can. J. Physiol. Pharmacol.*, **49**, 999 (1971).

416   E. E. Daniel and J. Crankshaw, *Blood Vessels*, **11**, 295, (1974).

417   F. Honda, S. Katsuki, J. T. Miyahara, and S. Shibata, *Br. J. Pharmacol.*, **60**, 529 (1977).

418   S. Harbon and H. Clauser, *Biochem. Biophys. Res. Commun.*, **44**, 1496 (1971).

419   M. F. Vesin and S. Harbon, *Mol. Pharmacol.*, **10**, 457 (1974).

420   S. Harbon, M. F. Vesin, and L. Do Khac (1975), in Ref. 151, pp. 83–100.

421   M. F. Vesin, L. Do Khac, and S. Harbon, *Mol. Pharmacol.*, **14**, 24 (1978).

422   J. Diamond and T. G. Holmes, *Can. J. Physiol. Pharmacol.*, **53**, 1099 (1975).

423   K. D. Meisheri, T. E. Tenner, and J. H. McNeill, *Eur. J. Pharmacol.*, **53**, 9 (1978).

424   N. I. A. Overweg and J. D. Schiff, *Eur. J. Pharmacol.*, **47**, 231 (1978).

425   T. P. Lee, J. F. Kuo, and P. Greengard, *Proc. Natl. Acad. Sci. U.S.A.*, **69**, 3287 (1972).

426   N. D. Goldberg, M. K. Haddox, S. E. Nicol, D. B. Glass, C. H. Sanford, F. A. Kuehl, and R. Estensen, *Adv. Cyclic Nucleotide Res.*, **5**, 307 (1975).

427   G. Schultz, J. G. Hardman, K. Schultz, C. E. Baird, and E. W. Sutherland, *Proc. Natl. Acad. Sci. U.S.A.*, **70**, 3889 (1973).

428   G. Schultz and J. G. Hardman, *Adv. Cyclic Nucleotide Res.*, **5**, 339 (1975).

429   G. Schultz, *Naunyn-Schmiedebergs Arch. Pharmacol.*, **297**, R11 (1977).

430   R. A. Janis and J. Diamond, *J. Pharmacol. Exp. Ther.*, **211**, 480 (1979).

431   F. Hirata, W. J. Strittmatter, and J. Axelrod, *Proc. Natl. Acad. Sci. U.S.A.*, **76**, 368 (1979).

432   J. H. Fleisch, in O. Carrier and S. Shibata, Eds., *Factors Influencing Vascular Reactivity*, Igaku-Shoin, Tokyo, 1977, pp. 78–95.

433   J. A. Bevan, *Circ. Res.*, **45**, 161 (1979).

434   M. E. Holman and A. M. Surprenant, *J. Physiol. (Lond.)*, **287**, 337 (1979).

434a   G. D. S. Hirst and T. O. Neild, *Nature (Lond.)*, **383**, 767 (1980).

435   E. Carlsson and A. Hedberg, *Acta Physiol. Scand., Suppl.*, **440**, 47 (1976).

436   E. J. Ariëns and A. M. Simonis, in Ref. 114, pp. 3–27.

437   J. H. Exton, *Am. J. Physiol.*, **238**, E3 (1980).

CHAPTER TWO

ADRENOCEPTORS AND ADRENERGIC MECHANISMS IN THE EMBRYONIC AND FETAL HEART

Achilles J. Pappano

Department of Pharmacology, The University of Connecticut Health Center, Farmington, Connecticut

Achilles J. Pappano is recipient of Research Career Development Award (HL-00027).

1 INTRODUCTION

The cardiovascular system is the first organ system to develop in embryonic vertebrates and its function is important for the growth and development of these
organisms. Intrinsic and extrinsic mechanisms regulate the operation of the cardiac
pump and assist the cardiovascular system in its function of maintaining homeo-
stasis. Intrinsic mechanisms include changes in impulse rate and end-diastolic fiber
length (Frank–Starling mechanism); these mechanisms operate quite early in the
development of the heart. For example, the Frank–Starling mechanism permits
heterometric regulation of cardiac performance in the embryonic chick as early as
the 3rd incubation day *in vivo* (1) and the 4th incubation day *in vitro* (2). Extrinsic
mechanisms include changes in heart rate and contractility produced by the parasympathetic (vagus) and sympathoadrenal divisions of the autonomic nervous system.

Sympathetic nerves to the heart consist primarily of adrenergic fibers that release
excitatory or stimulatory neurotransmitters (norepinephrine and/or epinephrine or
dopamine). The effects of these neurotransmitters on the heart muscle are due to
an action on transmitter receptors. This chapter considers the development and
operation of adrenergic receptors and adrenergic mechanisms on cardiac muscle
cells of embryonic, fetal, and neonatal vertebrates. The operation of the adrenergic
receptor and its cellular transduction mechanism is essential to allow sympathetic
adrenergic nerves to regulate cardiac performance. The adrenergic receptor and its
mechanism are present in some heart muscle cells before sympathetic adrenergic
innervation has occurred (reviewed in Ref. 3). Emphasis will be placed on the
mechanism by which activation of adrenergic receptor by neurotransmitter or hormone is transformed into a cellular signal that permits extrinsic regulation of cardiac
performance.

2 ADRENERGIC RECEPTORS IN EMBRYONIC AND FETAL CARDIAC MUSCLE

2.1 Detection by Physiological and Pharmacological Methods

The chick, which hatches 21 days after fertilization of the egg, displays a positive
chronotropic effect to epinephrine on the 2nd day after fertilization (4). Markowitz
concluded that the presence of a receptor permitted the catecholamines to accelerate
the heart very soon after spontaneous contractions had begun. This observation has
been confirmed and extended by many (5–7). Shideman's laboratory identified the
receptor in the avian heart responsible for the positive chronotropic and positive
inotropic effect of catecholamines as β-adrenergic (2). This conclusion has been
confirmed by others, who have used different adrenergic agonists and antagonists
in their experiments (8–11). Whereas it has been disputed that a positive inotropic
effect of catecholamines in the chick ventricle is demonstrable on the fourth incubation day (12), both laboratories agree that the β-adrenergic receptor-mediated
increase in adenosine 3′,5′-cyclic monophosphate (cyclic AMP) is present at this
time (see Section 3.2).

The mouse and the rat, which are born at 19½ and 22 days, respectively, after fertilization (13), display responses to catecholamines at a later time than the chick. For example, the rat heart is accelerated by epinephrine (14) and by isoproterenol (15) on day 10 to day 10½ of gestation, the time when spontaneous contractions develop. In this regard, the rat and chick are very similar because the onset of spontaneous contractions and of catecholamine-induced acceleration occur almost simultaneously in the chick at about 33–37 hr after fertilization, respectively (see Ref. 5). Strictly speaking, these adrenergic receptors for the positive chronotropic effect are probably present in the sinoatrial pacemaker; it is not known when the β-adrenergic receptor and its cellular transduction mechanisms are incorporated into the embryonic mammalian ventricle. Another similarity between these animals is the presence of the adrenergic receptor and its mechanism in heart cells before the heart has received sympathetic adrenergic innervation.

Myocardial contractions are first detected in the mouse at the 8th gestational day (13). Whereas norepinephrine and isoproterenol sometimes accelerated the sinoatrial pacemaker of the fetal mouse heart as early as 13–14 days of gestation, the positive chronotropic effect was not consistently observed until 15–16 days of gestation (16). It was concluded that the β-adrenergic receptor and its mechanism had differentiated before morphological innervation of the mouse heart occurred. However, the response to β-adrenergic agonists seemed poorly developed until the 15th–16th gestational day, when morphologic innervation occurs (16). The development of the β-adrenergic mechanism appears slower in the mouse heart than in the rat and chick heart. However, this view is not shared by others, who reported that adrenergic receptors were present on mouse heart cells *in vitro* obtained from hearts on the 9th gestational day (17). The reason for the divergent results is not known.

The earliest detection of an adrenergic response in the human fetal heart was reported by Gennser and Nilsson (18). Epinephrine had a positive inotropic effect on ventricular muscle from a 9-week fetus and a positive chronotropic effect on the sinoatrial pacemaker from a 10-week fetus. It is not known when the adrenergic mechanism is incorporated into human fetal hearts; the first contractions occur by the 3rd week of life *in utero* (13). The results obtained by Gennser and Nilsson extend those given earlier (19). The β-adrenergic nature of the adrenoceptor responsible for the positive inotropic effects in human hearts is supported by the observations that isoproterenol was more potent than epinephrine, which in turn was more potent than norepinephrine as a stimulant (19,20). Further, it has been reported that human fetal heart cells *in vitro* (53rd–71st day of gestational age) have adrenergic receptors sensitive to blockade by propranolol (21). However, epinephrine was more potent than isoproterenol, in contrast to the expected order of potency by β-adrenergic agonists. It should be noted that phentolamine had no effect by itself or on the positive chronotropic effects of the catecholamines (21). This observation is of interest with regard to the detection of α-adrenergic receptors in embryonic heart muscle (see below).

It is not clear from studies of other mammalian embryos when the β-adrenergic mechanism is incorporated into cardiac muscle. The fetal lamb heart is accelerated by sympathetic stimulation (stellate ganglion) as early as the 70th gestational day, and this requires the presence of adrenergic receptors (22). The β-adrenergic nature of these receptors was convincingly demonstrated by the elegant experiments of van Petten and Willes (23), who observed a surmountable blockade by propranolol of the pacemaker acceleration caused by isoproterenol in both ewe and fetus 100–120 gestational days). Neonatal dog (24,25) and rabbit (26) hearts are sensitive to β-adrenergic agonists, but little is known about the reactivity of hearts from dog and rabbit fetuses to catecholamines. There are several reasons to determine the earliest appearance of the β-adrenergic mechanisms in embryonic hearts. First, it is essential to know when extrinsic regulation of cardiac function by the β-adrenergic receptor is possible. Second, the study of the sensitivity of the heart to adrenergic agonists requires as complete a study as possible of the ontogenetic variations in drug action. Third, a study of ontogenetic appearance of β-adrenergic receptors by biochemical methods and the cellular transduction mechanism by pharmacological methods requires a systematic study from the first appearance of each component.

Attention has been directed toward the β-adrenergic receptor thus far. It is clear that the embryonic and neonatal heart has α-adrenergic receptors as well as β-adrenergic receptors. Cardiac Purkinje fibers from neonatal dogs (0–7 days after birth) displayed β-adrenergic effects insofar as automaticity was increased by isoproterenol, epinephrine, and phenylephrine (descending order of potency), and this stimulatory effect was antagonized by propranolol (27). The fibers also exhibited α-adrenergic effects insofar as automaticity was decreased by phenylephrine, epinephrine, and isoproterenol (descending order of potency). Phentolamine antagonized the inhibitory effects of these compounds. The contrasting effect of these compounds on automaticity is illustrated in the biphasic effect on frequency observed when the concentration was raised. The presence of α-adrenergic receptors has also been reported for mouse cardiac muscle cells *in vitro* (17). The positive chronotropic effect of isoproterenol, a relatively specific β-adrenergic agonist, was prevented by propranolol, whereas the positive chronotropic effect of phenylephrine, a relatively specific α-adrenergic agonist, was prevented by phentolamine. The positive chronotropic effects of epinephrine and norepinephrine, which can activate α- and β-adrenergic receptors, were blocked by a combination of propranolol and phentolamine (17). Others have concluded that the adrenergic mechanism in the embryonic heart includes only the β-adrenergic receptor (28). These seemingly divergent results cannot be compared until experiments are done under the same conditions. The presence of α-adrenergic receptors on embryonic heart cells from the mouse but not from the chicken need not be surprising. However, it would be helpful to study this matter of adrenergic receptor development and differentiation in a systematic fashion *in vivo* or *in vitro*. The temperature (29,30) and hormonal (31) status of the tissue are important variables to be considered. The failure to control the hormone levels of synthetic media is a disadvantage for the use of myocardial cell cultures.

2.2 Detection by Radioactive Ligand Binding

The study of hormone and neurotransmitter action has been enlarged by the introduction of radioactively labeled compounds (ligands) that react with receptors in target cells. The theoretical and experimental basis for the study of adrenergic receptors by radioactive ligand binding has been reviewed (31).

In the first experiments with embryonic and fetal hearts, the binding of radio-actively labeled agonists was used to identify the β-receptor. The binding of [³H]norepinephrine to a subcellular fraction was studied in fetal mouse hearts on the 13th gestational day (32). Experiments were done in a similar fashion with intact chick heart cells (12th–13th incubation day) maintained in monolayer culture conditions for 3 days (33,34). In both tissues, the binding of [³H]norepinephrine could be displaced by norepinephrine and by propranolol. Propranolol also inhibited the activation of adenylyl cyclase by norepinephrine in the subcellular fraction used for binding studies from the rat heart (32). It was concluded that the β-adrenergic receptor (and adenylate cyclase) were present in fetal rat heart cells before a physiologic effect of catecholamines could be demonstrated. This view conflicts with the observations of others (14,15), who detected an acceleratory action of catecholamines on the fetal rat heart as early as the 10th gestational day.

In these early experiments, it was not possible to conclude that the sites that bound [³H]norepinephrine were β-adrenergic receptors for several reasons. For example, the binding of radioactively labeled ligands did not exhibit stereospecificity, whereas the physiologic effects of β-agonists displayed such specificity. In addition, the concentration of propranolol needed to inhibit [³H]norepinephrine binding by 40% (10^{-4} M) is several orders of magnitude greater than that needed to antagonize the physiologic effects of β-agonists (34). Additional evidence against the reliable identification of β-adrenergic receptors by radioactively labeled agonist has been presented by Walker. Neonatal (0–5 days) rat heart cells in culture accumulated [³H]epinephrine (35), but most of the accumulated [³H]epinephrine was not associated with β-adrenergic receptors on cardiac muscle cells. Further, the accumulation could not be related to α-adrenergic receptors or to neuronal uptake (uptake I). It is of interest that the apparent dissociation constant (K_D) for dl-propranolol was about 1.3×10^{-8} M as determined by pharmacological tests (Schild plot) of antagonist interaction with the β-adrenergic receptor of the heart cells maintained in culture (35). By contrast, inhibition of [³H]epinephrine accumulation by propranolol was negligible (0%) at 10^{-4} M and complete (100%) at 10^{-3} M. At 10^{-3} M, propranolol also proved to be toxic to the cells.

The limitations of using radioactively labeled β-adrenergic agonists have been enumerated (see Ref. 31 for a review). More recently, radiolabeled β-adrenergic antagonists have been applied to identify the β-adrenergic receptor. There are few studies of β-adrenergic receptors in embryonic and neonatal hearts with radiolabeled β-adrenergic antagonists. [³H]Dihydroalprenolol was used in a study of β-adrenergic receptor development in the embryonic chick heart (36). As expected for a ligand that reacted with the β-adrenergic receptor, binding was not only saturable but also stereospecific. The specific activity determined with this ligand was 0.36 ± 0.04 pmol of β-adrenergic receptor per milligram of protein on days 4½–5½ and days 6½–7½. This activity decreased to 0.22 ± 0.02 pmol of β-adrenergic receptor per milligram of protein on day 9 and to 0.15 ± 0.01 and 0.16 ± 0.01 pmol on days 12–13 and 16–17, respectively. It was concluded that the density of β-adrenergic receptors decreased with age and specifically during development of cardiac adrenergic innervation. This interpretation is subject to at least two limitations. First, the amount of protein present in the muscle increases with age (37) and it is possible that the decrease in density results, at least in part, from an increase of protein content. Protein increases by 31% between the 6th–7th and 16th–17th incubation days in the chick heart. In addition, the measurements could be complicated by the development of β-adrenergic receptors on vascular smooth muscle cells. The coronary vascular bed develops rapidly between the 8th and 14th incubation days (38). These limitations would be eliminated if there were some stable marker of the cardiac muscle cell that could be used as a reference. Second, the implication that innervation changed the density of β-adrenergic receptors depends upon the time of innervation. This matter is discussed in detail in Section 4.1.

The β-adrenergic receptor has also been studied in the porcine heart from birth to 70 days later (39). The K_D for [^{3}H]alprenolol binding increased between birth (4.1 $\pm$ 0.09 nM) and 2 weeks (10.8 $\pm$ 2.1 nM); thereafter, it declined to an intermediate value at 40 (7.2 $\pm$ 0.7 nM) and 70 (7.7 $\pm$ 0.4 nM) days. Interestingly, the number of receptors (per gram of heart tissue) increased from birth (149 $\pm$ 19 fmol/g) to 1 week (500 $\pm$ 57 fmol/g). Afterward, the receptor number did not change significantly. The reversible increase of K_D is somewhat surprising, and this should be related to an increase of the apparent K_D for antagonist determined by pharmacological experiments (31). The significance of this observation remains to be explored. [^{3}H]Alprenolol has also been used to identify β-adrenergic receptors in heart cells obtained from embryonic mice (17). Whereas the binding of this ligand was competitively blocked by propranolol, it was concluded that the binding of [^{3}H]alprenolol was not sufficiently specific (30–40% nonspecific binding at saturation) to infer labeling only of β-adrenergic receptors. Others have reported that [^{3}H]dihydroalprenolol labeled β-adrenergic receptors (specific binding was 50% of total binding) in hearts from fetal, neonatal, and adult mice (39a). The apparent K_D for dihydroalprenolol binding was 0.33 $\pm$ 0.03 nM, and this did not change significantly during development. However, the maximum number of β-adrenergic binding sites increased significantly (see Section 4.1). Determination of apparent K_D from pharmacological experiments seems essential for evaluation of β-adrenergic receptor properties. In addition to the K_D for propranolol given previously, it has been reported that the β-antagonist 1-(6′-chloro-3′-methylphenoxy)-tert-(butylaminopropane-2-ol) (-)-KL255 had a K_D of 3.6×10^{-10} M in cultured heart cells from neonatal rats and cats (40). This value is consistent with the high affinity of antagonists for the β-adrenergic receptor.

3 MECHANISM OF THE β-ADRENERGIC RECEPTOR-INITIATED REACTION

3.1 Location of β-Adrenergic Receptors

It is generally assumed that the β-adrenergic receptors on heart cells are at the external aspect of the cell membrane. Evidence to support this position has been given by Venter et al. (41), who conducted experiments on embryonic chick hearts intact or in cell culture. Epinephrine or isoproterenol, covalently bound to glass beads or chips, accelerated the rate of beating of such preparations. Acceleration was observed when the glass beads with covalently bound catecholamine were applied to the cells, but it was not observed if the glass beads alone were applied. Placing glass beads with bound catecholamine in the bath but not in contact with the heart cells did not accelerate the rate of spontaneous contractions. These results are consistent with the external location of the β-adrenergic receptor on the plasma membrane. This conclusion has also been drawn by others (34), who used norepinephrine bound to agarose beads as a means of restricting the access of the catecholamine to the external surface of the plasma membrane in embryonic chick heart cells in culture. Norepinephrine bound to agarose prevented the binding of [³H]norepinephrine to the cells. In view of the difficulty encountered in identifying β-adrenergic receptors when radioactively labeled agonists are used (see Section 2.2), this experimental result is problematic. The experiments with catecholamines covalently bound to glass have also been criticized for the possibility that the drugs leaked from the glass surface; however, a test of this possibility gave negative results (42). Results obtained with Purkinje fibers from adult hearts indicate that the β-receptor is on the external surface of the plasma membrane. Iontophoretic application of isoproterenol changed membrane electrical properties when applied just outside the cell but not when applied intracellularly (43).

3.2 Adenylate Cyclase-Cyclic AMP

The participation of adenylate cyclase-cyclic AMP in the transduction of β-adrenergic receptor activity has received considerable attention in the embryonic hearts of the chick and of mammals. In this section, experiments that deal with stimulation of adenylate cyclase by β-adrenergic agonists will be considered.

Catecholamines caused cyclic AMP to accumulate in embryonic chick hearts as early as the 4th incubation day (44,45). The accumulation of cyclic AMP was directly related to the concentrations of norepinephrine and isoproterenol (44) and the elevation of cyclic AMP was maximal at 30 (44) to 180 sec (46) after addition of catecholamines. Propranolol prevented the accumulation of cyclic AMP by catecholamines, presumably by antagonizing β-receptor-mediated activation of adenylate cyclase (44,46). Analysis of the effect of catecholamines on cyclic AMP during development also showed that the cyclic nucleotide accumulated, but to a lesser extent, in hearts from older embryos (44–46). Results obtained in our laboratory showed that catecholamines increased cyclic AMP level in hearts from hatched chicks as well as those obtained late in development (Table 1).

Whereas the stimulatory effect of isoproterenol is in accordance with the role of adenylate cyclase at all stages examined, it is appropriate to consider apparent changes of the basal values of cyclic AMP in the heart at different stages. Basal levels of cyclic AMP were highest in hearts isolated from 4-day chick embryos and ranged from about 34 pmol/mg protein (44) to about 115 (46). The content diminished during development to 12 pmol/mg protein on the 16th incubation day (45). Measurements at the 18th incubation day and 1 week after hatching yielded results essentially the same as on the 16th incubation day (46). Qualitatively similar results

TABLE 1 Effect of Catecholamines on the Cyclic AMP Level in Chick Hearts

	18-Day Embryonic		7-Day Hatched	
	Cyclic AMP[a]	Twitch Tension[b]	Cyclic AMP	Twitch Tension
Control	0.43 A 0.06 (14)	100	0.64 A 0.05 (7)	100
Isoproterenol[c] (3C)	0.71 A 0.17 (7)	255	1.43 A 0.15 (5)	276 (2)
Isoproterenol (6C)	0.71 A 0.08 (5)	154	1.03 A 0.07 (4)	274 (2)

[a]Cyclic AMP expressed as picomol/mg wet weight.
[b]Twitch tension given as percent of control (z 100%).
[c]Concentration of 10^{w6} M.

have been reported by Hollman and Green (47), although data from Shideman's laboratory (44) showed no change between the 4th and 7th incubation days when others reported a very marked decrease of cyclic AMP. Consideration of these results is of importance in regard to the possibility that cyclic AMP may function chronically during development to regulate metabolically the availability of slow inward current channels for Ca^{2+} entry during cardiac excitation (see Section 3.3). It is not known how much of the change of cyclic AMP during ontogenesis is due to an alteration in cellular content of the cyclic nucleotide as compared to a change in cellular content of protein. The specific content of protein in the embryonic chick heart increases by 30% between the 6th and 18th incubation days (37). In a recent report (45), basal levels of cyclic AMP fell from 33.6 ± 2.2 to 11.7 ± 1.5 pmol/mg protein between the 4th and 16th incubation days. Therefore, the effect of a change in protein content on the specific activity of cyclic AMP need not be insignificant. Indeed, it has also been mentioned that the decline of cyclic AMP levels might result from a "dilution" of cardiac muscle cells by nonmuscle cells in older hearts (46).

Examination of basal cyclic AMP levels in our laboratory (48) showed that cyclic AMP increased significantly ($p < 0.05$) between the 18th incubation day and 1 week after hatching (Table 1). If the cyclic AMP content is expressed per milligram of protein, the values are 4.2 ± 0.41 and 4.8 ± 0.15 pmol/mg protein on the 18th incubation day and 1 week after hatching, respectively. There is no statistically significant difference when cyclic AMP is expressed as a function of protein, and this could simply be the result of cellular hypertrophy after hatching. The amount of protein per ventricle increases twofold over this time span, whereas the DNA content per ventricle (and therefore the number of cells per milligram of wet weight) changes very little (49). In our experiments, the ratio of protein to wet weight increased from 0.10 ± 0.003 on the 18th incubation day to 0.14 ± 0.003 at 1 week after hatching ($p < 0.01$). We concluded that the cyclic AMP content per cell is increased during this time (48). This increment has been implicated in the mechanism by which cyclic AMP influences autonomic neurotransmitter action (see Section 3.3).

Novak et al. (50) have studied the levels of cyclic AMP in the rat heart from a few days before birth until 50 days after birth. Cyclic AMP content (pmoles per milligram wet weight) rose from about 0.6 just before birth to about 1.4 at 10 days after birth. Thereafter, the cyclic AMP content declined slightly to about 1.2 pmol/ mg wet weight. These investigators also determined the protein content of the extracts (of 20,000g supernatant fraction) used at each age; the results showed essentially no change of protein to wet weight after birth. Protein kinase activity was greatest in hearts before birth and it fell afterward. On the other hand, cardiac muscle phosphorylase and phosphorylase kinase activities were low before birth and increased considerably by 30 days after birth. The nonactive forms of these two enzymes were involved in the postnatal increments. Norepinephrine stimulated adenylate cyclase in left ventricular muscle homogenates obtained from newborn (3 days after birth) rats (51). A concentration of 5×10^{-5} M norepinephrine was required in the newborn rat heart, whereas one of 5×10^{-6} M was sufficient to stimulate adenylate cyclase significantly in the adult rat. This stimulant effect of norepinephrine on adenylate cyclase in rat heart was confirmed in a more extensive series of experiments done by Kohrman (52). Basal enzyme activity was greatest on the day after birth (43 $\pm$ 10 pmol of cyclic AMP per milligram of protein per minute) and least at 7 (17 $\pm$ 2) and 14 (16 $\pm$ 1) days after birth. Because only the synthesis of cyclic AMP was measured, it was not possible to evaluate the mechanism for the maximal levels of cyclic AMP at 10 days after birth reported by others (50). The ability of isoproterenol to increase cyclic AMP in fetal rat heart cells maintained *in vitro* seems dependent upon the external concentrations ($[Ca^{2+}]_o$) of Ca^{2+} (53). At $[Ca^{2+}]_o$ less than 1.84 mM, isoproterenol had no significant effect on cyclic AMP. The increase of cyclic AMP by isoproterenol at 1.84 mM $[Ca^{2+}]_o$ was converted to a decrease when $[Ca^{2+}]_o$ was raised to 2.76 mM. At 2.76 mM $[Ca^{2+}]_o$, the basal cyclic AMP content was higher than at any lower concentration. Dissociation between the effects of isoproterenol on cyclic AMP and on heart rate is considered in Section 3.3.

In the pig heart, adenylate cyclase activity changed from 0.2 to 0.4 nmol cyclic AMP/10 min/g tissue between birth and 75 days after birth; there was no pattern to the fluctuations in enzyme activity during this time (54). At 150 days after birth, basal enzyme activity had diminished to less than 0.1. By contrast, cyclic AMP levels were rather constant from day 2 to day 150 after birth (1.4 to 1.5 nmol of cyclic AMP per gram wet weight). Phosphodiesterase activity in porcine heart muscle varied from about 1.5 to about 1.7 μmol cyclic AMP hydrolyzed/10 min/ g tissue between birth and 75 days later. The activity of this enzyme fell to about 1.0 at 150 day after birth. The physiological implications of the reduced activity of adenylate cyclase and of phosphodiesterase in the face of constant cyclic AMP levels are unknown (54).

Adenylate cyclase activity was also measured in homogenates of canine left ventricle at selected times (2 days; 1, 4, and 14 weeks; adult) after birth (55). Basal levels of this enzyme rose from 259.8 $\pm$ 9.4 pmol of cyclic AMP/mg protein/5 min in dogs 2 days after birth to 430.7 $\pm$ 13.5 in the 4-week-old group. Enzyme activity remained at this level through adulthood. Stimulation of adenylate cyclase by NaF (10 mM) and by epinephrine (100 μM) was maximal in the 4-week-old group. Stimulation of enzyme activity by epinephrine was antagonized by the β-adrenergic blocking agent, propranolol, but not by the α-adrenergic blocking drug, phentolamine. Whereas these results are not unexpected, it is of interest that the K_i (concentration for half-maximal inhibition) of propranolol was higher in the 4- and 14-week-old animals than at other ages. The significance of this observation has not been reported.

Most of the studies of adenylate cyclase-cyclic AMP in mammalian hearts have begun around birth and carried through adulthood. There are some instances in which observations have been made on embryonic and fetal hearts. In the human heart, adenylate cyclase activity was detected in tissues from embryos of 5 weeks gestational age (56). The enzyme could be activated by NaF but not by hormones at this stage. Catecholamine-induced activation of the enzyme was detected in hearts from the 6th gestational week; histamine also activated the enzyme at this time. A β-receptor was implicated because propranolol, but not phentolamine, prevented catecholamine-induced activation of the enzyme. By contrast, glucagon-induced activation of the enzyme appeared by the 8th–9th gestational weeks. It was concluded that additional experiments are required before the temporal sequence for development of hormone responsiveness (and hormone receptors) can be determined. These investigators also mentioned some of the difficulties encountered in measuring enzyme activity in a more purified preparation of fetal heart cells (57). In an earlier report, it had been noted that adenylate cyclase activity increased between the 12th and 22nd weeks of gestation in the human fetal heart (58). This trend was not observed in the experiments of Palmer and Dail (56) and there is no explanation given for the discrepancy.

It is of interest that experiments done with hearts from fetal (136 days; term = 147 days) and adult sheep also exhibited a temporal difference in the response of adenylate cyclase to hormones. Enzyme preparations from fetal and adult sheep hearts were activated by NaF (10 mM) and by epinephrine (10^{-6} M). However, only the enzyme from adult hearts was stimulated by glucagon (59). These results are consistent with the hypothesis that the temporal sequence of development of enzyme responsiveness to hormone depends upon the development of hormone receptors.

3.3 Physiological and Biochemical Effects of β-Receptor Activation: Relationship to Cyclic AMP

There is much written about the role of cyclic AMP as a mediator of the cardiac effects of β-adrenergic agonists (reviewed in Ref. 60 and by Venter in Chapter 7 of this volume). Divergent opinions have emerged among investigators who have studied the developing heart. Reconciliation of these divergent views will not be possible until experiments are done to examine critically all the criteria needed to accept (or reject) the hypothesis that cyclic AMP is an intracellular mediator of the β-adrenergic receptor.

In the embryonic chick ventricle, catecholamines increased cyclic AMP and increased contractility at the 4th incubation day (44). Sperelakis' laboratory concurred that cyclic AMP increased at this stage of development (45) but did not detect a positive inotropic effect (12). Experiments done in our laboratory (see Figure 1) showed that isoproterenol increased the force of contraction in ventricular muscle from the 4th incubation day. The muscle was adjusted to the optimal length on the length-tension relation and was driven at 3 Hz. The frequency of contractions did not change during exposure to isoproterenol; that is, the catecholamine did not induce pacemaker activity that would change the basic drive rate.

In addition to being associated with an increased force of contraction, stimulation by β-adrenergic agonists also increased membrane permeability to Ca^{2+} (60). Catecholamines increase the secondary inward current (i_{si}) carried by Ca^{2+} (and Na^+) during the plateau phase of the cardiac action potential. By this action, catecholamines restore calcium-dependent action potentials to cardiac cells whose early inward current (i_{Na}) is inactivated by K^+-induced deplorization or by tetrodotoxin (TTX) (61). Others have shown that β-adrenergic agonists restore or augment calcium-dependent action potentials to the embryonic chick heart (62). These observations have been confirmed and extended in a report from this laboratory (48). At the earliest stage studied, isoproterenol increased the amplitude of calcium-dependent action potentials in a potassium-depolarized ventricle from a 25–33 somite chick embryo (47–65 hr after fertilization). This is shown in Figure 2 (A and B), which also illustrates the stimulatory effect of isoproterenol on calcium-dependent action potentials on the 4th incubation day (Figure 2, C and D).

These results can be viewed as consistent with the hypothesis that cyclic AMP mediates the cellular actions of β-adrenergic agonists by increasing plasma membrane permeability to Ca^{2+} and thereby augmenting the force of contraction (reviewed in Ref. 60). Strictly speaking, it would be necessary to demonstrate that cyclic AMP increased at the earliest stages that displayed augmentation of calcium-dependent action potentials and contractions by catecholamines. This has not been done, although it is technically feasible. Polson et al. (44) have raised a more serious objection to the cyclic AMP hypothesis in the embryonic chick heart. Whereas these investigators found a relationship between β-adrenergic receptor stimulation and cyclic AMP accumulation, it appeared that increased cyclic AMP may not be related to the increased contractility. Although the increase of cyclic AMP preceded the increase of contractility at low concentrations of catecholamines,

Figure 1 Positive inotropic effect of isoproterenol (10^{-6} M) in embryonic chick ventricle from the 4th incubation day. Tissue was stimulated at 3 Hz with square-wave pulses; temperature 37°C. At arrow, superfusion with isoproterenol began. Time marks at 1-sec intervals show constant stimulus frequency in the absence and presence of isoproterenol. Recording speed reduced to 1/60 just before and after drug addition.

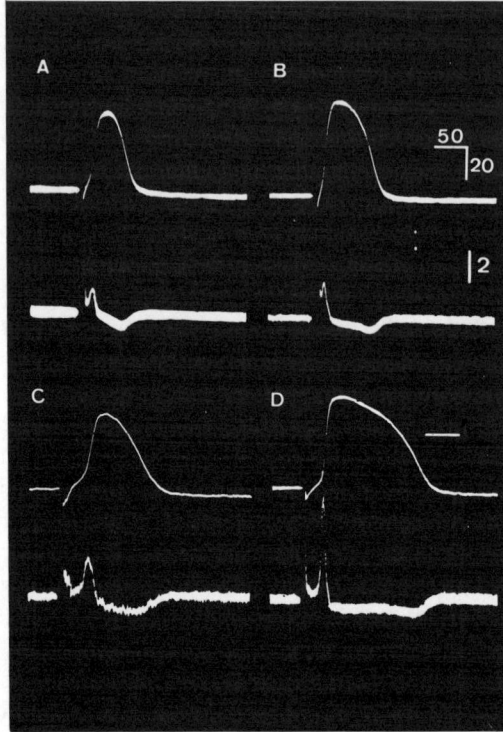

Figure 2 Ca^{2+}-dependent action potentials in young embryonic hearts. Vertical calibration for voltage (mV) and horizontal calibration for time (msec) refer to all records; zero potential marked by horizontal lines in curves B and D. Preparations were bathed in 25 mM K^+ to inactivate fast Na^+ conductance; only the Ca^{2+}-dependent action potential remained. Curves A and B are from one cell in a 25–33 somite (47–65 hr) embryo; C and D are from one cell in a 4-day embryo. Curves A and C, control; B and D, at 3 min in 10^{-6} M isoproterenol.

it was observed that cyclic AMP rose without a concomitant increase of the twitch at high concentrations. Further, twitch tension sometimes continued to rise at a time when cyclic AMP content was declining. These dissociations were viewed as apparent. Measurements of cyclic AMP in discrete cellular regions could help settle the matter of dissociations such as these. Results obtained in our laboratory are consistent with the cyclic AMP hypothesis for β-adrenergic agonist action on the embryonic avian heart, but they do not solve the apparent dissociations (see Section 5.1).

Methylxanthine phosphodiesterase inhibitors restored or augmented calcium-dependent action potentials to embryonic chick ventricular muscle (48,62). Inhibition of cyclic AMP phosphodiesterase, which allows cyclic AMP to accumulate, should have effects similar to addition of β-adrenergic agonists provided that the latter act by cyclic AMP. Wildenthal (16) has pointed out that the fetal mouse heart displayed a positive chronotropic effect to theophylline at a time when norepinephrine had minimal effects. This result can be interpreted as an example of the development of cellular responsiveness to cyclic AMP before that to norepinephrine. However, Wildenthal warned that some difficulties may be encountered with methylxanthine phosphodiesterase inhibitors such as theophylline because they had actions other than inhibition of phosphodiesterase. [The most appropriate methylxanthine for such experiments may be 1-methyl,3-isobutylxanthine, because this compound mimics the effects of isoproterenol on the twitch without stimulating β-adrenergic receptors (63). Neither theophylline nor papaverine reproduced all the characteristic effects of isoproterenol.]

The validity of the cyclic AMP hypothesis also depends upon exogenously applied cyclic AMP mimicking the effects of its intracellular counterpart. Shigenobu and Sperekalis (62,64) reported that cyclic AMP slowly augmented the plateau by increasing i_{si} carried by Ca^{2+}. An exposure of 10–30 min was needed in order for cyclic AMP to be effective, and it was also necessary to use theophylline to oppose hydrolysis of the cyclic nucleotide or to use dibutyryl cyclic AMP, which is resistant to hydrolysis. It had previously been reported that dibutyryl cyclic AMP (N^6,2'-O-dibutyryl-3',5'-adenosine monophosphate) but not cyclic AMP had a positive chronotropic effect on neonatal rat heart cells in culture (65). The greater lipid solubility and resistance to phosphodiesterase were considered the principal reasons for the effectiveness of dibutyryl cyclic AMP as compared to cyclic AMP (65). Additional support for the mimicry of endogenous by exogenous cyclic AMP came from experiments by Goshima on heart cells in culture obtained from fetal (13–15 day) and neonatal (1–3 day) mice (66).

Although the aforementioned results are encouraging, there are several observations that cast doubt on the role of cyclic AMP as the intracellular mediator of β-adrenergic agonists. Dibutyryl cyclic AMP (up to 10^{-2} M) had no acceleratory effect on pacemaker activity in fetal mouse hearts (13th gestational day to birth), even when the compound was present for 30 min (16,67). Fetal mouse hearts (19th gestational day) in organ culture displayed a negative chronotropic effect after a 90-min exposure to dibutyryl cyclic AMP (68). Significant deceleration occurred in 5×10^{-5} M dibutyryl cyclic AMP, and this effect became greater at higher concentrations. It was also reported that isoproterenol (10^{-5} M) had no effect on the frequency of contraction of fetal mouse hearts in organ culture when $[Ca^{2+}]_o$ was 0.92 and 1.38 mM (53). This result is consistent with the cyclic AMP hypothesis, because isoproterenol had no effect on cyclic AMP at these lower concentrations of Ca^{2+}. At 1.84 mM $[Ca^{2+}]_o$, isoproterenol increased cyclic AMP and the rate of spontaneous beating significantly, a finding also in accord with the cyclic AMP hypothesis. However, at 2.76 mM $[Ca^{2+}]_o$, addition of isoproterenol was accompanied by an increased rate of spontaneous contractions and a decreased level of cyclic AMP (53). Moreover, McN-2165 (ethyl-3-ethoxycarbonyl-4-hydroxy-2H-1,2-benzothiazine-2-acetate-1,1-dioxide) increased cyclic AMP and decreased the rate of spontaneous beating in this preparation. It is possible that these results represent a difference due to a particular animal, the mouse. These studies (16, 53,67,68) were not done under conditions that allowed the measurement of specific cellular pools of cyclic AMP. However, they represent a challenge to the cyclic AMP hypothesis that merits careful consideration.

Consideration should also be given to alternative models for the β-adrenergic mechanism. It is usually indicated that β-receptor stimulation increases cyclic AMP, which in turn increases membrane permeability to Ca^{2+}. Alternatively, it has been proposed that changes of intracellular Ca^{2+} (brought about by cyclic uptake and release from sarcoplasmic reticulum) control the intracellular content of cyclic AMP (69). A reduction of $[Ca^{2+}]_o$ to between 0.1 and 0.2 mM was accompanied by an increase of cyclic AMP. It may be erroneous to conclude that the "decrease in cellular Ca^{2+} increased the cyclic AMP" (70) because no measurement was made of cellular Ca^{2+} when $[Ca^{2+}]_o$ was varied. However, it is of interest that a calcium-dependent protein activator of cyclic AMP phosphodiesterase was reported in the sarcoplasmic reticulum of the rat heart (70), and this mechanism has been offered to explain the control of cellular cyclic AMP by Ca^{2+}.

A relationship among β-receptor, adenylate cyclase, and cell function can be glimpsed in studies of human heart. Isoproterenol increased the force of contraction in human embryonic hearts between the 12th and 22nd weeks of gestation (20). The ability of the embryonic heart to respond to isoproterenol increased during this time, although it is not possible to determine if cellular sensitivity to the catecholamine changed. The ability of catecholamine to stimulate adenylate cyclase is evident at 6–7 weeks (56). It is not known how soon after coupling of β-receptor to adenylate cyclase that catecholamines are able to increase the force of contraction. [It is difficult to reconcile the observations of Chang and Cumming (21) with the β-adrenergic receptor, cyclic AMP mechanism. These investigators found that positive chronotropic effect of catecholamines in human fetal hearts 53–71 days after fertilization was blocked by propranolol. However, epinephrine was a more potent stimulant of pacemaker activity than was isoproterenol, and this sequence does not agree with a pharmacological criterion for the β-adrenergic receptor.]

4 INTERCELLULAR REGULATION OF ADRENERGIC REACTIVITY DURING DEVELOPMENT

4.1 Adrenergic Innervation

Innervation of effector cells by voluntary and involuntary nerves has been related to changes in cell sensitivity to neurotransmitter (see the review in Ref. 3). It is reasonable to expect that the development of adrenergic innervation of the heart would be associated with altered sensitivity to catecholamines. On the one hand, the availability of neurons would provide a neuronal transport mechanism (uptake I) that would reduce the concentration of norepinephrine (and epinephrine) available for activation of postjunctional β-adrenergic receptors (71). On the other hand, continuous bombardment of postjunctional receptors by neurotransmitter diminishes receptor sensitivity to the transmitter, for example, by a reduction of the number of receptors (reviewed in Ref. 31).

Our laboratory had studied the development of sympathetic innervation to the sinoatrial pacemaker in the embryonic chick heart and the possibility that it is related to β-adrenergic receptor sensitivity (11,72,73). We have extended the experiments to the avian ventricle, and the results have provided a refined and somewhat revised model of the temporal relationship of innervation to β-adrenergic sensitivity.

Sympathetic adrenergic transmission to the right ventricle can be detected on the 16th incubation day when field stimulation is used to excite intracardiac nerves (Higgins and Pappano, unpublished results; see Table 2). After the 16th incubation day, the ability of nerve stimulation to augment the force of contraction increased until the response to nerve stimulation equaled that caused by exogenously applied isoproterenol. It was concluded that impulse transmission from adrenergic nerves to ventricular muscles began during the third week of life *in ovo*. [Experiments done with the sinoatrial pacemaker suggested that adrenergic innervation did not occur until the 21st incubation day (72). However, these experiments were done at 30°C. When the temperature was increased to 37°C, the same temperature used in experiments with ventricular muscle, adrenergic transmission to the sinoatrial pacemaker was detected at the 16th incubation day (Pappano and Skowronek, unpublished results). Because the change in temperature did not significantly alter pacemaker sensitivity to isoproterenol, it can be concluded that the release of transmitter was impaired at the lower temperature and that this prevented the detection of adrenergic transmission.]

It is helpful to consider the evidence in favor of the view that adrenergic transmission to the heart begins during the 3rd week of life *in ovo*. Others have concluded that such innervation takes place by the end of the first week of life in ovo, that is, between the 5th and 7th incubation days (see Refs. 2, 36 and 74). In untreated preparations, glyoxylic acid-induced fluorescence of adrenergic transmitter was not detected until the 12th and 14th incubation days, respectively, in the sinoatrium and in the ventricle (77). Fluorescent axons could be seen as early as the 11th

TABLE 2 Adrenergic Transmission and β-Adrenergic Sensitivity in Ventricular Muscle

		Isoproterenol
Age (incubation days)	$\Delta A/\Delta B^a$	EC_{50} ($\times$ 10^{-9} M)
8	-3 ± 2 (5)	5.2 ± 2.4 (3)
11	0 ± 0 (3)	3.2 ± 0.7 (4)
14	11 ± 7 (5)	5.3 ± 1.1 (7)
16	27 ± 1 (10)	10.2 ± 2.4 (5)
17	38 ± 18 (2)	38.2 ± 6.8 (6)
18	44 ± 4 (9)	35.2 ± 4.9 (12)
19	44 ± 13 (2)	30.3 ± 4.7 (4)
21 (hatching)	103 ± 19 (7)	8.0 ± 1.8 (7)
28 (1 week after hatching)	124 ± 6 (3)	2.7 ± 0.5 (3)

$^a\Delta A/\Delta B$ is the ratio of the positive inotropic effect caused by field stimulation to that caused by a maximal (10^{-6}M) concentration of isoproterenol. All measurements given as mean $\pm$ SE, with number of experiments in parentheses.

incubation day when the tissue was incubated in α-methylnorepinephrine, a compound taken up by adrenergic nerves but resistant to metabolism by monoamine oxidase. However, fluorescent axons were not detected before this time. These results agree with those given by others, who found fluorescent adrenergic nerves in the embryonic chick gut at the 11th–12th incubation days (76). Adrenergic nerves in the embryonic heart (75) and gut (76) seemed to develop an uptake I mechanism before they are able to synthesize sufficient norepinephrine in endogenous stores. The ability of cardiac preparations to accumulate [^{3}H]norepinephrine ([^{3}H]NE) increased in parallel with the appearance of fluorescent adrenergic axons. Before the 12th incubation day, there was no accumulation of [^{3}H]NE that could be attributed to uptake I (77). Release of [^{3}H]NE from adrenergic nerves by depolarizing stimuli (field stimulation, isotonic KCl) and by tyramine appeared between the 12th and 16th incubation days. The onset of transmitter secretion coincided temporally with the development of the propranolol-sensitive positive inotropic effect of field stimulation. Stimulation-induced overflow of [^{3}H]NE was inhibited by tetrodotoxin, removal of external Ca^{2+}, and by pretreatment with the adrenergic neuron blocking agents reserpine and 6-hydroxydopamine (77). Taken altogether, these results support the conclusion that adrenergic neuroeffector transmission appears for the first time in the chick heart (atrium and ventricle) during the 3rd week of life *in ovo*.

In light of this evidence, it is appropriate to consider the possibility that adrenergic innervation influences the sensitivity of adrenergic receptors on heart muscle. The results of tests of postjunctional sensitivity to isoproterenol are given in Table 2 as the EC_{50}, the concentration needed to produce a half-maximal increase in the force of contraction. Before adrenergic transmission begins, ventricular muscle cells are quite sensitive to isoproterenol. At the time of functional innervation, sensitivity decreases (the EC_{50} increases) to a low level from the 17th to the 19th incubation days. The subsensitivity is transient and the EC_{50} returns to values seen before innervation by the 21st incubation day (hatching) and remains there through the first week after hatching.

The occurrence of β-receptor subsensitivity at the time of innervation would not be unexpected (78); however, the transient character of the subsensitivity was puzzling. Experiments were done to interrupt the development of sympathetic adrenergic nerves by the administration *in ovo* of reserpine. Adrenergic transmission did not develop develop until after hatching in chicks treated on the 11th incubation day with reserpine. However, the transient subsensitivity to isoproterenol occurred at the same time in untreated preparations: between the 16th and 21st incubation days. These results are consistent with the hypothesis that the transient β-adrenergic subsensitivity is unrelated to the stores of norepinephrine (or epinephrine) in cardiac adrenergic nerves. [It has been concluded that both epinephrine and norepinephrine serve as transmitters of adrenergic nerves to the hearts of adult chickens (79). More recently, the results of biochemical (80) and pharmacological (81) experiments have suggested that both catecholamines were transmitters in the rectum of the chicken.] The possibility that the β-receptor subsensitivity was caused by catecholamines released from the adrenal medulla cannot be excluded. Moreover, other hormones, which might be released in large amounts toward the end of the incubation period in the egg, could also be considered as mediators of the transient subsensitivity.

Although it has not been possible to identify the mechanism of the subsensitivity, comment can be made regarding the relationship of sympathetic adrenergic innervation to the number of β-adrenergic receptors. It has been reported that the density of β-adrenergic receptors diminishes when the heart is innervated (36; see Section 2.2). It was assumed that sympathetic adrenergic innervation occurred by the 7th incubation day (37). However, it seems unlikely that sympathetic adrenergic innervation occurs this early in the chick (see above), and therefore it is not clear what the reason is for the reduction in β-receptor density with age. As mentioned in Section 2.4, the increase in protein content of muscle can explain about 30% of the suspected decrease in density. Perhaps the density of receptors is influenced by hormones (thyroid?) that become effective during this time.

Parenthetically, it should also be noted that the subsensitivity of the β-adrenergic mechanism observed in our experiments may not be explained by a reduction of the number of β-adrenergic receptors because density is unchanged at the time of subsensitivity (see Section 2.2). In the mouse heart, the density of β-adrenergic receptors (detected by [³H]dihydroalprenolol binding) increased late in fetal life in parallel with an increased sensitivity to isoproterenol (39a). This observation is in accordance with the need for β-adrenergic receptor development to permit the chronotropic effect of isoproterenol. The implications of the marked increase of ligand binding during the neonatal period (168% of adult values in 3-day neonate) remain to be studied. It is not known if a concomitant increase of sensitivity to isoproterenol occurs when ligand binding is elevated in the neonatal period.

Because of the difficulties encountered in the chemical measurement of β-adrenergic receptor density, it would be helpful to have a means to visualize the β-adrenergic receptor on the cell surface. The availability of a novel fluorescent marker of β-adrenergic receptors has been advanced in recent reports (82,83). The markers, 9-aminoacridine propranolol and a dansyl derivative, were reported to allow a fluorescent tagging of β-receptor binding sites on heart cells and in brain regions of the rat. The specificity of the binding of the fluorescent probes was supported by the observation that l-propranolol, but not d-propranolol, interfered with the binding. Experiments done in our laboratory with 9-aminoacridine propranolol and its dansyl derivative were not able to demonstrate clusters of fluorescent "tags" on heart cells from adult rats and from chicks. Negative results were obtained in each of two series of experiments. The failure to observe fluorescently tagged β-adrenergic receptors in our experiments with chick heart could result from a very low density of receptors. However, the negative results obtained with the rat heart cannot be explained by this mechanism since the positive reports (82) included the rat heart. It has recently been suggested that the fluorescent material observed after injection of 9-aminoacridine and its dansyl derivative may be β-adrenergic receptors, but it is not possible to distinguish them from autofluorescent granules, which are particularly prominent in large cells (84). Autofluorescent granules were demonstrated before injection of the probes and could also account for the inability to interfere with probe-induced fluorescence by treatment with *l*-propranolol (84). Improvements in the specificity of the reaction are needed before the technique can be used to identify the cellular distribution of β-adrenergic receptors.

The onset of functional adrenergic innervation of mammalian hearts has not been investigated systematically in some mammalian hearts. Functional innervation refers to transmission from nerve to muscle. This definition of innervation may be regarded as unsatisfactory because it neglects adrenergic neuronal functions such as uptake I and the synthesis and storage of transmitter. In many embryonic and fetal mammals, the uptake, synthesis, and storage of $[^3H]NE$ have been used as indices of adrenergic neuron properties (reviewed in Refs. 3 and 85). Considerable information has been obtained with several indices of adrenergic neuron function in the rat heart. In left atria from the rat, field stimulation evoked a positive inotropic effect that was very small at birth (10% of the maximal inotropic effect of norepinephrine) and that increased to almost adult values (6 weeks old) at 3 weeks after birth (86).

The positive inotropic effect of tyramine developed in parallel with that of field stimulation; these results are consistent with the progressive development of sympathetic adrenergic innervation of the rat left atrium after birth. Adrenergic innervation of the sinoatrial region may have developed to a considerable extent at birth because tyramine increased pacemaker frequency at this time. Therefore, the adrenergic innervation of the sinoatrial region seems to develop sooner than that of the left atrium, provided that transmission is the index of innervation. [Experiments done by others showed adrenergic transmission to the sinoatrial pacemaker at 1 week after birth but not in tissues from newborn rats (87). Whereas field stimulation was used to excite intracardiac nerves in these experiments, atropine was not present. Standen (86) used atropine in his experiments to prevent inhibitory cholinergic effects on norepinephrine release and on pacemaker cell function.] In light of this distinction, Standen evaluated the sensitivity of the two tissues to norepinephrine and to isoproterenol at birth and at 1, 3, and 6 weeks (adult) after birth. The sensitivity of right atrial preparations to norepinephrine and to isoproterenol did not change between birth and 6 weeks later. By contrast, the sensitivity of left atrial preparations to norepinephrine decreased after birth whereas sensitivity to isoproterenol did not change. These results indicated that the subsensitivity to norepinephrine was the result of increased uptake of this catecholamine by adrenergic nerves in older animals. Isoproterenol is not subject to neuronal uptake (88) and is a sensitive measure of postjunctional β-adrenergic sensitivity. These results are consistent with the view that adrenergic innervation of the sinoatrial region occurs earlier than that of the left atrium not only because of the earlier detection of transmission but also because of the earlier development of $uptake_1$. At birth, the EC_{50} for the positive chronotropic effect of norepinephrine was $1.5 \pm 0.3 \times 10^{-7}$ M and that of the positive inotropic effect was $3.0 \pm 0.5 \times 10^{-8}$ M. In the adult, the corresponding values for chronotropic and inotropic effects were $1.5 \pm 0.3 \times 10^{-7}$ M and $1.54 \pm 0.19 \times 10^{-7}$ M, respectively. Therefore, uptake I may have been as well developed in adrenergic nerves of the sinoatrial region at birth as at 6 weeks later. This was not the case for adrenergic nerves of the left atrium (86). Moreover, it was suggested that functional innervation may develop later in the ventricles than in the atria (86). This supposition is consistent with the finding that ventricular uptake of norepinephrine lagged behind atrial uptake until 2 weeks after birth (89). Observations such as these provide a note of caution about the interpretation of experimental results obtained from a biochemical or physiological analysis of the entire heart.

Slotkin's laboratory has used drug-induced changes of ornithine decarboxylase activity of the rat heart to ascertain the development of sympathetic adrenergic innervation. Nicotine, which releases catecholamines from adrenergic nerve endings, increased ornithine decarboxylase activity in adult rat hearts (90). Because this effect of nicotine was absent in neonatal rats less than 8 days old, it was concluded that the nicotine-induced change in enzyme activity depended upon the presence of sympathetic adrenergic fibers that had innervated the heart. This position was strengthened by the finding that reserpine treatment (1 day after birth) delayed the development of the nicotine-induced increase of ornithine decarboxylase activity. Indeed, reserpine treatment delayed heart growth. However, the effect of reserpine was not attributed specifically to an interruption of sympathetic input to tissue growth (see Ref. 91 for another view). Because of the significant temporal differences noted for adrenergic innervation of left and right atria (86) and of atria and ventricles (89), it might be that the nicotine-induced changes of ornithine decarboxylase activity are only atrial in origin. The inclusion of the ventricles might dilute the effects of nicotine with enzyme activity not subject to sympathetic control. In view of this limitation, it is of interest that the sensitivity of the β-adrenergic receptor increased between 11 and 16 days after birth. The age-dependent supersensitivity (the EC_{50} was measured) was considered independent of the development of cardiac adrenergic innervation (92). A puzzling aspect in these experiments was that chlorisondamine, a long-acting autonomic ganglion blocking drug, significantly reduced heart rate when injected into rats 16 days after birth and older. The negative chronotropic effect of chlorisondamine was considered due to blockade of transmission through sympathetic ganglia because administration of atropine did not allow acceleration of heart at any age (in the absence of chlorisondamine). It is imprudent to conclude that a component of the autonomic nervous system is present simply because the heart rate changes in the presence of an autonomic blocking drug. One might have expected heart rate to increase in atropine because the sinoatrial pacemaker is innervated by vagal inhibitory fibers. Hower, the lack of pacemaker acceleration in the presence of atropine, even in adult animals, does not mean that vagal inhibitory innervation is lacking. (A general caution about the interpretation of experiments with autonomic agonists and antagonists has been given by Woods et al. (93). On the basis of experiments done in fetal and neonatal lambs and in sheep, it appeared that sensitivity of the cardiovascular system to adrenergic agonists increases soon after birth. However, the authors noted that the change of sensitivity to the pressor effect of norepinephrine could have resulted simply from the closure of vascular shunts after birth.) Further, if it can be concluded that the negative chronotropic effect in chlorisondamine indicates sympathetic adrenergic regulation of the cardiac pacemaker, then these results differ from another index of adrenergic innervation. The use of nicotine-induced changes of ornithine decarboxylase activity suggested that sympathetic adrenergic innervation occurred between the 5th and 8th postnatal days (90).

A progressive increase of the maximum positive chronotropic effect of norepinephrine and isoproterenol occurred in rat atria isolated from neonatal (1–2 day), 8–9 day, 18–19 day, and adult rats (94). Whereas measurements such as these indicate changes in the maximum responsiveness of the tissue (perhaps because of an increase of myofibrillar content and organization), it is not possible to discern changes in sensitivity of the β-adrenergic mechanism when results are presented in this fashion. It is preferable to present results as "percent of maximum drug-induced effect" (see Refs. 3 and 31). In this way, one can ascertain the changes in β-adrenergic sensitivity (isoproterenol) from those due to uptake by adrenergic nerves (norepinephrine). This distinction has been considered in the work done on canine hearts. Sympathetic nerve stimulation evoked stimulatory effects in the dog heart that were similar in animals 10 days after birth and in adults (24,25). The coincidence was attributed to several factors (24). Transmitter reuptake and release are less in the young canine heart as compared to the adult. It would be expected that the postjunctional β-adrenergic mechanism would be "supersensitive" in the young heart because of the smaller amounts of transmitter released (24). In the adult animal, the heart would be "subsensitive" to norepinephrine because of the changes produced postjunctionally by large amounts of transmitter and because of the greater development of uptake I that reduced the amount of norepinephrine in the synaptic space (24). These results indicate that the acquisition of secretory properties does not signal the end of adrenergic neuron development (25). Taken altogether, the experimental evidence in this section indicates that a test of neuronal influence on adrenergic sensitivity is probably best done on a selected portion of the heart rather than on the whole organ. The conclusion that sympathetic adrenergic innervation of the canine heart occurs at different times in parts of the organ is reinforced by measurements of sympathetic nerve-induced changes in refractory period in young (1- to 6-week-old) dogs (95).

4.2 Hormonal Effects

The effects of hormones on adrenergic sensitivity have not received much attention in the embryonic and fetal heart. An interesting experiment from Slotkin's laboratory (96) has directed attention to thyroid hormone and its effect on cardiac sympathetic innervation. Sympathetic transmission to the heart was tested indirectly by measuring the increase of ornithine decarboxylase activity caused by insulin-induced hypoglycemia. By this index, sympathetic transmission in untreated animals occurred between the 4th and 6th day after birth. Sympathetic transmission was detected as early as the 2nd day after birth in rats that received triiodothyronine from the day of birth. It would be of interest to know if the biochemical (89) and histochemical development of sympathetic adrenergic nerves is also accelerated by treatment with thyroid hormone. Administration of triiodothyronine to neonatal (1 day) and adult rats for 3 consecutive days was not associated with a change in the stimulatory effect of norepinephrine on cardiac adenylate cyclase (51). Hormone administration was associated with an increased heart weight and rate of beating. Although it has been concluded that neuronal influence on cardiac development may be slight (96), it would be of interest to repeat the experiments in the presence of propranolol. The study by Brus and Hess (51) can be viewed as evidence in favor of a presynaptic (i.e., neuronal) effect of thyroid hormone treatment.

By contrast, experiments done in the fetal mouse heart in organ culture showed that treatment with triiodothyronine increased pacemaker sensitivity to isoproterenol, but not to acetylcholine, theophylline, or glucagon (67). The selectivity of the action of triiodothyronine prompted the conclusion that the hormone altered the density or affinity of β-adrenergic receptors and/or the intracellular mechanism activated by the β-receptor. The hypothesis regarding the properties of the β-adrenergic receptor has been tested in adult animals where thyroid hormone increased the density of receptors (reviewed in Ref. 31). Although treatment with thyroid hormone might have had a neuronal component to allow β-adrenergic "supersensitivity" (see Ref. 97), the demonstration of supersensitivity to isoproterenol suggests a significant contribution of the postjunctional component. The discrepancies among the reported targets of thyroid hormone in fetal and neonatal animals remain to be solved.

Experiments done with chick embryo heart cells in culture suggest that a "hormonal" factor may assist in the development of β-adrenergic receptors (98). Heart cells dissociated from chicks 2.5–4.5 days after fertilization did not accelerate when exposed to epinephrine, even when allowed to grow *in vitro* for several days. Cells prepared from hearts at 5–6 days after fertilization responded to epinephrine with acceleration. Interestingly, incubation of heart cells from the 2.5 to 4.5-day group in medium containing 10% chick embryo extract permitted acceleration of beating by epinephrine. It appeared that some component in the chick embryo extract induced the development (or unmasking) of β-adrenergic receptors (98). If this is the case, the development of cardiac β-adrenergic receptors *in ovo* would depend upon a factor that circulates in the blood. Blood flow begins at the 16-somite stage (45–49 hr after fertilization) in the chick. Because β-adrenergic effects are observed at this time and earlier, there may be other factors, not circulating (or humoral), that influence the differentiation of the β-adrenergic receptor.

5   INTRACELLULAR REGULATION OF ADRENERGIC REACTIVITY DURING DEVELOPMENT

5.1   Muscarinic Inhibition of the β-Adrenergic Mechanism

Activation of the β-adrenergic receptor increases g_{si} and the force of contraction in the embryonic heart (see Section 3.3). The reaction sequence between the β-adrenergic receptor on the one hand and the increase of membrane permeability to calcium and the force of contraction on the other hand can be interrupted by acetylcholine. The inhibitory effect of acetylcholine is antagonized by atropine, hence the designation as muscarinic. Interestingly, the mechanism of the muscarinic inhibition of g_{si} and of contractions changed during development (48). Before hatching, acetylcholine per se had no effect on calcium-dependent action potentials and on contractions (Figure 3, A–C; see also Ref. 62). However, acetylcholine inhibited calcium-dependent action potentials and contractions that had been augmented by catecholamines. This "indirect" inhibitory effect of acetylcholine is shown in Figure 3 (D–F). The specificity of the inhibitory effect of acetylcholine is evident from several aspects. Muscarinic inhibition did not occur if calcium-dependent action potentials were generated by an increase of external Ca^{2+} (rather than by addition of catecholamines). The inhibition did not occur if calcium-dependent action potentials originated from a resting potential of -70 to -80 mV (in the presence of tetrodotoxin). This result precludes the possibility that the inability to observe inhibition by acetylcholine before hatching was the result of inactivation of the calcium-dependent component of the action potential. Muscarinic inhibition did occur when calcium-dependent action potentials were augmented by drugs (catecholamines, histamine) that stimulate adenylate cyclase. The inhibitory effect of acetylcholine on calcium-dependent action potentials induced by isoproterenol was also associated with a reduction of cyclic AMP accumulation (see Table 3). Acetylcholine per se had no effect on cyclic AMP content on the 18th incubation day. By contrast, acetylcholine per se reduced cyclic AMP content in hearts 1 week after hatching (Table 3). Concomitantly, acetylcholine reduced the amplitude of calcium-dependent action potentials (Figure 4) and contractions (Table 3) in the absence of added catecholamines. Indeed, muscarinic inhibition after hatching oc-

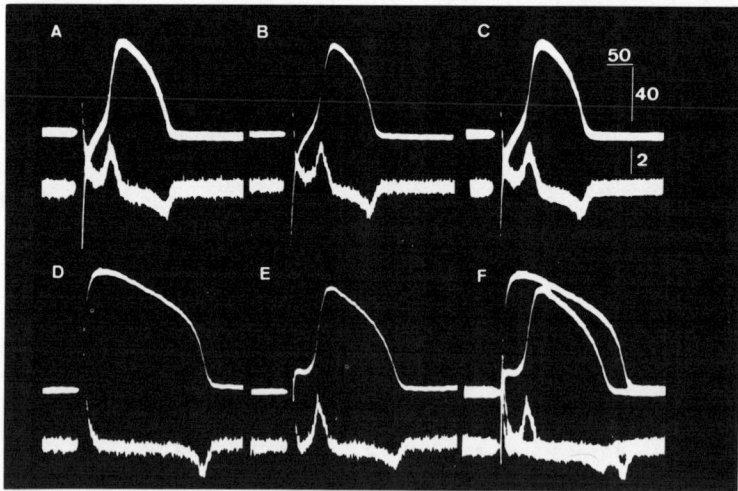

Figure 3 "Indirect" muscarinic inhibition. All records are taken from one cell in an 18-day embryonic ventricle in 25 mM K^+. Calibration format as in Figure 2. Upper vertical trace is membrane voltage (mV) and lower vertical trace is rate of rise (V/sec). Curve A: control. Curve B: 4 min after 10^{-5} M acetylcholine (ACH). Curve C: A and B superimposed to illustrate lack of effect of ACh. Curve D: in 10^{-6} M isoproterenol, the Ca^{2+}-dependent action potential has increased in amplitude duration, and maximum rate of rise. Curve E: in the presence of 10^{-5} M Ach (5 min) and isoproterenol. Curve F: D and E superimposed. From Ref. 48; reproduced with permission from the American Heart Association.

TABLE 3 Effect of Acetylcholine on the Cyclic AMP Level in Chick Hearts

	18-Day Embryonic		7-Day Hatched	
	Cyclic AMP[a]	Twitch Tension[b]	Cyclic AMP	Twitch Tension
Control	0.4 A 0.04 (18)	100	0.6 A 0.05 (7)	100
Acetylcholine (3C)	0.4 A 0.05 (4)	95 A 7 (4)	0.4 A 0.08 (5)	76 A 9 (6)
Isoproterenol (6C)	0.7 A 0.08 (5)	154 A 29 (3)	1.0 A 0.07 (4)	228 A 51 (3)
Acetylcholine v isoproterenol	0.5 A 0.08 (4)	107 A 4 (3)	0.8 A 0.03 (3)	90 A 7 (2)

[a]Cyclic AMP expressed as picomol/mg wet weight.
[b]Twitch tension given as percent of control (z 100%).
[c]Acetylcholine (10^{w6} M) was administered for 3 min by itself or during the last 3 min of a 6-min exposure to isoproterenol (10^{w6} M).

curred in preparations treated with 6-hydroxydopamine (100 mg/kg i.v. 24 hr before experiment) and propranolol (3×10^{-7} M), so that it is reasonable to conclude that acetylcholine "directly" inhibited g_{si} and contractions. The indirect inhibitory effect of acetylcholine persisted after hatching; this is shown in the summary of experiments with isoproterenol in Table 3.

A model has been proposed for these results (48) and it is shown in Figure 5. In this model, guanosine triphosphate (GTP) links the β-adrenergic receptor with adenylate cyclase and cyclic AMP modulates membrane permeability to Ca^{2+} (99–104). It is proposed that "indirect" muscarinic inhibition arises from interruption of GTP-dependent regulation of β-adrenergic receptor/adenylate cyclase interaction and that "direct" muscarinic inhibition arises from interruption of GTP-dependent activation of adenylate cyclase. The possibility that "direct" muscarinic inhibition depended simply upon an increased cellular level of cyclic AMP seems remote. Cyclic AMP averaged 0.52 ± 0.03 pmo/lmg wet weight in ventricles from the 12th incubation day ($N = 5$). "Direct" muscarinic inhibition did not occur at this age

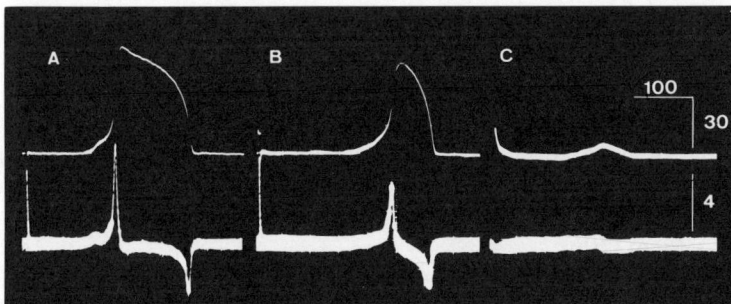

Figure 4: "Direct" muscarinic inhibition. All records taken from one cell in a ventricle, 1 week after hatching (pretreated with 6-hydroxydopamine). Tissue superfused with 25 mM K^+ and propranolol (3×10^{-7} M) is present. Curve A: control (no isoproterenol present); curve B: 2.5 min and curve C, 4 min after 10^{-5} M acetylcholine. From Ref. 48; reproduced with permission from the American Heart Association.

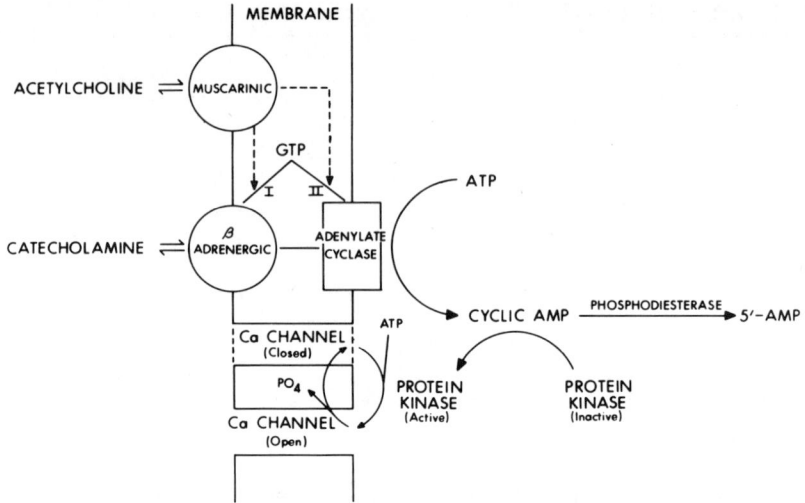

Figure 5 Model of "indirect" and "direct" inhibition of Ca^{2+}-dependent action potentials by acetylcholine. The model depicts the muscarinic and β-adrenergic receptors, the GTP binding protein, and the enzyme adenylate cyclase as plasma membrane components. Interaction between β-adrenergic receptor and adenylate cyclase is regulated by GTP (pathway I); GTP has a regulatory protein (not indicated) that is required to permit stimulation of adenylate cyclase by GTP itself (pathway II). Cyclic AMP activates a protein kinase that phosphorylates a membrane site and increases the number of Ca^{2+} conductance sites. Addition of a substance that increases cellular cyclic AMP will increase the number of g_{xi} channels activated at a given membrane voltage and thereby increase the calcium-dependent action potential. Dephosphorylation of the membrane site decreases the calcium conductance. Muscarinic agents affect adenylate cyclase activation by pathways that indicate the possible mechanisms for indirect (I) and direct (II) cholinergic inhibition. The pathway linking the muscarinic receptor to mechanism II operates only after hatching, whereas mechanism I operates before and after hatching. Adapted from Ref. 48; reproduced with permission from the American Heart Association.

or earlier (4th–7th incubation days) when others have reported very high cyclic AMP levels (44–47). This consideration does not exclude the possibility that "direct" inhibition by acetylcholine depends upon a particular cellular store of cyclic AMP whose concentration is not revealed by our measurements.

The "indirect" but not the "direct" pathway for muscarinic inhibition has been observed in the mammalian ventricle (reviewed in Ref. 105). The development of muscarinic inhibition of calcium-dependent action potentials and contractions has not been studied systematically in the mammalian heart. Interestingly, the inability of acetylcholine to inhibit ventricular contractility in the newborn lamb heart (1–7 days after birth) was attributed to an incomplete development of the sympathetic adrenergic system (106).

In preliminary experiments, acetylcholine also inhibited the calcium-dependent action potentials and contractions that had been augmented by 3-isobutyl-1-methylxanthine (IBMX), an inhibitor of phosphodiesterase (Biegon and Pappano, unpublished results). These results are consistent with the cyclic AMP hypothesis for the regulation of Ca^{2+} permeability and contractions. Because the inhibitory effect of acetylcholine was not associated with a reduction in cyclic AMP content in the presence of IBMX, it appears that acetylcholine inhibits at a site beyond cyclic AMP in the proposed model for muscarinic inhibition (see Figure 5). This view does not exclude the possibility that muscarinic inhibition of the effects of IBMX is achieved by a reduction of a small but critically located pool of cyclic AMP that is masked by a larger cellular pool of the cyclic nucleotide.

5.2   Cell Growth and Differentiation

In the neonatal rat heart, β-adrenergic agonists or dibutyryl cyclic AMP inhibited the incorporation of [³H]thymidine into the DNA of cardiac muscle cells (91). The effect is reversible and it is blocked by propranolol. It is not initiated by an α-adrenergic agonist, phenylephrine. Therefore, the inhibition of thymidine incorporation by catecholamines is consistent with an action on β-adrenergic receptors. The synthesis of DNA by rat heart cells diminishes progressively after birth and ceases during the third week after birth (107). Because adrenergic nerves and the content of norepinephrine and cyclic AMP progressively increased during this time, it was tentatively concluded that adrenergic innervation might regulate (inhibit) cell proliferation and differentiation. This interesting hypothesis has not been tested by experiments done in sympathectomized animals. Because sympathetic adrenergic innervation of the rat heart may occur at different times for atria and ventricles (see Section 4.1), it would be helpful to analyze the hypothesis in a particular cardiac tissue (e.g., ventricles). Further, it has been noted that sympathetic adrenergic innervation of the rat heart, as measured by norepinephrine content (108), cate-cholamine histochemistry (109), and retention of [³H]norepinephrine (110) reached adult levels by the 6th week after birth. These results are consistent with a later maturation of the cardiac adrenergic nerves in the rat than that assumed by Claycomb (91).

Another feature of catecholamine action on the cardiovascular system is the ability to produce anomalous development of the heart and aortic arch (111). The cardiovascular anomalies were produced by β-adrenergic agonists with isoproterenol the most potent and they were antagonized by the β-adrenergic antagonist propranolol. These results indicate that cardiovascular anomalies (aortic arch defects, ventricular septal defect, dual outlet right ventricle, aortic hypoplasia) can be induced by activation of β-adrenergic receptors. It was suggested that the anomalies were produced by changes in cardiovascular hemodynamics caused by the β-adrenergic agonists.

The ability of these compounds to produce cardiovascular anomalies is reinforced by the work of others, who have studied the effects of isoproterenol at different times during embryogenesis of the chick (112). The anomalies produced by isoproterenol varied with the time when the drug was given. Thus transposition of the aorta and pulmonary artery was observed from the 5th to the 7th incubation days, whereas degeneration of cardiac muscle was observed from the 10th to the 12th incubation days. It was concluded that the anomalous changes in cardiovascular development caused by isoproterenol were caused specifically during limited critical periods (112).

6 SUMMARY

This chapter has considered the development of β-adrenergic receptors and the β-adrenergic mechanism in embryonic, fetal, and neonatal heart cells. This is essentially a description of the β-adrenergic system, although there are some mechanistic aspects that have emerged, especially in the area of intercellular and intracellular regulation of adrenergic reactivity.

Emphasis has been given to the possible role of nerves and hormones on the properties of the β-receptor and its intracellular mechanism. The possible role of the β-receptor on the properties of the sympathetic adrenergic nerves has not been considered. This aspect of neuromuscular development in the autonomic nervous system is the focus of a rapidly developing area of neurobiology. Over 30 years ago, Yntema and Hammond had suggested that the development of postganglionic autonomic neurons from the neural crest was regulated by effector cells (113). They wrote: "In the case of the autonomic nerons it is possible that the relation of the crest cells to blood vessels or to the wall of the gut activates their differentiation." The experimental verification of this hypothesis has been obtained quite recently (see Ref. 114 for a summary). Development of sympathetic neurons is influenced to a large extent by extrinsic factors (reviewed in Ref. 115), and the choice of neuronal transmitter is subject to control by signals that arise from effector cells. The ability of the environment to influence transmitter choice *in vivo* has been elegantly demonstrated in the experiments done by N. LeDouarin and Teillet (116). Research in this area will provide a stimulating challenge to traditional ideas of neuromuscular relationships and will inevitably increase our understanding of regulatory mechanisms in cell biology.

ACKNOWLEDGMENTS

Work done in the author's laboratory was supported by the National Institutes of Health (HL-13339), the Connecticut Heart Association, and the University of Connecticut Research Foundation.

Thanks are due Dr. Rebecca Biegon and Dr. Dennis Higgins for providing information and records from their experiments and for criticizing the paper. Frances DiBattista and Carol Skowronek helped to prepare the manuscript.

REFERENCES

1   J. J. Faber, *Am. J. Physiol.*, **214**, 475 (1968).

2   L. P. McCarty, W. C. Lee, and F. E. Shideman, *J. Pharmacol. Exp. Ther.*, **129**, 315 (1960).

3   A. J. Pappano, *Pharmacol. Rev.*, **29**, 3 (1977).

4   C. Markowitz, *Am. J. Physiol.*, **97**, 271 (1931).

5   F.-Y. Hsu, *Chin. J. Physiol.*, **7**, 243 (1933).

6 A. Barry, *Circulation,* **1,** 1362 (1950).

7 E. Fingl, L. A. Woodbury and H. H. Hecht, *J. Pharmacol. Exp. Ther.,* **104,** 103 (1952).

8 T. B. Bolton, *Br. J. Pharmacol.,* **31,** 253 (1967).

9 O. C. Jaffee, *Teratology,* **5,** 153 (1972).

10 F. Michal, F. Emmett, and R. H. Thorp, *Comp. Biochem. Physiol.,* **22,** 563 (1967).

11 K. Löffelholz and A. J. Pappano, *J. Pharmacol. Exp. Ther.,* **191,** 479 (1974).

12 Letters to the Editor, *Circ. Res.,* **34,** 268 (1974).

13 N. J. Sissman, *Am. J. Cardiol.,* **25,** 141 (1970).

14 E. K. Hall, *J. Cell. Comp. Physiol.,* **49,** 187 (1957).

15 M. A. Robkin, T. H. Shepard, and D. C. Dyer, *Proc. Soc. Exp. Biol. Med.,* **151,** 799 (1976).

16 K. Wildenthal, *J. Clin. Invest.,* **52,** 2250 (1973).

17 M. A. Lane, A. Sastre, M. Law, and M. Salpeter, *Dev. Biol.,* **57,** 254 (1977).

18 G. Gennser and E. Nilsson, *Experientia,* **26,** 1105 (1970).

19 J. B. E. Baker, *J. Physiol. (Lond.)* **120,** 122 (1953).

20 D. J. Coltart and B. A. Spilker, *Experientia,* **28,** 525 (1972).

21 T. D. Chang and G. R. Cumming, *Circ. Res.,* **30,** 628 (1972).

22 G. S. Dawes, *Foetal and Neonatal Physiology,* Year Book Medical Publishers, Chicago, 1968, p. 184.

23 G. R. Van Petten and R. F. Willes, *Br. J. Pharmacol.,* **38,** 572 (1970).

24 W. P. Geis, C. J. Tatooles, D. V. Priola, and W. F. Friedman, *Am. J. Physiol.,* **228,** 1685 (1975).

25 P. Gauthier, R. A. Nadeau, and J. de Champlain, *Can. J. Physiol. Pharmacol.,* **53,** 763 (1975).

26 R. Brus and D. Jacobowitz, *Arch. Int. Pharmacodyn.,* **200,** 266 (1972).

27 M. R. Rosen, A. J. Hordof, J. P. Ilvento, and P. Danilo, Jr., *Circ. Res.,* **40,** 390 (1977).

28 R. J. Ertel, D. E. Clarke, J. C. Chao, and F. R. Franke, *J. Pharmacol. Exp. Ther.,* **178,** 73 (1971).

29 B. G. Benfey, *Fed. Proc.,* **36,** 2575 (1977).

30 M. Nickerson and G. Kunos, *Fed. Proc.,* **36,** 2580 (1977).

31 L. T. Williams and R. J. Lefkowitz, *Receptor Binding Studies in Adrenergic Pharmacology,* Raven Press, New York, 1978, Chap. 9.

32 S. Martin, B. A. Levey, and G. S. Levey, *Biochem. Biophys. Res. Commun.,* **54,** 949 (1973).

33 R. J. Lefkowitz, D. S. O'Hara, and J. Warshaw, *Nature New Biol. (Lond.),* **244,** 79 (1973).

34 R. J. Lefkowitz, D. O'Hara, and J. B. Warshaw, *Biochim. Biophys. Acta,* **332,** 317 (1974).

35 M. J. A. Walker, *Br. J. Pharmacol.,* **62,** 185 (1978).

36 R. W. Alexander, J. B. Galper, E. J. Neer, and T. W. Smith, *Circulation,* **57** and **58,** Suppl. II-21 (1978).

37 A. L. Romanoff, *Biochemistry of the Avian Embryo,* Interscience, New York, 1967.

38 Z. Rychter and R. Jelinek, *Physiol. Bohemoslov.,* **20,** 131 (1971).

39 H. C. Stanton and H. J. Mersmann, *Fed. Proc.,* **38,** 361 Abs. (1979).

39a F.-C. M. Chen, H. I. Yamamura, and W. R. Roeske, *Eur. J. Pharmacol.,* **58,** 255 (1979).

40 A. J. Kaumann and R. Wittmann, *Naunyn-Schmiedebergs Arch. Pharmacol.,* **287,** 23 (1975).

41 J. C. Venter, J. E. Dixon, P. R. Maroko, and N. O. Kaplan, *Proc. Natl. Acad. Sci. U.S.A.,* **69,** 1141 (1972).

42 W. R. Ingebretsen, Jr., E. Becker, W. F. Friedman, and S. E. Mayer, *Circ. Res.,* **40,** 474 (1977).

43 H. Reuter, *J. Physiol. (Lond.),* **242,** 429 (1974).

44 J. B. Polson, N. D. Goldberg, and F. E. Shideman, *J. Pharmacol. Exp. Ther.,* **200,** 630 (1977).

45 J.-F. Renaud, N. Sperelakis, and G. LeDouarin, *J. Mol. Cell. Cardiol.*, **10**, 281 (1978).

46 M. J. McLean, R. A. Lapsley, K. Shigenobu, F. Murad, and N. Sperelakis, *Dev. Biol.*, **42**, 196 (1975).

47 M. Hollman and R. D. Green, *Fed. Proc.*, **32**, 711, abstr. (1973).

48 R. L. Biegon and A. J. Pappano, *Circ. Res.*, **46**, 353 (1980).

49 C. M. Doyle, R. Zak, and D. A. Fischman, *Dev. Biol.*, **37**, 133 (1974).

50 E. Novak, G. I. Drummond, J. Skala, and P. Hahn, *Arch. Biochem. Biophys.* **150**, 511 (1972).

51 R. Brus and M. E. Hess, *Endocrinology*, **93**, 982 (1973).

52 A. F. Kohrman, *Pediatr. Res.*, **7**, 575 (1973).

53 C. L. Eyer and W. E. Johnson, *Arch. Int. Pharmacodyn.*, **237**, 119 (1979).

54 H. J. Mersmann, G. Phinney, L. J. Brown, and D. G. Steffen, *Biol. Neonate*, **32**, 266 (1977).

55 Y. Vulliemoz, M. Verosky, M. Rosen, and L. Triner, *Fed. Proc.*, **36**, 318, abstr. (1977).

56 G. C. Palmer and W. G. Dail, Jr., *Pediatr. Res.*, **9**, 98 (1975).

57 W. G. Dail, Jr., and G. C. Palmer, *Anat. Rec.*, **177**, 265 (1973).

58 D. J. Coltart, G. M. Davies, I. M. Gillibrand, and J. Hamer, *J. Physiol. (Lond.)*, **225**, 38P (1972).

59 G. Ahumada, B. E. Sobel, and W. F. Friedman, *Am. J. Physiol.*, **230**, 1590 (1976).

60 R. W. Tsien, *Adv. Cyclic Nucleotide Res.*, **8**, 363–420 (1977).

61 A. J. Pappano, *Circ. Res.*, **27**, 379 (1970).

62 K. Shigenobu and N. Sperelakis, *Circ. Res.*, **31**, 932 (1972).

63 M. Korth, *Naunyn-Schmiedebergs Arch. Pharmacol.*, **302**, 77 (1978).

64 K. Shigenobu and N. Sperelakis, *Jap. J. Pharmacol.*, **25**, 481 (1975).

65 E.-G. Krause, W. Halle, E. Kallabis, and A. Wollenberger, *J. Mol. Cell. Cardiol.*, **1**, 1 (1970).

66 K. Goshima, *J. Mol. Cell. Cardiol.*, **8**, 713 (1976).

67 K. Wildenthal, *J. Pharmacol. Exp. Ther.*, **190**, 272 (1974).

68 C. L. Eyer and W. E. Johnson, *Eur. J. Pharmacol.*, **51**, 423 (1978).

69 I. Harary, J.-F. Renaud, E. Sato, and G. A. Wallace, *Nature*, **261**, 60 (1976).

70 I. Harary and G. A. Wallace, *Recent Adv. Stud. Card. Struct. Metab.*, **12**, 635–643 (1978).

71 L. L. Iversen, *The Uptake and Storage of Noradrenaline in Sympathetic Nerves*, Cambridge University Press, Cambridge, 1967.

72 A. J. Pappano and K. Löffelholz, *J. Pharmacol. Exp. Ther.*, **191**, 468 (1974).

73 A. J. Pappano, *J. Pharmacol. Exp. Ther.*, **196**, 676 (1976).

74 N. G. Culver and D. A. Fischman, *Am. J. Physiol.*, **232**, R116 (1977).

75 D. Higgins and A. J. Pappano, *J. Mol. Cell. Cardiol.*, **11**, 661 (1979).

76 M. L. Epstein and M. D. Gershon, *Soc. Neurosci.*, **4**, 271 (1978).

77 D. Higgins and A. Pappano, *Fed. Proc.*, **38**, 1396 Abs. (1979).

78 T. Deguchi and J. Axelrod, *Mol. Pharmacol.*, **9**, 612 (1973).

79 V. P. DeSantis, W. Langsfeld, R. Lindmar, and K. Löffelholz, *Br. J. Pharmacol.*, **55**, 343 (1975).

80 S. Konaka, H. Ohashi, T. Okada, and T. Takewaki, *Br. J. Pharmacol.*, **65**, 257 (1979).

81 S. Komori, H. Ohashi, T. Okada, and T. Takewaki, *Br. J. Pharmacol.*, **65**, 261 (1979).

82 E. Melamed, M. Lahav, and D. Atlas, *Experientia*, **32**, 1387 (1976).

83 D. Atlas and A. Levitzki, *Proc. Natl. Acad. Sci. U.S.A.*, **74**, 5290 (1977).

84 A. Hess, *Brain Res.*, **160**, 533 (1979).

85 W. F. Friedman, *Prog. Cardiovasc. Dis.*, **15**, 87 (1972).

86 N. B. Standen, *Br. J. Pharmacol.*, **64**, 83 (1978).

87 J. Vlk and F. F. Vincenzi, *Biol. Neonate*, **31**, 19 (1977).

88 G. Hertting, *Biochem. Pharmacol.*, **13**, 1119 (1964).

89 G. F. Atwood and N. Kirschner, *Dev. Biol.*, **49**, 532 (1976).

90 J. Bartolome, C. Lau, and T. A. Slotkin, *J. Pharmacol. Exp. Ther.*, **202**, 510 (1977).

91 W. C. Claycomb, *J. Biol. Chem.*, **251**, 6082 (1976).

92 F. J. Seidler and T. A. Slotkin, *Br. J. Pharmacol.*, **65**, 431 (1979).

93 J. R. Woods, Jr., A. Dandavino, K. Murayama, C. R. Brinkmann III, and N. Assali, *Circ. Res.*, **40**, 401 (1977).

94 A. Nukari-Siltovuori, *Experientia*, **33**, 1611 (1977).

95 F. A. Kralios and C. K. Millar, *Cardiovasc. Res.*, **12**, 547 (1978).

96 C. Lau and T. A. Slotkin, *J. Pharmacol. Exp. Ther.*, **208**, 485 (1979).

97 K. Wildenthal, *J. Clin. Invest.*, **51**, 2702 (1972).

98 S. Lipshultz, J. Shanfeld, and S. Chacko, *J. Cell Biol.*, **67**, 244a (1975).

99 R. Niedergerke and S. Page, *Proc. R. Soc. Lond. Ser. B*, **197**, 333 (1977).

100 A. J. Pappano and E. E. Carmeliet, *Pflügers Arch.*, **382**, 17 (1979).

101 H. Reuter and H. Scholz, *J. Physiol. (Lond.)*, **264**, 49 (1977).

102 E. M. Ross, M. E. Maguire, T. W. Sturgill, R. L. Biltonen, and A. G. Gilman, *J. Biol. Chem.*, **252**, 5761 (1977).

103 N. Sperelakis and J. A. Schneider, *Am. J. Cardiol.*, **37**, 1079 (1976).

104 A. M. Watanabe, M. M. McConnaughey, R. A. Strawbridge, J. W. Fleming, L. R. Jones, and H. R. Besch, Jr., *J. Biol. Chem.*, **253**, 4833 (1978).

105 M. N. Levy, in W. C. Randall, Ed., *Neural Regulation of the Heart*, Oxford University Press, New York, 1977, pp. 95–129.

106 S. E. Downing, J. C. Lee, and J. F. N. Taylor, *Am. J. Physiol.*, **233**, H451 (1977).

107 W. C. Claycomb, *J. Biol. Chem.*, **250**, 3229 (1975).

108 L. L. Iversen, J. deChamplain, J. Glowinski, and J. Axelrod, *J. Pharmacol. Exp. Ther.*, **157**, 509 (1967).

109 T. H. Schiebler and R. Heene, *Histochemie*, **14**, 328 (1968).

110 J. Glowinski, J. Axelrod, I. J. Kopin, and R. J. Wurtman, *J. Pharmacol. Exp. Ther.*, **146**, 48 (1964).

111 R. J. Hodach, A. E. Hodach, J. F. Fallon, J. D. Folts, H. J. Bruyere, and E. F. Gilbert, *Teratology*, **12**, 33 (1975).

112 B. Ostadal, Z. Rychter, and V. Rychterova, *J. Mol. Cell. Cardiol.*, **8**, 533 (1976).

113 C. L. Yntema and W. S. Hammond, *Biol. Rev.*, **22**, 344 (1947).

114 E. J. Furshpan, P. R. MacLeish, P. H. O'Lague, and D. D. Potter, *Proc. Natl. Acad. Sci. U.S.A.*, **73**, 4225 (1976).

115 G. Burnstock, *Prog. Neurobiol.*, **11**, 205 (1978).

116 N. M. LeDouarin and M-A. M. Teillet, *Dev. Biol.*, **41**, 162 (1974).

CHAPTER THREE

ADRENOCEPTORS IN GANGLIA

David A. Brown and Malcolm P. Caulfield*

Department of Pharmacology, The School of Pharmacy, University of London, London, England

Catecholamines (CA) can both facilitate and depress sympathetic ganglionic transmission [reviewed by De Groat and Volle (1), Haefely (2), Kosterlitz and Lees (3), and Volle (4)]. The former appears to be a β-effect, the latter an α-effect (1). Depression of transmission is more frequently observed and appears to be the main effect seen *in vitro*.

1 GANGLIONIC DEPRESSION: PRE- AND POSTSYNAPTIC COMPONENTS

Two components of depressant action have been reported: inhibition of acetylcholine release (5) and postsynaptic hyperpolarization (1). However, transmission failure in rabbit (6,7) and guinea pig (8) ganglia is unequivocally *pre*synaptic in origin: depression of the excitable postsynaptic potential (EPSP) and a fall in its quantal content *m* can be detected without postsynaptic hyperpolarization and without any

---

*Present addess: Glaxo Group Laboratories, Ltd, Greenford, Middx. UB6 OHE, England

reduction in postsynaptic excitability or in the response to iontophoretic acetylcholine. A presynaptic effect is also probably the primary mechanism in other species,
such as the rat. Thus 1 μM (-)adrenaline reduces the stimulus-evoked release of
[³H]acetylcholine from isolated rat ganglia by 68 ± 5% (mean ± SE; $n = 5$) at
a stimulus frequency of 1 Hz (Figure 1); at the same concentration the compound
postganglionic action potential was reduced by about 35%, in spite of an almost
negligible (~0.1 mV) ganglionic hyperpolarization (Figure 2). Lower concentrations could reduce the spike amplitude without hyperpolarization (Figure 3).

1.1 Presynaptic Receptors

The presynaptic depressant effect of adrenaline in rabbit ganglia is α-mediated (6).
Recent experiments on rat ganglia (9) suggest that the receptors concerned belong
to the "α_2"-subclass of α-receptors (see 10). Thus the depressant effect of adrenaline
on transmission through the rat ganglion—almost certainly presynaptic in origin

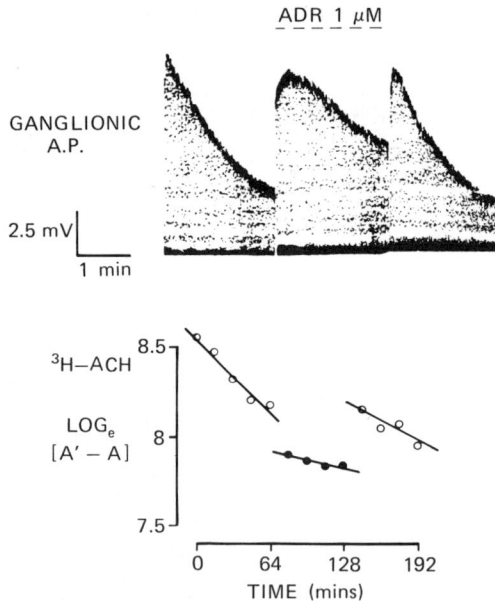

Figure 1   Effect of (-)adrenaline (1 μM) on the release of radiolabeled acetylcholine (ACh) from an
isolated rat superior cervical ganglion. The ACh stores were labeled by preincubation with [³H]choline,
and the [³H]ACh formed was subsequently released by stimulating the preganglionic nerve at 1 Hz for
2-min periods at 16-min intervals in the presence of 50 μM neostigmine (see Ref. 87). The upper record
shows the amplitudes of the compound ganglion action potential recorded during three such stimulus
periods: before, during, and after adrenaline perfusion, displayed on a chart recorder using a peak-height
detector (see Ref. 88). The graph shows the amount of tritium released by each train of stimuli (resting
release subtracted) expressed on a logarithmic scale against time. Evoked tritium release (= [³H]ACh:
see Ref. 89 and A. J. Higgins, unpublished data) declines with successive stimulus bursts as the pool
of labeled ACh diminishes; this decline is a first-order process whose rate coefficient is given by the
slope of the line. Adrenaline reduced this rate coefficient from 0.051 to 0.011 min⁻¹: the mean reduction
in five experiments (± SE) was 68 ± 8%. From M. P. Caulfield and A. J. Higgins, unpublished data.

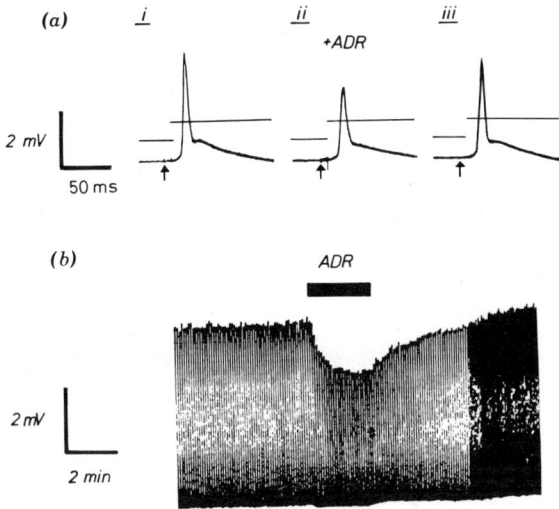

Figure 2   Effect of adrenaline (1 μM) on the compound action potential in an isolated rat superior cervical ganglion following single supramaximal preganglionic stimuli at 0.2 Hz, recorded extracellularly across an insulated partition with respect to an indifferent electrode on the postganglionic trunk (see Ref. 88). The upper record (a) shows oscilloscope records of the action potentials recorded (i) before, (ii) during, and (iii) after adrenaline application (arrows mark stimulus artifacts). The second oscilloscope beam (b) monitors the sampling period of a peak-height detector (see Ref. 90), used to "hold" the spike deflexion for display on a dc chart recorder (lower trace). Note that the compound-action-potential height is reduced by 1.6 mV, whereas the resting demarcation potential was increased by only ~0.1 mV; note also the reduced amplitude of the synaptic potential in upper record (ii). From M. P. Caulfield, unpublished data.

(see above)—is readily blocked by yohimbine but not by the peripheral α_1-selective blocking agent prazosin (Figure 4). Similarly, the spectrum of agonist potencies as inhibitors of transmission accords with an α_2-receptor (Table 1; note the high potency of clonidine relative to methoxamine or phenylephrine; see also 11). Confirmation for this view is provided by recent measurements of [^{3}H]dihydro-ergocryptine binding to isolated rat ganglia (12): the binding, which was strongly reduced by α-agonists, including clonidine ($K_i \sim 1.5 \times 10^{-6}$ M), was reduced by about half on preganglionic denervation.

In dog paravertebral lumbar ganglia, separate presynaptic dopamine receptors also seem to be present, which are activated by dopamine and apomorphine and inhibited by haloperidol and pimozide (13). However, the presynaptic depressant action of dopamine on rabbit ganglia is blocked by phenoxybenzamine (7) and hence is probably mediated by an α-receptor. Similarly, there do not seem to be any presynaptic dopamine receptors in rat ganglia, since apomorphine is inactive.

1.2   Mechanism of Presynaptic Depression

The type of receptor involved (α_2) suggests that the presynaptic depressant action of CA on the cholinergic nerves in sympathetic ganglia is merely another example of the widely distributed presynaptic "modulatory" action of CA on peripheral

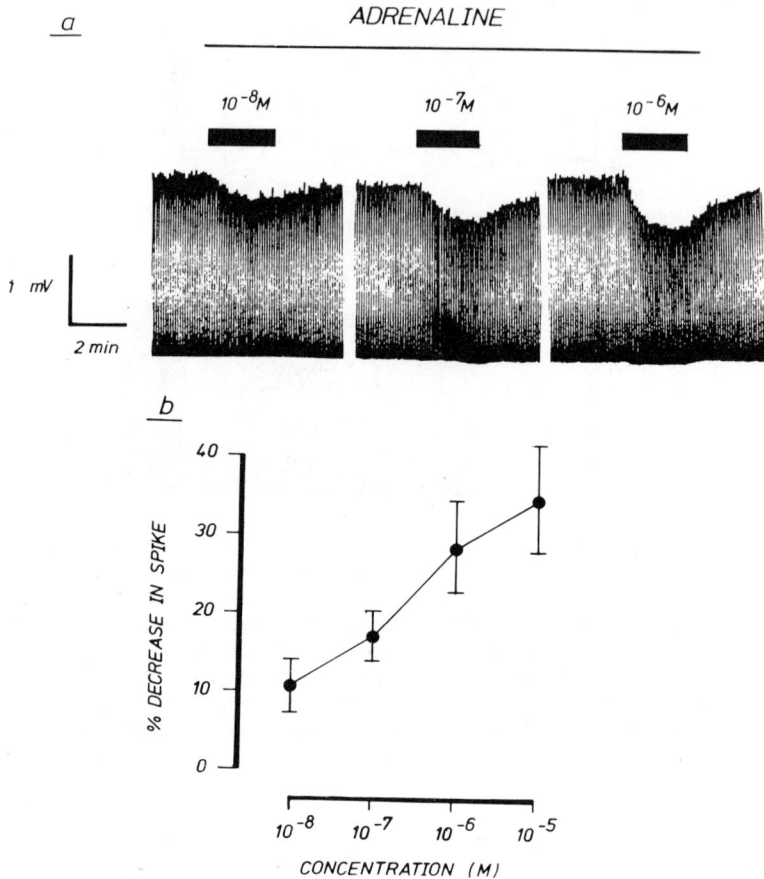

Figure 3 *(a)* Effect of three concentrations of adrenaline on the ganglionic action potentials recorded and displayed as in Figure 2. Note the absence of any baseline ganglionic hyperpolarization at 10^{-8} M and the small hyperpolarization at 10^{-7} and 10^{-6} M. *(b)* Dose–response curve for percent reduction in ganglionic action potential (mean ± SE of five experiments). From M. P. Caulfield, unpublished data.

autonomic nerves, including both adrenergic (14,15) and cholinergic (16) nerves. The currently favored view regarding the mechanism of depression is that CA inhibit inward Ca^{2+} currents (see, e.g., 15,17–19). However, the evidence for this is essentially indirect, depending primarily on *(a)* restriction of inhibition to Ca^{2+}-dependent neurosecretion and *(b)* apparent antagonism by Ca^{2+} ions. (More direct evidence for an effect on Ca^{2+} currents is available from recent observations on the *post*synaptic effect of CA on sympathetic neurones; see below.)

Christ and Nishi (20) analyzed the presynaptic action of adrenaline on rabbit superior cervical ganglia using electrophysiological techniques. They could obtain no evidence for any effect on the membrane potential or excitability of the terminal or preterminal fibers, implying an action on excitation–secretion coupling. The reduction in the EPSP was diminished from 68% in 2 mM $[Ca^{2+}]$ to 16% in 10 mM $[Ca^{2+}]$, in accord with the general "antagonistic" action between the two.

$+PRAZOSIN$ 10^{-5} M $+YOHIMBINE$ 10^{-6} M

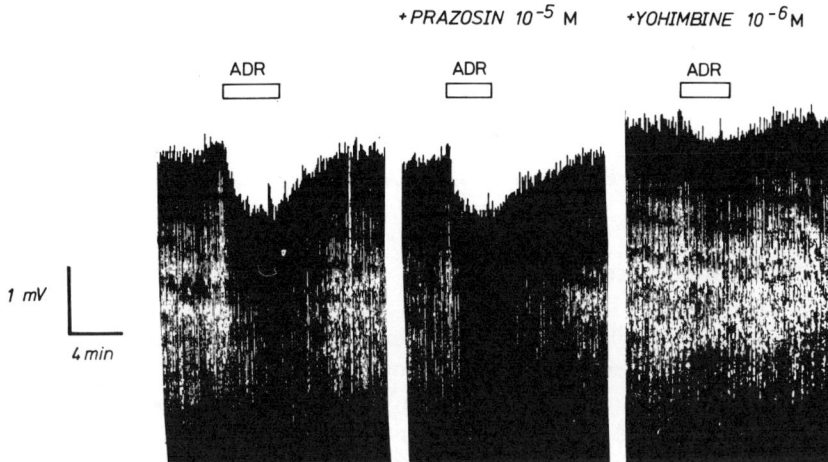

Figure 4 Effects of prazosin (10 μM) and yohimbine (1 μM) on the depression of the orthodromically generated ganglionic action potential produced by 1 μM (-)adrenaline, recorded as in the lower record of Figure 2. From M. P. Caulfield, unpublished data.

However, an attempted measurement of the quantal release parameters m (= quantal content), n (= number of quanta available for release), and p (= probability of single quantal release: $m = np$) suggested differential effects of adrenaline and Ca^{2+}: the reduction of m produced by 10 μM adrenaline (-35%) appeared to result from a reduced quantal *store (n)* rather than a reduced probability of release *(p)*, whereas raising Ca^{2+} increased p, not n. The force of this distinction is diminished somewhat by the fact that *(a)* the comparison was with *raised* Ca^{2+} rather than *reduced* Ca^{2+}, and *(b)* the quantal analysis was based on the assumption of a Poisson distribution of EPSP amplitudes, whereas binomial statistics may be more appropriate (21–23, but see Ref. 24). Bearing in mind the complexities of quantal analysis

TABLE 1 Inhibition of Transmission through the Isolated Rat Superior Cervical Ganglion[a]

Agonist	EC_{50} (M)	Maximum Depression of Spike (%)
(−) Adrenaline	4.1×10^{-8}	47
Clonidine	2.2×10^{-6}	60
(−) Noradrenaline	7.4×10^{-5}	40
Dopamine	7.9×10^{-5}	70
Phenylephrine	1.5×10^{-4}	40
Methoxamine	2.4×10^{-4}	100
(±) Isoprenaline	2.5×10^{-4}	100

[a]Measured from the depression of the compound postganglionic action potential (see Figure 2).

at this site, further electrophysiological analyses are necessary, although it is doubtful whether they would provide any real insight into the detailed mechanisms of CA depression.

2 POSTSYNAPTIC HYPERPOLARIZATION

2.1 Postsynaptic Receptors

De Groat and Volle (1) classified the hyperpolarizing receptors in cat ganglia as α. This has recently been confirmed in isolated rat ganglia (25), with the added refinement that the receptors concerned are again of the α_2-subclass (i.e., those more usually associated with sympathetic nerve terminals) (see Refs. 14 and 15). This classification was based on the observations that *(a)* selective α_1-agonists, such as amidephrine, were ineffective hyperpolarizing agents (Table 2); *(b)* the α_2-agonists clonidine, oxymetazoline, and ergometrine were potent partial agonists; and *(c)* responses to all agonists were blocked by phentolamine and yohimbine, but not by prazosin. The binding characteristics of [^{3}H]dihydroergocryptine to isolated rat ganglia also accord with this classification (12; Table 2). The hyperpolarizing CA receptor in the rabbit ganglion is also α (6), probably of the α_2-type (26).

The presence of a "presynaptic" type of receptor on the "postsynaptic" neurone may seem rather peculiar. However, it is now clear that the pharmacological classification into α_1 and α_2 cannot be strictly equated with pre- and postjunctional receptors. In particular, there is a strong analogy with the central nervous system,

TABLE 2   Effect of Some Sympathetic Agonists as Hyperpolarizing Agents (25) and Inhibitors of [^{3}H] Dihydroergocryptine Binding (12) in Isolated Rat Superior Cervical Ganglia[a]

Agonist	EC_{50} for Hyperpolarization (μM)	K_1 for Inhibition of Binding (μM)
(−) Adrenaline	0.17 ± 0.03 (4)	0.4
(−) Noradrenaline	1.7 ± 0.6 (5)	1.6
(±) Isoprenaline	4.1 ± 0.8 (6)	1.5
(−) Phenylephrine	4.2 ± 0.4 (4)	—
(±) Amidephrine	1100 ± 310 (3)	—
ADTN	6.2 ± 0.4 (3)	—
Dopamine	17 ± 5 (4)	60
Clonidine[b]		1.5

[a]Values are means ± SE (*n* in parentheses).
[b]Partial agonist: estimated EC_{50}.

where postsynaptic (inhibitory) α_2-receptors have been demonstrated both by electrophysiological (27) and radiolabeled ligand binding (28,29) methods.

The additional presence of specific "dopamine" receptors on sympathetic neurones has sometimes been suggested, or implied (see, e.g., Refs. 30 and 31), but this seems unjustified on the evidence so far available. Thus Libet (30) described dopamine as "equally effective" with noradrenaline in hyperpolarizing the rabbit ganglion, without defining their relative potencies. Further, antagonism of dopamine hyperpolarization in this species by haloperidol (32) is inconclusive in view of haloperidol's powerful α-blocking action (33). Further, the "dopamine-sensitive" adenylate cyclase reported by (34) was not antagonized by conventional dopamine-receptor blocking agents. Dun and Nishi (7) had previously reported that the hyperpolarizing action of dopamine on the rabbit ganglion was blocked by phenoxybenzamine, suggesting an effect on α-receptors. This has recently been confirmed in rat ganglia (25): although both dopamine and the rigid analog 2-amino-6,7-dihydroxy-1,2,3,4-tetrahydronaphthalene (ADTN) were quite effective hyperpolarizing agents, apomorphine was ineffective; further, the action of dopamine was blocked by α-receptor antagonists such as phentolamine or yohimbine, but not by the dopamine-receptor blockers haloperidol, fluphenazine, or α-flupenthixol. It appears that, in the rat and rabbit, both pre- and postsynaptic effects of dopamine are mediated via α_2-receptors, not dopamine receptors.

2.2   Mechanism of the Postsynaptic Effects of CA

2.2.1   *Electrophysiology*

On intracellular recording the CA-induced hyperpolarization is small ($\leqslant 5$ mV), rather variable, and accompanied by no change or a small ($\leqslant 10\%$) increase in cell input resistance measured under current clamp (6,7,35–37). Figure 5 illustrates an effect of noradrenaline on a rat ganglion cell; there was a small (~ 1 mV) hyperpolarization, with no clear change in the voltage–current relationship of the cell. The most noticeable electrophysiological response to a catecholamine is, in fact, not the change in membrane potential or resistance, but a change in *spike configuration*. In particular, the *after-negativity following the spike is strikingly reduced* (38,39*b*; see also Figure 5). Since the after-negativity appears to result—in part, at least—from an increased G_K consequent upon Ca^{2+} entry during the spike (39*a*,40), this may presage a primary effect of CA on the inward Ca^{2+} current. In accord with this, Minota and Koketsu (38) and Horn and McAfee (39*b*,39*c*) found that CAs reduce the amplitude of those components of the action potential in frog and rat sympathetic neurones attributable to the inward Ca^{2+} current, in both normal (high-Na^+) and low Na^+/high Ca^{2+} solution. Such effects were annulled in a low-Ca^{2+} solution.

This action of catecholamines is obviously very interesting in view of the abovementioned conclusion that the ganglionic receptor is, in fact, a "presynaptic" type of receptor, because it forms the most direct evidence yet available that the *presynaptic action of CA might also result from an inhibition of the inward Ca^{2+}

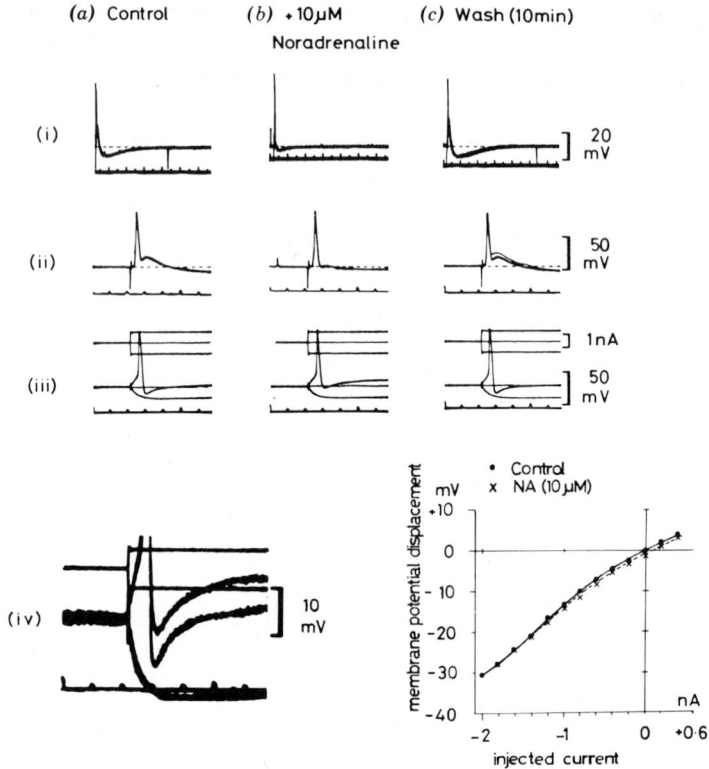

Figure 5   Effect of noradrenaline (10 μM, 2 min application) on the electrical properties of a rat superior cervical ganglion cell recorded with an intracellular microelectrode in the manner described by Brown and Constanti (91). Oscilloscope records show: (i) and (ii) responses to single preganglionic stimuli; and (iii) superimposed responses to the injection of depolarizing and hyperpolarizing current pulses through the bridge-balanced recording microelectrode. Note that noradrenaline greatly reduced the amplitude of the synaptic potential in (i) and (ii) and reduced the amplitude of the after-negativity following the spike in (iii), without changing spike amplitude or input resistance as measured by the electrotonic response to the hyperpolarizing current injection in (iii). The change in spike after-negativity is shown in the enlarged superimposed sweeps in (iv). The graph shows the steady-state voltage–current curves before (●) and during (×) this application of noradrenaline: the membrane potential was increased by 1 mV but the voltage–current relationship was unchanged. From A. Constanti and D. A. Brown, unpublished data.

current [a point made by Minota and Koketsu (38) before the pharmacological analogy had become fully apparent]. However, the *post*synaptic consequences of CA-mediated inhibition of inward Ca^{2+} currents are not yet clear. For example, does it cause any modification of the repetitive firing behavior of the neurone? [This might be anticipated both from studies on the role of the Ca^{2+}-mediated K^+ current ($G_{K(Ca)}$) in other neurones (see, e.g., Ref. 41) and from previous observations on the effects of divalent cation inhibition of the Ca^{2+} current in autonomic neurones (39,42).] Second, is the hyperpolarization itself due to a reduced inward Ca^{2+} current at rest potential? The potential (and resistance) change resulting when the calcium current is reduced may depend greatly on the relationship between the voltage sensitivity of the Ca^{2+} current itself ($G_{Ca(V)}$) and that of the Ca^{2+}-activated K^+ current ($G_{K(Ca)}$). In neuroblastoma cells there is quite a large separation (43), so that inhibiting $G_{Ca(V)}$ at membrane potentials more negative than -20 mV should induce a net outward current and hyperpolarization. Rat sympathetic neurones may behave similarly, since a reduction in $[Ca^{2+}]_o$ from 2.5 to 1.25 mM also produces a small hyperpolarization, whereas increasing $[Ca^{2+}]_o$ to 10 mM produces a depolarization (42). On the other hand, in myenteric neurones there appears to be a large Ca^{2+}-triggered component of G_K at *rest* potential, so that inhibition of the inward Ca^{2+} current at rest produces a *de*polarization accompanied by a large increase in cell input resistance (44).

2.2.2 Electrogenic Effects

An alternative interpretation of the CA-induced hyperpolarization is that of an "electrogenic" origin, that is, that it results from the direct activation of an electrogenic outward Na^+ pump or some other metabolically dependent nonneutral pump (36,45). Similar views have been advanced to explain the synaptically mediated "slow inhibitory postsynaptic potential (IPSP)" (46,47), although the evidence that this is mediated by catecholamines must now be regarded as suspect (see below).

The principal evidence for pump stimulation stems from experiments on frog ganglia. Initial evidence was essentially indirect, relying on the changes in adrenaline-induced hyperpolarization observed with extracellular (sucrose-gap) recording caused by removing extracellular Na^+ or by other procedures (such as cooling or application of ouabain) expected to inhibit the Na^+ pump (36). Since no control test for pumping activity was used, it is difficult to know how far the observed effects were primary or secondary consequences of pump inhibition or membrane potential change. Akasu and Koketsu (48a,b) obtained stronger and more direct evidence for an acceleration of electrogenic sodium pumping in frog ganglia by adrenaline, taking the form of (a) an increased hyperpolarization on adding K^+ to a K^+-free solution (see Ref. 49) and (b) an increased rate of Na^+extrusion/K^+ recapture after injecting NaCl intracellularly (measured from the change in spike configuration). This suggests that CA can accelerate Na^+ pumping in frog ganglia but leaves open the question of whether this accounts for the hyperpolarization, since the contribution of electrogenic Na^+ extrusion to the *resting* potential was not determined. (A striking feature of these experiments was the delayed and prolonged action of adrenaline. No effect was seen at 5 min and the effect increased progressively over a 35-min period; in contrast, the hyperpolarization is immediate.)

A great deal more is known about the properties of the Na^+ pump in *mammalian* ganglia, from direct measurements of total (50) and unidirectional (51) Na^+ fluxes in rat sympathetic ganglia, and from the electrogenic response to Na^+ extrusion after Na^+ loading in rat (52) and rabbit (53) ganglia. High concentrations of a catecholamine (dopamine 100 μM) failed to accelerate ^{24}Na efflux (54); nor did dopamine modify the electrogenic response to Na^+ pumping, measured as the after-hyperpolarization following washout of a nicotinic agonist (D. A. Brown, unpublished experiments; see Ref. 52). Further, inhibition of this electrogenic effect by a combination of ouabain and K^+ removal did not reduce the hyperpolarizing response to noradrenaline (Figure 6): indeed, the latter was increased, probably because of membrane depolarization. A comparable distinction between electrogenic and CA-induced hyperpolarization has been reported in rabbit ganglia (55).

Thus in mammalian cells, where, by virtue of their high pumping rate and small size, the electrogenic effects of Na^+ extrusion are very pronounced, CA do *not* activate the Na^+ pump and the CA hyperpolarization is clearly *not* electrogenic in nature. The slowly developing effect reported in frog ganglia might be some secondary consequence of prolonged CA application, unrelated to the initial hyperpolarization.

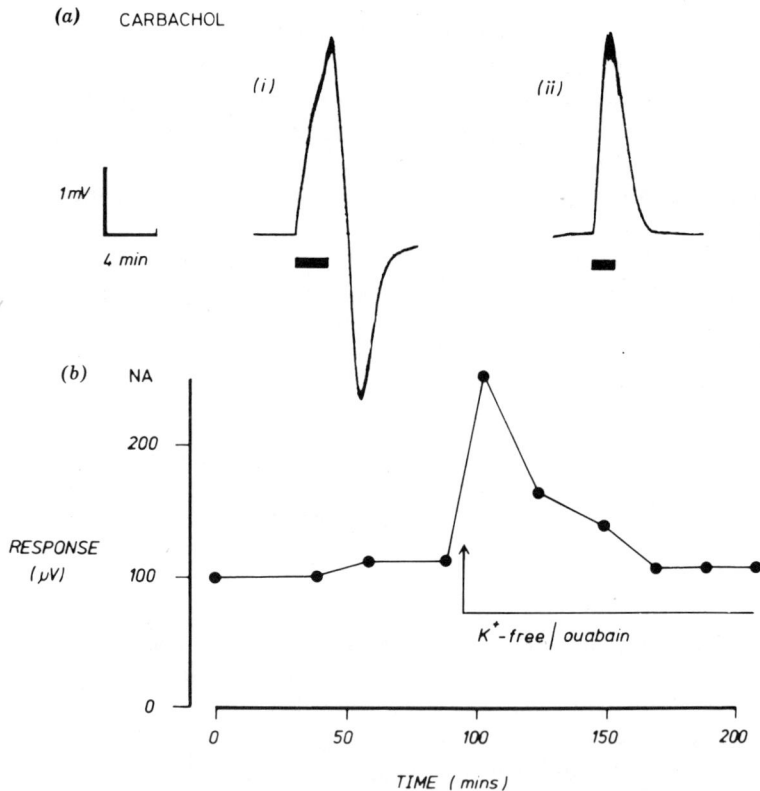

Figure 6 Effect of Na^+-pump inhibition on *(a)* electrogenic Na^+ hyperpolarization and *(b)* noradrenaline-induced hyperpolarization. In *(a)* an isolated rat ganglion was "Na^+-loaded" by applying carbachol (100 μM for 2 min, at bar), which depolarizes the ganglion; on washing out the carbachol, the electrogenic extrusion of accumulated Na^+ ion produces a large after-hyperpolarization (see Ref. 52). In (ii) this after-hyperpolarization is totally inhibited by perfusing the ganglion with a K^+-free solution containing 1 mM ouabain, even though the preceding depolarization (and hence the amount of Na^+ loading) was unchanged. In *(b)* application of the same solution *increased,* instead of reducing, the hyperpolarizations produced by 10 μM noradrenaline, applied for 1 min at 10-min intervals. From M. P. Caulfield, unpublished data.

2.2.3 The Slow IPSP

The slow IPSP is a prolonged ganglion hyperpolarization generated by repetitive orthodromic stimulation, which can be detected when nicotinic transmission is blocked with curare, but which is blocked by atropine (56,57). It has been suggested (30,57) that the slow IPSP is mediated by a catecholamine liberated from inter-neurones via a muscarinic action of the transmitter. The principal evidence for this disynaptic process was that (a) the slow IPSP was blocked by dibenamine (57), phenoxybenzamine (58), or dihydroergotamine (59), or by procedures depleting the ganglion of catecholamines, especially dopamine (60), and (b) a muscarinic agonist, bethanechol, produced an initial hyperpolarization that was blocked in a high Mg^{2+}/low Ca^{2+} solution (30,59). However, these studies predated our current knowledge of the pharmacology of the ganglionic adrenoceptors, and it is now clear that the slow IPSP is *not* inhibited by selective adrenoceptor blockade (8a,61): the previously reported effects of dibenamine and phenoxybenzamine may be attributed to their known activity on muscarinic receptors. Further, there is good evidence for direct hyperpolarizing actions of muscarinic agonists on frog ganglia (62,63; but see 64) and hamster parasympathetic neurones (37).

Thus the relationship between the synaptically evoked slow IPSP and the post-synaptic action of catecholamines is far from clear (65 provides a useful review). Nevertheless, the slow IPSP *has* been subjected to considerable study and the results—even by analogy—might be relevant to concepts of catecholamine action.

At least four ionic mechanisms have been proposed:

1 Nishi and Koketsu (46) proposed that the slow IPSP in frog ganglia was generated by activating an electrogenic sodium pump. Their evidence is essentially the same type of indirect evidence (from pump-inhibition procedures) as that discussed earlier, with the added problems caused by possible effects on transmitter release. Smith and Weight (66) have shown fairly convincingly that the slow IPSP in frog ganglia is much more resistant to pump inhibition than is the electrogenic extrusion of Na^+ from Na^+-loaded ganglia.

2 Kobayashi and Libet (47), while excluding electrogenic Na^+ pumping per se, postulated some other metabolically dependent process for the slow IPSP in rabbit ganglia, which has so far defied definition.

3 Weight and Padjen (62,67) suggested that the slow IPSP in frog ganglia resulted from a reduced Na^+ conductance, G_{Na}, on the basis that (a) the slow IPSP was accompanied by an *increased* input resistance; (b) it was increased in amplitude by membrane hyperpolarization, presumably reflecting an increased driving force $(E_M - E_{Na})$; and (c) the analogous hyperpolarization produced by acetylcholine was reduced on reducing external $[Na^+]$, in spite of a membrane hyperpolarization. The properties of the Ca-induced hyperpolarization are, on the whole, compatible with a similar mechanism, as an alternative to the effect on G_{Ca} (see above); this proposal warrants further study, particularly to identify the type of "Na^+ channel" affected.

4 Hartzell et al. (63) identified the slow IPSP and the hyperpolarizing response of frog *para*sympathetic neurones to bethanechol as resulting from a voltage-sensitive *increase* in G_K. Although the voltage sensitivity might explain some of the anomalous changes in the slow IPSP to applied changes in membrane voltage (see Ref. 46), this mechanism is clearly not applicable to either the slow IPSP or the CA-induced hyperpolarization of sympathetic neurones, since the latter are not accompanied by an appropriate fall in input resistance even at hyperpolarized membrane potentials (see Figure 5).

2.2.4 *Interaction of CA With Muscarinic Agonists*

Muscarinic agonists produce a "slow" depolarization mediated by a decrease in K^+ conductance (68–71). If CA hyperpolarization resulted from a reduced G_{Na}, such as that postulated for the slow IPSP by Weight and Padjen (62), then the two agonists might effectively "cancel" each other—that is, the CA-induced hyperpolarization would be enhanced during a muscarinic depolarization, and vice versa. There are some indications that this might be so. Thus De Groat and Volle (72) noted that the muscarinic depolarization and asynchronous discharge in cat ganglia was very easily inhibited by CA, and we (54) have reported a striking potentiation of the CA hyperpolarization in isolated rat ganglia during an ongoing muscarinic depolarization. [A long-lasting potentiation of the depolarization of rabbit ganglia by muscarinic agonists following application of dopamine has been reported by Libet et al. (73), but this is probably an unrelated phenomenon since it far outlasts the dopamine hyperpolarization.]

  Although initially attractive, recent experiments on the mechanism of action of muscarinic agonists and CA suggest that this conceptual approach to the interaction is far too simplistic, since the two agonists affect *separate* voltage-sensitive components of membrane conductance. Thus muscarinic agonists inhibit a voltage-sensitive component of G_K (the M current) operating (in frogs) over the range of -60 to -10 mV (71; a comparable voltage-sensitive current, operating at perhaps a somewhat more negative membrane potential, may be the primary target in rat ganglion cells: 42), whereas the only current so far identified as peculiarly sensitive to CA is a voltage-sensitive Ca^{2+} current (see above). When the two agonists are added in combination (Figure 7), the effect of each on the respective currents can still be discerned and appears unchanged. Since both the muscarine-sensitive M current and the CA-sensitive Ca^{2+}-activated K^+ current may be concerned primarily with the control of repetitive neuronal discharge rates, interaction of the two agonists may best be sought at this level. This type of "dynamic" interaction may also apply to brain cells.

2.2.5 *Biochemical Mechanisms: Adenylate Cyclase*

Greengard and co-workers (31,34,74,75) have presented an emphatic argument in favor of adenosine 3',5'-cyclic monophosphate (cyclic AMP) as the mediator of both the CA hyperpolarization and the slow IPSP in ganglia. The evidence regarding CA-induced hyperpolarization is essentially as follows: (*a*) dopamine and (less effectively) noradrenaline activate adenylate cyclase in chopped bovine ganglia via a phentolamine-sensitive (i.e., α) receptor (75); (*b*) cyclic AMP and its mono- and dibutyryl derivatives hyperpolarized isolated rabbit ganglia (74); and (*c*) theophylline augmented dopamine hyperpolarization of the rabbit ganglia (74).

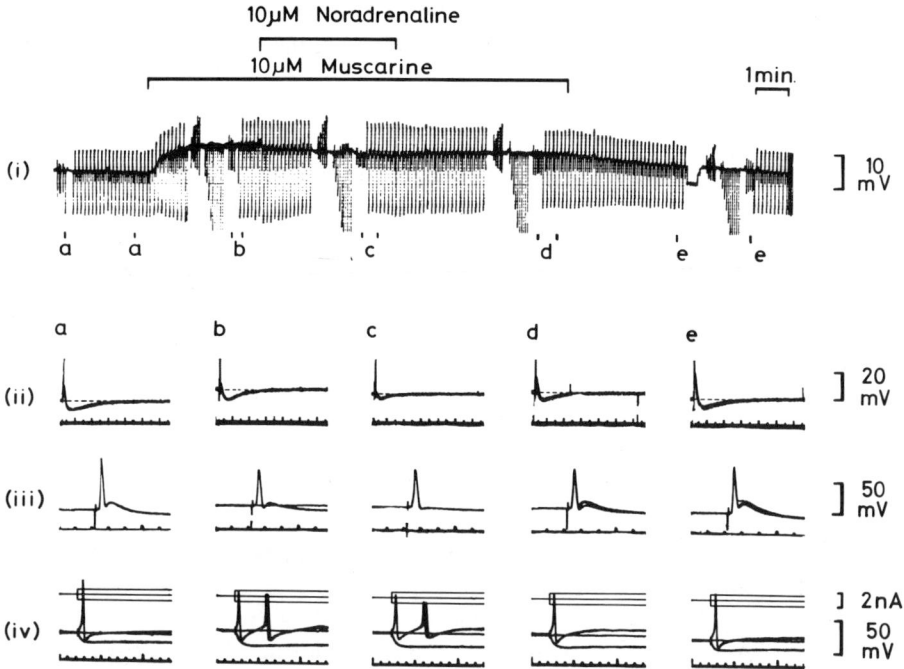

Figure 7   Effect of 10 μM noradrenaline applied in the presence of 10 μM muscarine on the same cell as that illustrated in Figure 5. The upper record is a continuous dc record of membrane potential and of the peak voltage deflections produced by constant hyperpolarizing current injections (-0.8 nA, 520-msec duration: downward deflections). Lower records (ii) to (iv) are oscilloscope records of responses to orthodromic stimuli [(ii) and (iii)] and depolarizing and hyperpolarizing current injections (iv) obtained at the times corresponding to the letters a–e in record (i). Muscarine depolarized the cell by 8 mV, increased its input resistance from 33 to 47 MΩ, and induced repetitive-firing capability during depolarizing current injections (42), probably by inhibiting a voltage-sensitive outward K$^+$ current (71). Superimposition of noradrenaline now hyperpolarized the cell by 3 mV (as opposed to 1 mV in the absence of muscarine: see Figure 5) but did not change the high input resistance or repetitive firing induced by the muscarine; conversely, the adjuvant presence of muscarine has not affected the depressant actions of noradrenaline on the synaptic potential [column c, rows (ii) and (iii)] or on the spike after-negativity [column c, row (iv)]; see Figure 5. From A. Constanti and D. A. Brown, unpublished data.

However, more recent studies have shown that this hypothesis is untenable for rat and rabbit ganglia (if not for cow ganglia, in which correlative electrophysiological studies are impossible). *(a)* In rat ganglia there is a total contradiction between the pharmacological properties of the CA receptors linked to adenylate cyclase and to the hyperpolarization: whereas the latter is α-mediated (see above), cyclic AMP is only increased via β-receptors and not at all via α-receptors (11, 76–79). *(b)* Intracellular application of cyclic AMP fails to replicate the action of CA (80–82). In most studies externally applied cyclic nucleotides were also ineffective (32,82–84); in other experiments (on rats) where adenosine compounds did produce a hyperpolarization, this appeared to result from activation of an *external* adenosine receptor, which was blocked by theophylline (78). *(c)* The original report that CA-induced hyperpolarization is potentiated by phosphodiesterase inhibitors

(74) has not been confirmed in several subsequent studies (32,79,82,83). Thus it appears most unlikely that cyclic AMP is involved in the hyperpolarizing response to CA. (Equally negative conclusions can be deduced regarding the slow IPSP, although, since this is of uncertain relevance to CA effects, further elaboration seems pointless.)

The increase in cyclic AMP produced by β-agonists is, however, very striking (see, e.g., Ref. 79). Its consequences remain to be determined: they may not be directly electrogenic in nature. Kobayashi et al. (81) and Libet (85) have presented a case for a second "modulatory" effect on subsequent muscarinically mediated responses as the direct, long-term response to cyclicAMP, but it is not clear that these effects are mediated via β-receptors. In fact, much of the β-activated adenylate cyclase may be localized to glial cells (75,86).

3 CONCLUSIONS

In a complex situation it is helpful to attempt some form of simplification, if only to focus the most appropriate lines of future experimentation. With a little extrapolation we might arrive at the following schema:

1   The primary action of CA is to reduce voltage-sensitive Ca^{2+} currents in both pre- and postsynaptic elements. In mammals this is mediated by α_2-receptors.

2   Presynaptically, this reduces transmitter release. This is the main factor in depressing ganglionic transmission.

3   Postsynaptically, the main effect might be a modification of the discharge-frequency characteristics, resulting from a reduction of the Ca^{2+}-driven component of K^+ conductance. This may be particularly significant under conditions in which another frequency-determining current, the M current, is suppressed by muscarinic agonists or by peptides (see Refs. 71, 92 and 93). There is also a small hyperpolarization, perhaps reflecting a reduction of G_{Ca} at rest potential; this is of minor significance under normal circumstances, but may become more important in determining firing rate during an ongoing depolarization.

4   Although there may be some form of indirect linkage between the α-receptor and current inhibition, this is not cyclic AMP.

5   β-Agonists produce a large increase in cyclic AMP. This may be related to the facilitation of transmission seen *in vivo* with β-agonists. However, since much of the β-mediated adenylate cyclase activation may be localized to glial cells rather than neurones, it might mediate long-term metabolic effects rather than immediate electrogenic effects.

NOTES ADDED IN PROOF

1   Ivanov and Skok (94) confirm that NA-hyperpolarization of rabbit superior cervical ganglion cells is not accompanied by a clear reduction in input conductance. In contrast, the slow IPSP was accompanied by an *increased* conductance (to K^+ ions) in 5 out of 11 cells tested.

2 P. R. Adams & M. Galvan (personal communication) have now confirmed that 10 μM NA substantially and reversibly reduces the voltage-dependent Ca^{2+}-current in voltage-clamped rat sympathetic neurones.

3 P. M. Dunn (personal communication) has observed that CAs can produce a *de*polarization of freshly-dissected and isolated rat superior cervical ganglia through activation of β_2-receptors.

4 Variable actions of CA (depolarization and hyperpolarization) have also been described on parasympathetic ganglia (95, 96).

REFERENCES

1 W. C. De Groat and R. L. Volle, *J. Pharmacol. Exp. Ther.*, **154**, 1–13 (1966).

2 W. E. Haefely, *Prog. Brain Res.*, **31**, 61–72 (1969).

3 H. W. Kosterlitz and G. M. Lees, *Catecholamines*, in H. Blaschko and E. Muscholl, Eds., Handbook of Experimental Pharmacology, Vol. 33, Springer-Verlag, Berlin, 1972, pp. 762–812.

4 R. L. Volle, in D. A. Kharkevich, Ed., *Pharmacology of Ganglionic Transmission*, Handbook of Pharmacology, Vol. 53, Springer-Verlag, Berlin, 1980, pp. 385–410.

5 W. D. M. Paton and J. W. Thompson, *Abstr. XIX Int. Physiol. Congr.*, pp. 664–665 (1953).

6 D. D. Christ and S. Nishi, *J. Physiol. (Lond.)*, **213**, 107–117 (1971).

7 N. Dun and S. Nishi, *J. Physiol. (Lond.)*, **239**, 155–164 (1974).

8 N. Dun and A. G. Karczmar, *J. Pharmacol. Exp. Ther.*, **200**, 328–335 (1977).

8a N. J. Dun and A. G. Karczmar, *Neurosci. Abstr.*, **5**, 739 (1979).

9 M. P. Caulfield, Pharmacological Characteristics of Catecholamine Receptors in Rat Sympathetic Ganglia, Ph.D. thesis, University of London, 1978.

10 S. Berthelsen and W. A. Pettinger, *Life Sci.*, **21**, 595–606 (1977).

11 T. Lindl, *Neuropharmacology*, **18**, 227–235 (1979).

12 M. S. Kafka and N. B. Thoa, *Biochem. Pharmacol.*, **28**, 2485–2489 (1979).

13 J. L. Willens, *Arch. Pharmacol.*, **279**, 115–126 (1973).

14 S. Z. Langer, *Biochem. Pharmacol.*, **23**, 1793–1800 (1974).

15 K. Starke, *Rev. Physiol. Biochem. Pharmacol.*, **77**, 1–124 (1977).

16 J. E. S. Wikberg, *Nature (Lond.)*, **273**, 164–166 (1978).

17 L. Stjarne, *Arch. Pharmacol.*, **278**, 323–327 (1973).

18 K. Dismukes, A. A. De Boer, and A. H. Mulder, *Arch. Pharmacol.*, **299**, 115–122 (1977).

19 M. Gothert, I.-M. Pohl, and E. Wehking, *Arch. Pharmacol.*, **307**, 21–27 (1979).

20 D. D. Christ and S. Nishi, *Br. J. Pharmacol.*, **41**, 331–338 (1971).

21 J. G. Blackman and R. D. Purves, *J. Physiol. (Lond.)*, **203**, 173–198 (1969).

22 E. M. McLachlan, *J. Physiol. (Lond.)*, **245**, 447–466 (1975).

23 T. Bennett, T. Florin, and A. G. Pettigrew, *J. Physiol. (Lond.)*, **257**, 597–620 (1976).

24 O. Sacchi and V. Perri, *Pflügers Arch.*, **329**, 207–219 (1971).

25 D. A. Brown and M. P. Caulfield, *Br. J. Pharmacol.*, **65**, 435–445 (1979).

26 A. E. Cole and P. Shinnick-Gallagher, *Neurosci. Abstr.*, **5**, 738 (1979).

27 J. M. Cedarbaum and G. K. Aghajanian, *Eur. J. Pharmacol.*, **44**, 375–385 (1977).

28 D. C. U'Prichard and S. H. Snyder, *Life Sci.*, **24**, 79–88 (1979).

29 W. S. Young III and M. J. Kuhar, *Eur. J. Pharmacol.*, **59**, 317–319 (1979).

30 B. Libet, *Fed. Proc.*, **29**, 1945–1956 (1970).

31 P. Greengard and J. W. Kebabian, *Nature (Lond.)*, **260**, 101–108 (1974).

32 N. J. Dun, K. Kaibara, and A. G. Karczmar, *Science (N.Y.)*, **197**, 778–780 (1977).

33 M. Gothert, H.-J. Lox, and J.-M. Rieckesmann, *Arch. Pharmacol.*, **300**, 255–265 (1977).

34 J. W. Kebabian and P. Greengard, *Science (N.Y.)*, **190**, 157–159 (1971).

35 H. Kobayashi and B. Libet, *J. Physiol. (Lond.)*, **208**, 353–372 (1970).

36 K. Koketsu and M. Nakamura, *Jap. J. Physiol.*, **26**, 63–77 (1976).

37 T. Suzuki and R. L. Volle, *Arch. Pharmacol.*, **304**, 15–20 (1978).

38 S. Minota and K. Koketsu, *Jap. J. Physiol.*, **27**, 353–366 (1977).

39 T. Suzuki and R. L. Volle, *Life Sci.*, **24**, 79–88 (1979).

39a T. Suzuki and K. Kusano, *J. Neurobiol.*, **9**, 367–392 (1978).

39b J. P. Horn and D. A. McAfee, *Science (N.Y.)*, **204**, 1233–1235 (1979).

39c J. P. Horn and D. A. McAfee, *J. Physiol. (Lond.)*, **301**, 191–204 (1980).

40 D. A. McAfee and P. J. Yarowsky, *J. Physiol. (Lond.)*, **290**, 507–523 (1979).

41 E. F. Barrett and J. N. Barrett, *J. Physiol. (Lond.)*, **255**, 737–774 (1976).

42 D. A. Brown and A. Constanti, *Br. J. Pharmacol.*, **70**, 593–608 (1980).

43 W. H. Moolenar and I. Spector, *J. Physiol. (Lond.)*, **292**, 307–323 (1979).

44 J. D. Wood and C. J. Meyer, *J. Neurophysiol.*, **42**, 569–581 (1979).

45 M. Nakamura and K. Koketsu, *Life Sci.*, **11**, 1165–1174 (1972).

46 S. Nishi and K. Koketsu, *J. Neurophysiol.*, **31**, 717–728 (1968).

47 H. Kobayashi and B. Libet, *Proc. Natl. Acad. Sci. U.S.A.*, **60**, 1304–1311 (1968).

48a T. Akasu and K. Koketsu, *Experientia (Basel)*, **32**, 57–59 (1976).

48b T. Akasu and K. Koketsu, *Jap. J. Physiol.*, **26**, 289–301 (1976).

49 H. P. Rang and J. M. Ritchie, *J. Physiol. (Lond.)*, **196**, 183–221 (1968).

50 D. A. Brown and C. N. Scholfield, *J. Physiol. (Lond.)*, **242**, 307–319 (1974).

51 D. A. Brown and C. N. Scholfield, *J. Physiol. (Lond.)*, **242**, 320–351 (1974).

52 D. A. Brown, M. J. Brownstein, and C. N. Scholfield, *Br. J. Pharmacol.*, **44**, 651–670 (1972).

53 G. M. Lees and D. I. Wallis, *Br. J. Pharmacol.*, **50**, 79–93 (1974).

54 D. A. Brown and M. P. Caulfield, in E. Szabadi, C. M. Bradshaw, and P. Bevan, Eds., *Recent Advances in the Pharmacology of Adrenoceptors*, Elsevier/North-Holland, Amsterdam, 1978, pp. 57–66.

55 B. Libet, T. Tanaka, and T. Tosaka, *Life Sci.*, **20**, 1863–1870 (1977).

56 R. M. Eccles, *J. Physiol. (Lond.)*, **117**, 196–217 (1952).

57 R. M. Eccles and B. Libet, *J. Physiol. (Lond.)*, **157**, 484–503 (1961).

58 B. Libet and T. Tosaka, *Proc. Natl. Acad. Sci. U.S.A.*, **67**, 667–673 (1970).

59 B. Libet and H. Kobayashi, *J. Neurophysiol.*, **37**, 805–814 (1974).

60 B. Libet and Ch. Owman, *J. Physiol. (Lond.)*, **237**, 635–662 (1974).

61 A. E. Cole and P. Shinnick-Gallagher, *Brain Res.*, **187**, 226–230 (1980).

62 F. F. Weight and A. Padjen, *Brain Res.*, **55**, 219–224 (1973).

63 H. C. Hartzell, S. W. Kuffler, R. Stickgold, and D. Yoshikami, *J. Physiol. (Lond.)*, **271**, 817–846 (1977).

64 N. J. Dun and A. G. Karczmar, *Proc. Natl. Acad. Sci. U.S.A.*, **75**, 4029–4032 (1978).

65 J. P. Gallagher, P. Shinnick-Gallagher, A. E. Cole, W. H. Griffith III, and B. J. Williams, *First Galveston Symposium on Neurosciences, in press*.

66 P. A. Smith and F. F. Weight, *Nature (Lond.)*, **267**, 68–70 (1977).

67 F. F. Weight and A. Padjen, *Brain Res.*, **55**, 225–228 (1973).

68 F. F. Weight and Z. Votava, *Science (N.Y.)*, **170**, 755–758 (1970).

69 K. Kuba and K. Koketsu, *Jap. J. Physiol.*, **26,** 651–669 (1976).

70 K. Kuba and K. Koketsu, *Jap. J. Physiol.*, **26,** 703–716 (1976).

71 D. A. Brown and P. R. Adams, *Nature (Lond.)*, **283,** 673–676 (1980).

72 W. C. De Groat and R. L. Volle, *J. Pharmacol. Exp. Ther.*, **154,** 200–215 (1966).

73 B. Libet, H. Kobayashi, and T. Tanaka, *Nature (Lond.)*, **258,** 155–157 (1965).

74 D. A. McAfee and P. Greengard, *Science (N.Y.)*, **178,** 310–312 (1972).

75 P. Greengard and J. W. Kebabian, *Fed. Proc.*, **33,** 1059–1067 (1974).

76 H. Cramer, D. G. Johnson, I. Hanbauer, S. D. Silverstein, and I. J. Kopin, *Brain Res.*, **53,** 97–105 (1973).

77 T. Lindl and H. Cramer, *Biochem. Biophys. Res. Commun.*, **65,** 731–739 (1975).

78 D. A. Brown, M. P. Caulfield, and P. J. Kirby, *J. Physiol. (Lond.)*, **290,** 441–451 (1979).

79 L. Quenzer, D. Yahn, K. Alkhadi, and R. L. Volle, *J. Pharmacol. Exp. Ther.*, **208,** 31–36 (1979).

80 J. P. Gallagher and P. Shinnick-Gallagher, *Science (N.Y.)*, **189,** 851–853 (1977).

81 H. Kobayashi, T. Hashiguchi, and N. S. Ushiyama, *Nature (Lond.)*, **271,** 268–270 (1978).

82 N. A. Busis, F. F. Weight, and P. A. Smith, *Science (N.Y.)*, **200,** 1079–1081 (1978).

83 T. Akasu and K. Koketsu, *Br. J. Pharmacol.*, **60,** 331–336 (1977).

84 N. J. Dun and A. G. Karczmar, *J. Pharmacol. Exp. Ther.*, **202,** 89–96 (1977).

85 B. Libet, *Life Sci.*, **24,** 1043–1058 (1979).

86 A. G. Gilman and M. Nirenberg, *Proc. Natl. Acad. Sci. U.S.A.*, **68,** 2165–2168 (1971).

87 D. A. Brown and A. J. Higgins, *Br. J. Pharmacol.*, **66,** 108–109P (1979).

88 D. A. Brown and S. Marsh, *Brain Res.*, **156,** 187–191 (1978).

89 D. A. Brown, K. B. Jones, J. V. Halliwell, and J. P. Quilliam, *Nature (Lond.)*, **226,** 958–959 (1970).

90 C. J. Courtice, *J. Physiol. (Lond.)*, **268,** 1–2P (1977).

91 D. A. Brown and A. Constanti, *Br. J. Pharmacol.*, **63,** 217–224 (1978).

92 P. R. Adams and D. A. Brown, *Br. J. Pharmacol.*, **68,** 353–355 (1980).

93 D. A. Brown, A. Constanti, and S. Marsh, *Brain Res.*, **193,** 614 (1980).

94 A. Y. Ivanov and V. I. Skok, *J. aut. nerv. Syst.* **1,** 255 (1980).

95 W. C. DeGroat and A. M. Booth, *Fed. Proc.* **39,** 2990 (1980).

96 W. H. Griffith, J. P. Gallagher and P. Shinnick-Gallagher, *Fed. Proc.* **38,** 276 (1978).

CHAPTER FOUR

MECHANISMS INVOLVED IN α-ADRENERGIC EFFECTS OF CATECHOLAMINES

John H. Exton

Howard Hughes Medical Institute and Department of Physiology, Vanderbilt University School of Medicine, Nashville, Tennessee

1 INTRODUCTION

Epinephrine and norepinephrine play a major role in the control of most body functions and in the responses of organisms to external stimuli, particularly those that are threatening to life. They act on the cells of their target tissues through specific binding sites termed adrenergic receptors, which are divided into two main classes, α and β, based on differences in their sensitivities to various agonists and antagonists (Table 1). In some tissues, the physiological responses mediated by α-receptors are different or opposite to those mediated by β-receptors, but in other tissues the responses are similar. The major α-adrenergic responses are contraction of most smooth muscles, relaxation of intestinal smooth muscle, inhibition of insulin secretion, stimulation of salivary K^+ and water secretion, aggregation of platelets, and stimulation of glycogen breakdown in liver in some species.

   α-Receptors have been further subdivided into postsynaptic α_1-receptors located on the target cells, which mediate the majority of α-adrenergic responses, and

TABLE 1   Characteristics of α- and β-Adrenergic Receptors

Characteristics	*α-Receptors*	*β-Receptors*
Sensitivity to agonists[a]	Epi > Norepi > Phenyl > Iso	Iso > Epi > Norepi > Phenyl
Typical antagonists	Phentolamine	Propranolol
	Phenoxybenzamine	Dichloroisoproterenol
	Ergot alkaloids	Alprenolol
Typical responses		
Most smooth muscles	Contraction	Relaxation
Intestinal smooth muscle	Relaxation	Relaxation
Salivary glands	K^+ and H_2O secretion	Amylase secretion
Cardiac muscle	Increased contractility	Increased contractility and heart rate
Platelets[b]	Aggregation	Inhibition of aggregation
Pancreas	Inhibition of insulin secretion	Stimulation of insulin secretion
Adipose tissue[b]	Inhibition of lipolysis	Lipolysis
Liver[b]	Glycogenolysis	Glycogenolysis
Skeletal muscle		Glycogenolysis

[a]Abbreviations: Epi, epinephrine; Norepi, norepinephrine; Phenyl, phenylephrine; Iso, isoproterenol.
[b]Species variations observed.

presynaptic α_2-receptors located on the sympathetic nerve endings, which are involved in the feedback inhibition of catecholamine secretion from these endings (1). However, it is becoming clear that the α-receptors of some target cells exhibit the properties of α_2-receptors; that is, this subclass of α-receptors is not always located presynaptically (2). This chapter will be concerned mainly with postsynaptic α_1-adrenergic responses and will explore the evidence that these are due to an increase in cytosolic Ca^{2+} concentration.

2   IDENTIFICATION OF α-RECEPTORS

In comparison to the β-adrenergic system, the mechanisms involved in the α-adrenergic actions of catecholamines are not well understood. Only recently have methods become available to identify the α-receptor. These have involved the use of the α-antagonists [3H]dihydroergocryptine and [3H]WB-4101 and the labeled agonist [3H]clonidine as radioligands (3–12). In addition, [3H]epinephrine and [3H]norepinephrine have been used in combination with propranolol, catechol, and ascorbate (4,5). However, it is clear from several studies that these radioligands bind preferentially to different sites (4–6,10). Thus in brain membranes, [3H]epinephrine, [3H]norepinephrine, and [3H]clonidine bind to stereoselective sites that have the characteristics of α-receptors, whereas [³H]WB-4101 binds to other sites which also have the features of α-receptors, but which have greater affinity for antagonists and lesser affinity for agonists (4,6). In these tissues, [³H]dihydroergocryptine behaves as a mixed agonist–antagonist and binds to both types of α-receptor (7). These results can be attributed partly to the presence of both α_1-and α_2-receptors in brain (10).

In rat liver, where cell heterogeneity is much less of a problem than in brain, there is evidence that [³H]catecholamines bind preferentially to high-affinity plasma membrane sites which are α_1 in type and probably represent the physiological α-receptor (5,12). These sites show a stereoselectivity and potency series for agonists and antagonists which are essentially identical to those seen for two characteristic physiological α-adrenergic responses in intact hepatocytes: stimulation of calcium release and activation of phosphorylase (5) (Figure 1). On the other hand, [³H]dihydroergocryptine binds preferentially to sites which are also of the α_1 type, but are very different in many respects from those which preferentially bind catecholamines (5,12). These sites are of unknown function and exhibit higher affinity for α-antagonists and lower affinity for α-agonists than seen for the sites corresponding to the physiological α-receptor (5,8,9).

Kunos et al. (11) have also obtained data indicating that [³H]dihydroergocryptine binds predominantly to sites in uterine smooth muscle which are different from the α-receptors mediating the physiological contractile response. These findings indicate that the use of labeled ergot alkaloids to identify the physiological α-receptor may be misleading unless binding is correlated with relevant physiological parameters.

In most tissues, the cellular location of the membranes containing α-receptor binding sites has not been identified. In liver, however, sites have been located on plasma membranes, but not on microsomes or mitochondria (5,8,9). Furthermore, epinephrine immobilized by covalent linkage to a large polymer (13,000 daltons) potently elicits α-responses in isolated hepatocytes under conditions under where it is not released from the polymer and therefore probably cannot penetrate the cell membrane (13).

3 α-ADRENERGIC RECEPTORS AND CYCLIC NUCLEOTIDES

In all tissues examined except for certain regions of the brain, activation of α-receptors does not cause accumulation of adenosine 3′,5′-cyclic monophosphate (cyclic AMP) or activation of cyclicAMP-dependent protein kinase. In cerebral cortex of several species, the rise in cyclic AMP induced by catecholamines is partly blocked by α-antagonists and thus appears to be mediated in part by α-receptors (for references, see 14). It has also been found that in liver cells depleted of calcium by treatment with ethyleneglycol-bis (β-aminoethylether) N, N-tetraacetic acid (EGTA), α-adrenergic stimulation promotes cyclic AMP accumulation due to activation of adenylate cyclase (15), whereas in normal liver cells, this does not happen (16).

In platelets, human adipose tissue, pancreatic islets, and melanophores, on the

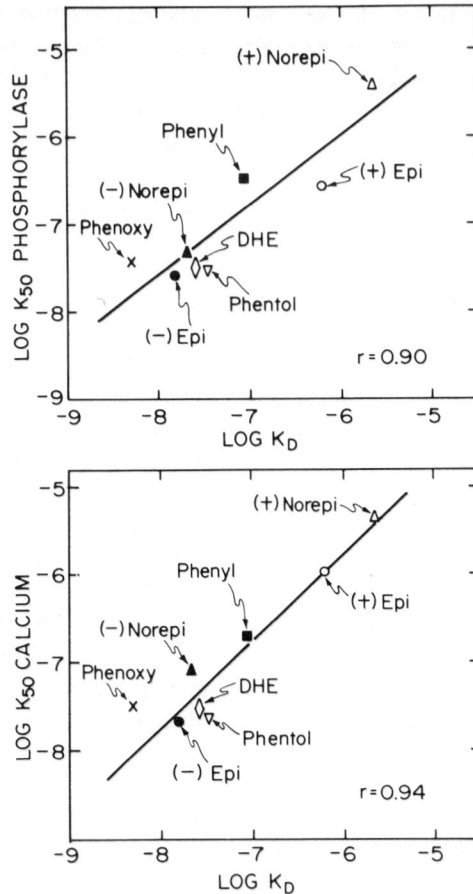

Figure 1   Correlations between the effects of α-agonists and antagonists on [^{3}H]epinephrine binding to rat liver plasma membranes and on phosphorylase activation and calcium efflux in isolated rat hepatocytes. The concentrations of α-adrenergic agents half-maximally activating phosphorylase or calcium release (α-agonists), or half-maximally inhibiting phosphorylase activation or calcium release induced by 40 nM epinephrine (α-antagonists) are plotted logarithmically as K_{50}'s versus the K_D's (dissociation constants) calculated from the [^{3}H]epinephrine binding data. From Ref. 5.

other hand, α-agonists have been reported to decrease cyclic AMP when this is raised by other agents (for references, see 17). Jakobs et al. (18,19) have shown that in human platelets, the effect is due to inhibition of adenylate cyclase and is related to the ability of catecholamines to cause aggregation. In summary, there is evidence that α-agonists can inhibit, stimulate, or be without effect on adenylate cyclase depending on the tissue or conditions. The inhibitory effects appear to be mediated by α_2-receptors.

Stimulation of α-receptors in certain smooth muscles, parotid, platelets, cerebellum, and pineal gland has been reported to increase the level of guanosine 3',5'-cyclic monophosphate (cyclic GMP) (20–23). The effect is dependent upon the presence of Ca^{2+} ions. Since α-agonists have not been shown to activate guanylate cyclase directly, the effect may be secondary to a rise in intracellular Ca^{2+}. Although one group reported that catecholamines acting through α-receptors increased cyclic GMP in liver (24), they also claimed that carbachol, insulin, A23187 ionophore, and glucagon had the same effect. We and others (25) have been unable to confirm these findings. The role of cyclic GMP in α-adrenergic actions in liver and other tissues is uncertain, although it has been suggested that the nucleotide may exert a negative feedback on processes regulating the intracellular level of Ca^{2+} (26).

4 ROLE OF Ca^{2+} IONS IN α_1-ADRENERGIC RESPONSES

By analogy with the β-receptor/cyclic AMP system, there is probably a primary intracellular second messenger for the α_1-system which influences a variety of cell processes. There is much evidence that this putative mediator influences cell calcium metabolism resulting in a rise in cytosolic Ca^{2+} concentration. The evidence for this includes (a) the dependence upon calcium for many α-adrenergic responses (14,20,27–41); (b) the stimulatory effect of α-adrenergic agonists on calcium fluxes in several tissues (26–28,38–50); (c) the calcium-dependent mimickry of the effects of α-adrenergic activation by agents such as vasopressin, angiotensin II, and the ionophore A23187, which alter Ca^{2+} fluxes in many tissues (24,26–28,36,49,51,52); (d) the inhibition by calcium antagonists of many α-responses (33,35,53); and (e) the known effects of Ca^{2+} ions on processes influenced by α-adrenergic stimulation (26,27,29,54–58).

The role of Ca^{2+} ions in α_1-adrenergic mechanisms is well illustrated for glycogenolysis in rat liver and K^+ release in rat parotid gland. The sustained effects of α-agonists on both processes are inhibited by the chelator EGTA (27,30,35,36, 38,39,41). The ionophore A23187 in the presence of Ca^{2+} reproduces the effects of α-agonists on both processes (27,28,35,36,49,51,52). α-Adrenergic stimulation of calcium fluxes in liver (Figure 2) and parotid glands has been observed (27,28, 39–42,45–51). One of the key enzymes involved in α-adrenergic activation of liver phosphorylase activation, namely, phosphorylase b kinase, is known to be stimulated by concentrations of Ca^{2+} ions within the probable intracellular range (for references, see Ref. 27). Ca^{2+} ions have also been postulated to play a key role in K^+ release from salivary and lacrimal gland by increasing the permeability of the plasma membrane to K^+ (35,38).

Ca^{2+} ions play a fundamental role in the contraction of striated muscle through their release from sarcoplasmic reticulum and binding to troponin C. This binding decreases the inhibition of the interaction between actin and myosin exerted by the troponin–tropomyosin complex. An analogous system may also operate in the smooth muscle of higher vertebrates (54–56,58). However, in the smooth muscle of most species, Ca^{2+} ions appear to act additionally by stimulating a specific kinase which phosphorylates the light chain of myosin, thereby altering its interaction with actin (58). Recent studies have shown that a calcium-dependent regulator protein (CDR) is required for interaction of Ca^{2+} ions with the light-chain myosin kinase and that this protein is very similar to troponin C and probably identical with the Ca^{2+}-binding proteins which stimulate adenylate cyclase, cyclic nucleotide phos-

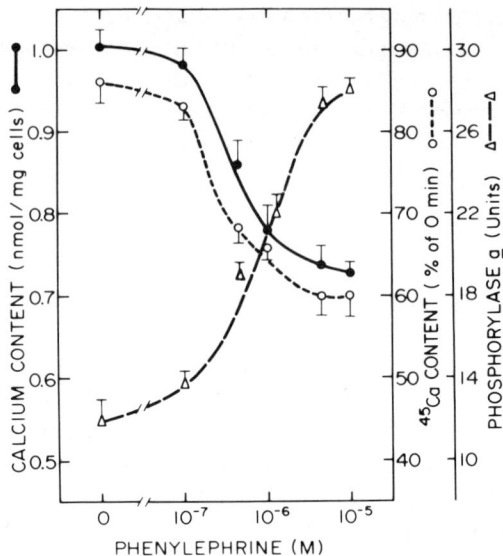

Figure 2 Dose–response curves for the effects of the specific α-agonist phenylephrine on phosphorylase activation, calcium content, and ^{45}Ca content in rat hepatocytes previously incubated with ^{45}Ca. From Ref. 49.

phodiesterase, and Ca^{2+}-stimulated adenosine triphosphata⸱⸱ (ATPase) in certain tissues (59). Whatever the mechanism by which α-adrenergic contraction occurs in smooth muscle, it is generally agreed that the primary intracellular signal is a rise in cytosolic Ca^{2+} (54–56,58).

The two mechanisms by which $α_1$-adrenergic activation increases cytosolic Ca^{2+} are by causing the mobilization of Ca^{2+} from intracellular stores and by promoting the influx of extracellular Ca^{2+} (or Ca^{2+} loosely bound to superficial sites) (Figure 3). In some tissues (e.g., salivary glands) the first mechanism appears to be mainly involved since most of the α-responses are dependent upon external Ca^{2+} (26, 30,33–36,38,39,46,47). In others (e.g., liver) mobilization of internal Ca^{2+} is the major mechanism (41,45,49,50) and the actions of α-agonists persist when extracellular Ca^{2+} is removed in *short-term* experiments (41,49). In smooth muscle, the intracellular calcium store mobilized during contraction is probably the sarcoplasmic reticulum (54–56,60), although mitochondria and calcium bound to the inner surface of the plasma membrane may also be involved (61,62) (Figure 3). In liver, α-adrenergic stimulation can release up to 30% of total cell calcium (49), indicating that the cation is coming from a major intracellular source (45,49). Recent experiments (50) have shown that the mitochondria are the major source of the calcium mobilized in liver by α-agonists (Figure 4), but release from other organelles and membranous structures has not been totally excluded. The rise in cytosolic Ca^{2+} concentration is postulated to stimulate calcium-sensitive enzymes and also result in the expulsion of Ca^{2+} from liver cells by a pump presumed to be located in the plasma membrane or adjacent endoplasmic reticulum (50).

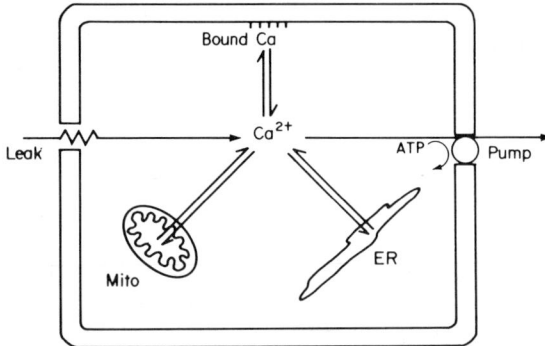

Figure 3 Factors regulating the cytosolic concentration of Ca^{2+} in liver cells. Leak refers to the penetration of Ca^{2+} into the cell due to the very high extracellulal cytosolic Ca^{2+} concentration gradient. Pump refers to the energy-dependent Ca^{2+} extrusion mechanism. Mito and ER refer to the mitochondria and endoplasmic reticulum, which contain a high concentration of calcium. Bound Ca refers to calcium bound to specific sites on plasma and other cell membranes.

The mechanisms by which α_1-adrenergic activation affects cell calcium are unknown. It is generally considered that, in those tissues in which calcium influx occurs, there is opening of the Ca^{2+} "gates" across the plasma membrane rather than inhibition of the putative pump which expels Ca^{2+} (Figure 3). The calcium channels controlled by the α-receptor appear to be the same as those controlled by muscarinic cholinergic receptors in certain tissues (35). Since the α-receptor could be connected to the "gating" mechanism in the plasma membrane, it is not necessary to postulate the existence of a second messenger for this effect. In those tissues,

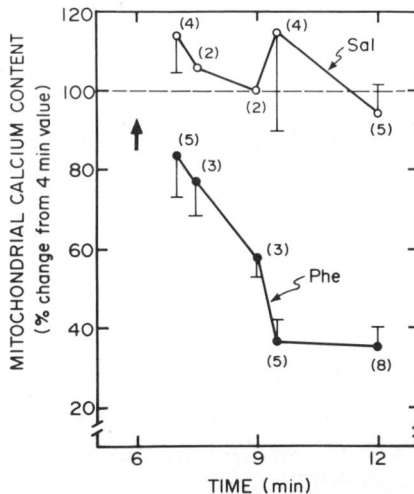

Figure 4 Mobilization of mitochondrial Ca^{2+} by phenylephrine in the perfused rat liver as a function of time. Data are plotted as a percent of mitochondrial calcium measured at 6 min (i.e., immediately prior to infusion of saline or 10^{-5} M phenylephrine). From Ref. 50.

such as liver, where the major release of internal calcium is from mitochondria (Figures 3 and 4) and the α-receptor is localized to the plasma membrane, some primary intracellular messenger or coupling of these organelles to the plasma membrane is needed.

Many α-adrenergic responses can be attributed to changes in cell calcium. As indicated above, there is much evidence that the effects of α-agonists on salivary K^+ release, glycogen breakdown in liver, and smooth muscle contraction involve a rise in cytosolic Ca^{2+}. There are also data indicating a role for Ca^{2+} ions in the following actions of α-agonists: (a) stimulation of hepatic gluconeogenesis (63); (64,65); (b) inhibition of hepatic glycogen synthesis (63); (c) glycogenolysis in other tissues (64,65); (d) platelet aggregation (26); (e) cyclic GMP accumulation in several tissues (20,21,24); (f) cyclic AMP accumulation in cerebral cortex (14); (g) hyperpolarization and relaxation of intestinal smooth muscle (31); (h) glucose metabolism and amylase release in salivary gland (32–34); (i) peroxidase secretion from lacrimal gland (38–40); and (j) K^+ fluxes in liver (42,50,52). However, it has not been determined whether or not these ions are important in α-adrenergic stimulation of hepatic amino acid transport (66,67); oxygen consumption and lactate release (68); glucose uptake to adipose tissue, diaphragm, and heart (69–71); inotropism and cyclic GMP accumulation in heart (70,72); and α-adrenergic inhibition of insulin secretion (73) and of cyclic AMP accumulation in several tissues (17).

It is possible that the putative primary messenger of the α-adrenergic system may affect some cellular processes more directly than via changes in cell calcium.

Garrison and co-workers (63,74) have incubated isolated hepatocytes in medium containing $^{32}PO_4^{3-}$ and examined the effects of glucagon, cyclic AMP, catecholamines, vasopressin, and angiotensin II on the labeling of cytosolic proteins separated by polyacrylamide gel electrophoresis. They found that stimulation of α-receptors by catecholamines in the presence of β-blockade or exposure to vasopressin and angiotensin II increased the phosphorylation of most, if not all, of the 12 protein bands whose labeling was increased by glucagon or cyclic AMP. Several of these bands were later identified (63) as enzymes known to be phosphorylated by cyclic AMP-dependent protein kinase (phosphorylase, glycogen synthase, and pyruvate kinase). This indicates that either α-adrenergic activation increases the activity of a protein kinase with specificity similar to cyclic AMP-dependent protein kinase, or decreases the activity of a phosphoprotein phosphatase. The possibility that the protein phosphorylation induced by α-adrenergic activation is due to activation of a kinase modulated by Ca^{2+} is supported by the findings (63) that the phosphorylation of liver cell proteins induced by vasopressin, angiotensin II, and A23187 was abolished in Ca^{2+}-free medium.

5 POSSIBLE ROLE OF PHOSPHATIDYLINOSITOL BREAKDOWN IN α_1-ADRENERGIC RESPONSES

Michell (75–79) and others (80–83) have observed that α-adrenergic stimulation increases the incorporation of ^{32}P into phosphatidylinositol in liver, parotid gland, and other tissues, and it has been postulated that altered phosphatidylinositol metabolism is a primary response to activation of α_1-receptors (75,78,84). Current evidence indicates that the earliest event is a stimulation of phosphatidylinositol breakdown to 1,2-diacylglycerol (79) and that the increased ^{32}P incorporation occurs during resynthesis of phosphatidylinositol stimulated by increased levels of 1,2-diacylglycerol.

Whatever the mechanism(s) by which phosphatidylinositol breakdown may mediate α_1-adrenergic responses, there is much evidence suggesting that this breakdown is coupled in some way to alterations in cell calcium metabolism (78,84). For example, labeling of phosphatidylinositol with ^{32}P is stimulated in many tissues by compounds (muscarinic cholinergic agents, vasopressin, angiotensin, 5-hydroxytryptamine, histamine, substance P, pancreozymin, plant lectins) that alter Ca^{2+} fluxes (75,78,84). The possibility that the change in phosphatidylinositol metabolism is secondary to the increase in Ca^{2+} is excluded by the observations that it is not affected by removal of extracellular Ca^{2+} and is not mimicked by the ionophore A23187 in a Ca^{2+}-dependent manner (76–80).

The link, if any, between phosphatidylinositol breakdown and the alterations in cell calcium caused by α_1-adrenergic activation is unknown. It is possible that the P-lipid change could alter the binding of Ca^{2+} to the plasma membrane. Alternatively, the change may release a natural ionophore that promotes Ca^{2+} influx into the cell and/or releases Ca^{2+} from intracellular stores. It must be stressed, however, that the relationship(s) between the P lipid and the calcium alterations is far from clear and that both events could arise separately from a primary response to α_1-adrenergic activation.

Another change in phosphatidylinositol metabolism induced by α-adrenergic stimulation in one tissue is increased breakdown of triphosphoinositide (phosphatidylinositol 4,5-bisphosphate) (84,85). This compound has a high affinity for Ca^{2+} and appears to be located predominantly in plasma membranes (84). It has been proposed that α-adrenergic stimulation could release "trigger" amounts of Ca^{2+} which would stimulate the Ca^{2+}-sensitive plasma membrane phosphodiesterase and phosphomonoesterase catalyzing the breakdown of triphosphoinositide (84,85), and that this could result in altered membrane permeability to ions (84,85) and perhaps other effects.

6 CONCLUDING COMMENTS

Clearly, much more work is needed to define the mechanisms involved in α-adrenergic phenomena. To date, α-receptors have been located only on the plasma membranes of target cells, although their possible presence on intracellular structures cannot yet be excluded. By analogy with the adenylate cyclase system, it may be speculated that activation of α_1-receptors increases the activity of a plasma membrane-bound enzyme, which generates a primary intracellular α-signal that promotes the net release of mitochondrial calcium and/or increased influx of extracellular calcium and other effects (Figure 5). There is much evidence that phosphatidylinositol breakdown is an early response to stimulation of α-adrenergic or muscarinic cholinergic receptors and is not secondary to a change in cell calcium. However,

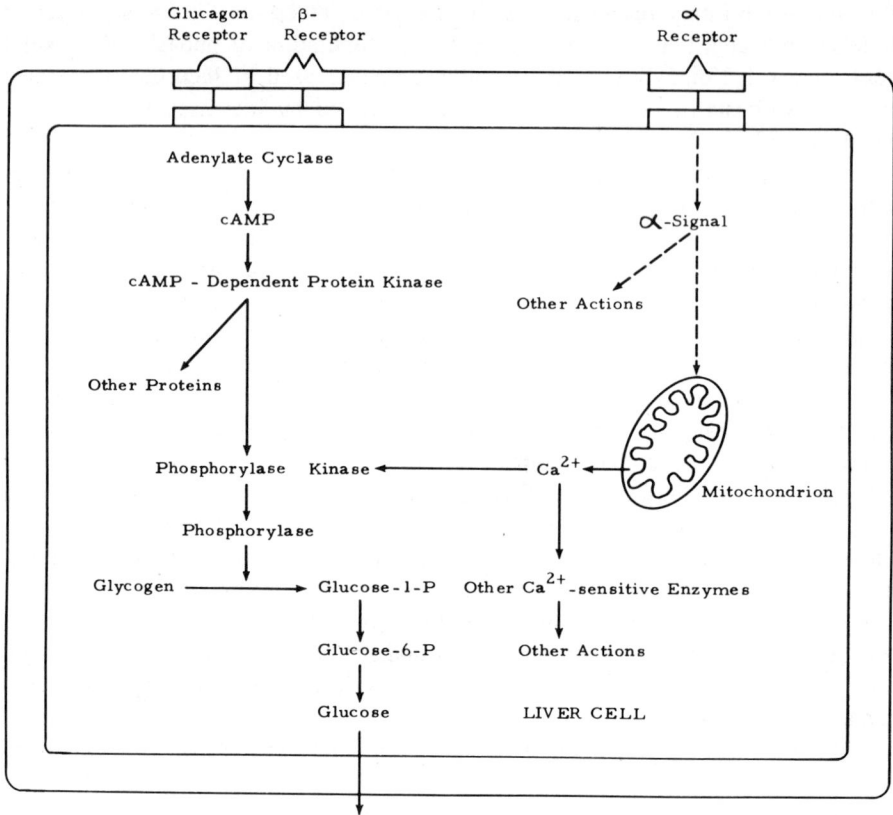

Figure 5 Postulated mechanisms by which activation of glucagon and β-adrenergic and α-adrenergic receptors leads to glycogenolysis and other responses in the liver cell. Dashed lines refer to mechanisms for which no evidence currently exists.

it is not known whether this phospholipid change is responsible for the alteration(s) in cell calcium or that both changes result separately from a more primary response.

It seems probable that a rise in cytosolic Ca^{2+} is responsible for most of the changes induced by α_1-receptor activation. In view of the ubiquity of calcium-dependent regulator protein in tissues (59,86) and the growing evidence that this mediates many intracellular effects of Ca^{2+}, it is tempting to suggest that it is also involved in α-adrenergic responses. In addition, the discovery of protein kinases activated by calcium-dependent regulator protein plus Ca^{2+} (87) opens the possibility that the α-adrenergic system could act partly by stimulating the phosphorylation of specific proteins (Figure 5). Such a system would explain the increased labeling of specific cytosolic proteins in α-adrenergically stimulated hepatocytes incubated with $^{32}P_i$ (74) and would show many parallels to the β-receptor–cyclic AMP–cyclic AMP-dependent protein kinase system. However, it is possible that some α_1-adrenergic responses could also result from a more direct action of the putative primary intracellular α-signal on cell proteins (Figure 5).

7 SUMMARY

Epinephrine and norepinephrine binding sites with the physiological characteristics of α_1-adrenergic receptors have been identified in the plasma membranes of liver and other cells. Interaction of catecholamines with the receptors causes a mobilization of Ca^{2+} ions from mitochondria and perhaps other intracellular stores in liver cells. In other cells, there may also be influx of extracellular Ca^{2+} ions. Evidence is presented in support of the hypothesis that the rise in cytosolic calcium ions resulting from these changes is responsible for many of the α_1-adrenergic actions of catecholamines. Possible mechanisms by which activation of α_1-adrenergic receptors causes changes in calcium and other aspects of cellular metabolism are discussed. These include the possible role of phosphatidylinositol breakdown in early α_1-adrenergic events, the generation at the plasma membrane of an intracellular second messenger which causes release of mitochondrial Ca^{2+}, and the possible role of a Ca^{2+}- and CDR-dependent protein kinase in α_1-adrenergic-induced enzyme changes.

REFERENCES

1   S. Z. Langer, *Biochem. Pharmacol.*, **23**, 1793 (1974).

2   S. Berthelsen and W. A. Pettinger, *Life Sci.*, **21**, 595 (1977).

3   R. J. Lefkowitz, *Fed. Proc.*, **37**, 123 (1978).

4   D. C. U'Prichard and S. H. Snyder, *J. Biol. Chem.*, **252**, 6450 (1977).

5   M. F. El-Refai, P. F. Blackmore, and J. H. Exton, *J. Biol. Chem.*, **254**, 4375 (1979).

6   D. C. U'Prichard, D. A. Greenberg, and S. H. Snyder, *Mol. Pharmacol.*, **13**, 454 (1977).

7   D. A. Greenberg and S. H. Snyder, *Mol. Pharmacol.*, **14**, 38 (1978).

8   G. Guellaen, M. Yates-Aggerbeck, G. Vauquelin, D. Strosberg, and J. Hanoune, *J. Biol. Chem.*, **253**, 1114 (1978).

9   W. R. Clarke, L. R. Jones, and R. J. Lefkowitz, *J. Biol. Chem.*, **253**, 5975 (1978).

10   D. C. U'Prichard, M. E. Charness, D. Robertson, and S. H. Snyder, *Eur. J. Pharmacol.*, **50**, 87 (1978).

11   G. Kunos, B. Hoffman, Y. N. Kwok, W. H. Kan, and L. Mucci, *Nature (Lond.)*, **278**, 254 (1979).

12   M. F. El-Refai and J. H. Exton, *Eur. J. Pharmacol.*, **62**, 201 (1980).

13   J.-P. Dehaye, P. F. Blackmore, J. C. Venter, and J. H. Exton, *J. Biol. Chem.*, **255**, 3905 (1980).

14   U. Schwabe and J. W. Daly, *J. Pharmacol. Exp. Ther.*, **202**, 134 (1977).

15   T. M. Chan and J. H. Exton, *J. Biol. Chem.*, **252**, 8645 (1977).

16   A. D. Cherrington, F. D. Assimacopoulos, S. C. Harper, J. D. Corbin, C. R. Park, and J. H. Exton, *J. Biol. Chem.*, **251**, 5209 (1976).

17   G. A. Robison, R. W. Butcher, and E. W. Sutherland, *Cyclic AMP*, Academic Press, New York, 1971.

18   K. H. Jakobs, W. Saur, and G. Schultz, *J. Cyclic Nucleotide Res.*, **2**, 381 (1976).

19   K. H. Jakobs, W. Saur, and G. Schultz, *Naunym-Schmiedebergs Arch. Pharmacol.*, **302**, 285 (1978).

20   G. Schultz, K. Schultz, and J. G. Hardman, *Metabolism*, **24**, 429 (1975).

21   R. F. O'Dea and M. Zatz, *Proc. Natl. Acad. Sci. U.S.A.*, **73**, 3398 (1976).

22   F. R. Butcher, R. Lynn, C. Emler, and M. Nmerovski, *Mol. Pharmacol.*, **12**, 862 (1976).

23   J. A. Ferendelli, D. A. Kinscherf, and M. M. Chang, *Brain Res.*, **84**, 63 (1975).

24   R. H. Pointer, F. R. Butcher, and J. N. Fain, *J. Biol. Chem.*, **251**, 2987 (1976).

25   D. A. Hems, C. J. Davies, and K. Siddle, *FEBS Lett.*, **87**, 196 (1978).

26   M. J. Berridge, *Adv. Cyclic Nucleotide Res.*, **6**, 1 (1975).

27   F. D. Assimacopoulos-Jeannet, P. F. Blackmore, and J. H. Exton, *J. Biol. Chem.*, **252**, 2662 (1977).

28   S. Keppens, J. R. Vandenheede, and H. deWulf, *Biochim. Biophys. Acta*, **496**, 448 (1977).

29   G. Van de Werve, L. Hue, and H-G. Hers, *Biochem. J.*, **162**, 135 (1977).

30   Z. Selinger, S. Batzri, S. Eimerl, and M. Schramm, *J. Biol. Chem.*, **248**, 369 (1973).

31   E. Bulbring and T. Tomita, *Proc. R. Soc. Lond. Ser. B*, **197**, 271 (1977).

32   M. P. Thompson and D. H. Williamson, *Biochem. J.*, **160**, 597 (1976).

33   B. A. Leslie, J. W. Putney, Jr., and J. M. Sherman, *J. Physiol.*, **260**, 351 (1976).

34   O. H. Peterson, N. Ueda, R. A. Hall, and T. A. Gray, *Pflügers Arch. Eur. J. Physiol.*, **372**, 231 (1977).

35   S. H. Marier, J. W. Putney, Jr., and C. M. Van de Walle, *J. Physiol.*, **279**, 141 (1978).

36   J. W. Putney, Jr., B. A. Leslie, and S. H. Marier, *Am. J. Physiol.*, **235**, C128 (1978).

37   W. W. Douglas and A. M. Poisner, *J. Physiol.*, **105**, 528 (1963).

38   R. J. Parod and J. W. Putney, Jr., *J. Physiol.*, **281**, 359, 371 (1978).

39   J. W. Putney, Jr., C. M. Van deWalle, and B. A. Leslie, *Am. J. Physiol.*, **235**, C188 (1978).

40   G. Herman, S. Busson, L. Ovtracht, C. Maurs, and B. Rossignol, *Biol. Cell.*, **31**, 255 (1978).

41   S. J. Weiss and J. W. Putney, Jr., *J. Pharmacol. Exp. Ther.*, **207**, 699 (1979).

42   D. G. Haylett, *Br. J. Pharmacol.*, **57**, 158 (1976).

43   R. C. Deth and C. Van Breemen, *Pflügers Arch. Eur. J. Physiol.*, **348**, 13 (1974).

44   R. C. Deth and C. Van Breemen, *J. Membr. Biol.*, **30**, 363 (1977).

45   J.-L. J. Chen, F. Babcock, and H. A. Lardy, *Proc. Natl. Acad. Sci. U.S.A.*, **75**, 2234 (1978).

46   P. Kanagasuntheram and P. J. Randle, *Biochem. J.*, **160**, 547 (1976).

47   B. E. Miller and D. L. Nelson, *J. Biol. Chem.*, **252**, 3629 (1977).

48   S. Foden and P. J. Randle, *Biochem. J.*, **170**, 615 (1978).

49   P. F. Blackmore, J. L. Marks, F. T. Brumley, and J. H. Exton, *J. Biol. Chem.*, **253**, 4851 (1978).

50   P. F. Blackmore, J.-P. Dehaye, and J. H. Exton, *J. Biol. Chem.*, **254**, 6945 (1979).

51   Z. Selinger, S. Eimerl, and M. Schramm, *Proc. Natl. Acad. Sci. U.S.A.*, **71**, 128 (1974).

52   G. M. Burgess, M. Claret, and D. H. Jenkinson, *Nature (Lond.)*, **279**, 544 (1979).

53   C. Van Breemen, B. R. Farinas, R. Casteels, P. Gerba, F. Wuytack, and R. Deth, *Philos. R. Soc. Lond. Ser. B*, **265**, 57 (1973).

54   A. P. Somlyo and A. V. Somlyo, *Pharmacol. Rev.*, **20**, 197 (1968); **22**, 249 (1970).

55   L. Hurwitz and A. Suria, *Annu. Rev. Pharmacol.*, **11**, 303 (1971).

56   D. J. Triggle, *Annu. Rev. Pharmacol.*, **12**, 185 (1972).

57   R. P. Rubin, *Calcium and the Secretory Process*, Plenum Press, New York, 1974.

58   S. V. Perry, R. S. Adelstein, S. Chacko, B. Barylko, S. P. Scordilis, and M. A. Contri, in S. V. Perry, A. Margreth, and R. S. Adelstein, Eds., *Contractile Systems in Non-muscle Tissues*, Elsevier/North-Holland, Amsterdam, 1976, p. 141.

59   T. C. Vanaman, F. Sharief, and D. M. Watterson in R. H. Wasserman, R. A. Corradino, E. Carafoli, R. H. Kretsinger, D. H. Maclennan, and F. L. Siegel, Eds., *Calcium Binding Proteins and Calcium Function*, Elsevier/North-Holland, Amsterdam, 1977, p. 107.

60   C. E. Devine, A. V. Somlyo, and A. P. Somlyo, *J. Cell Biol.*, **52**, 690 (1972).

61 A. P. Somlyo, A. V. Somlyo, C. E. Devine, P. D. Peters, and T. A. Hall, *J. Cell Biol.*, **61,** 723 (1974).

62 G. Debbas, L. Hoffman, E. J. Landon, and L. Hurwitz, *Anat. Rec.*, **182,** 47 (1975).

63 J. C. Garrison, M. K. Borland, V. A. Florio, and D. A. Twible, *J. Biol. Chem.*, **254,** *7147 (1979).*

64 G. Herman and B. Rossignol, *Eur. J. Biochem.*, **55,** 105 (1975).

65 J. C. Lawrence, Jr., and J. Larner, *Mol. Pharmacol.*, **13,** 1060 (1977).

66 A. LeCam and P. Freychet, *Endocrinology,* **102,** 379 (1978).

67 M. W. Pariza, F. R. Butcher, J. E. Becker, and V. R. Potter, *Proc. Natl. Acad. Sci. U.S.A.,* **74,** 234 (1977).

68 A. Jakob and S. Diem, *Biochim. Biophys. Acta,* **404,** 57 (1975).

69 Y. Saitoh, K. Itaya, and M. Ui, *Biochim. Biophys. Acta,* **343,** 492 (1974).

70 S. L. Keely, J. D. Corbin, and T. Lincoln, *Mol. Pharmacol.*, **13,** 965 (1977).

71 J. P. Luzio, R. C. Jones, K. Siddle, and C. N. Hales, *Biochim. Biophys. Acta,* **362,** 29 (1974).

72 W. C. Grovier, *J. Pharmacol. Exp. Ther.*, **159,** 82 (1968).

73 D. Porte, Jr., *J. Clin. Invest.*, **46,** 86 (1967).

74 J. C. Garrison, *J. Biol. Chem.*, **253,** 7091 (1978).

75 R. H. Michell, *Biochim. Biophys. Acta,* **415,** 81 (1975).

76 R. H. Michell and L. M. Jones, *Biochem. J.*, **138,** 47, 52 (1974).

77 L. M. Jones and R. H. Michell, *Biochem. J.*, **148,** 479 (1975).

78 L. M. Jones and R. H. Michell, *Biochem. Soc. Trans.*, **6,** 673 (1978).

79 M. M. Billah and R. H. Michell, *Biochem. Soc. Trans.*, **6,** 1033 (1978).

80 Y. Orom, M. Lowe, and Z. Selinger, *Mol. Pharmacol.*, **11,** 79 (1975).

81 A. A. Abdel-Latif, M. P. Owen, and J. L. Mathey, *Biochem. Pharmacol.*, **25,** 461 (1975).

82 C. J. Kirk, T. R. Verrinder, and R. A. Hems, *FEBS Lett.*, **83,** 267 (1977).

83 J. M. Stein, and C. N. Hales, *Biochem. J.*, **128,** 532 (1972).

84 R. H. Michell, *TIBS*, **4,** 128 (1979).

85 R. A. Akhtar and A. A. Abdel-Latif, *J. Pharmacol. Exp. Ther.*, **204,** 655 (1978).

86 W. Y. Cheung, T. J. Lynch, and R. W. Wallace, *Adv. Cyclic Nucleotide Res.*, **9,** 233 (1978).

87 D. M. Waisman, T. J. Singh, and J. H. Wang, *J. Biol. Chem.*, **253,** 3387 (1978).

CHAPTER FIVE

DIRECT BINDING STUDIES OF α-ADRENOCEPTORS

David C. U'Prichard

Department of Pharmacology Northwestern University Medical School Chicago, Illinois

1 INTRODUCTION

The study of receptors for neurotransmitters and hormones has received a tremendous impetus over the last decade with the development of high-specific-activity radioactive ligands as receptor probes (1). In particular, the identification and description of the molecular characteristics of adrenergic receptors, whose physio-logical substrates are the sympathetic and central neurotransmitter norepinephrine (NE), and the adrenal neurohormone and putative central transmitter epinephrine (EPI), has shown recent extremely rapid development. At the present time, the state of knowledge of the molecular and cellular biochemistry of β-adrenergic receptors (β-receptors) is much further advanced than our corresponding knowledge of α-adrenergic receptors (α-receptors) (2,3), primarily because specific β-receptor ra-dioligands were developed about 2 years prior to α-receptor ligands (4–6), and because of the well-established connection between β-receptors and adenylate cy-clase, an interaction that has been particularly amenable to exhaustive study in cultured cell systems (3). The corresponding molecular effector mechanisms as-sociated with α-receptors are by no means so well understood, and this has led to a general difficulty in correlating α-receptor binding constants with pharmacological responses, other than at the gross organ level. In addition, α-receptor binding studies have been complicated by the fact that in very many tissues, two subtypes of α-receptor can be labeled, the α_1- and α_2-receptors (see below), which, unlike the subclasses of β-receptor, are probably very distinct with respect to their associated cellular mechanisms, and their location at central synapses and peripheral neu-roeffector junctions.

Over the last 2 years, nevertheless, considerable progress has been made in discriminating and selectively labeling α-receptor subtypes, and the recent advances in the understanding of the connection between α_2-receptors and adenylate cyclase in simple cellular systems suggests that for this type of α-receptor especially, a combined approach of binding studies and cellular response measurements may soon yield information of a quantity and accuracy comparable to our knowledge of β-receptor function. The present review briefly outlines some current concepts of α-receptor function which are important to a discussion of binding studies, before moving on to recent progress on identification of α-receptors by radioligand-binding techniques in the brain and various peripheral organs, with a particular focus on the distinction in binding studies between α_1- and α_2-receptors, and the information imparted in these studies about the junctional location of α-receptor subtypes. Studies on allosteric regulation of affinity constants of α-receptor binding sites, and agonist-induced receptor regulation, are discussed from the viewpoint of the in-formation that these data can give with regard to α-receptor coupling mechanisms. Finally, preliminary results and future prospects for α-receptor isolation and pu-rification are examined.

2 THE PHARMACOLOGICAL SUBDIVISION OF ADRENERGIC RECEPTORS

The classical demonstration by Dale (7) that ergot preparations reversed the pressor response to EPI was an early indication that catecholamines interact at functionally distinct receptor sites in sympathetically innervated tissues. However it was more than 40 years later than Ahlquist (8) classified these sites as α- and β-receptors, on the basis of the differential response to six sympathomimetic amines in several peripheral organ systems. In one group of tissue responses, the potency order among catecholamine drugs was EPI ≥ NE ≥ isoproterenol (ISO), whereas in the other group the potency order was ISO > EPI ≥ NE. Ahlquist classified the former group of responses as α-receptor-mediated. At the time of Ahlquist's experiments, the known adrenolytic drugs, such as the irreversibly acting haloalkylamines, dibenamine and phenoxybenzamine, various ergot derivatives, and the imidazoline antagonists, phentolamine and tolazoline, appeared to specifically block α-receptor responses, whereas specific β-receptor antagonists such as pronethalol and propranolol were not developed for another 10 years. In general, the selectivity of the foregoing antagonists at adrenergic receptor subtypes is much greater than that of the phenylethylamine agonists, and so α- and β-receptor classification in various tissues has been based on differential antagonist potencies, and to a lesser degree on differential catecholamine structure–activity relationships, both in pharmacological measurements of receptor-mediated response and in receptor binding studies.

While Lands et al. (9) established a subclassification of β-receptors into β_1- and β_2-types, the subdivision of α-receptors has taken a different route. Stemming from initial observations in the isolated, perfused cat spleen that the α-blocker phenoxybenzamine increased the overflow of NE into the perfusate (10), it became apparent from studies in perfused peripheral organs and brain slices that NE and other catecholamine and imidazoline agonists acted at an adrenergic receptor to inhibit the stimulated release of NE stored in central and peripheral noradrenergic terminals, while α-antagonist drugs prevented this inhibitory effect (for reviews, see Refs. 11–14). Although the release-modulating receptor has general pharmacological characteristics much more closely akin to α- than to β-receptors, differences emerged between this receptor and the postjunctional α-receptor. The imidazoline agonists, such as clonidine, tramazoline, and oxymetazoline, the catecholamine agonist α-methyl-NE, and the antagonists yohimbine, piperoxan, and tolazoline, are more potent at the release-modulating receptor, whereas the agonists phenylephrine and methoxamine and the antagonist phenoxybenzamine are selectively potent at the classical postjunctional α-receptor (12). Two newer α-adrenolytic drugs, prazosin and the benzodioxane derivative WB-4101, are very potent at, and extremely selective for, the postjunctional α-receptor (15–19). These pharmacological differences have led Langer (11) to postulate the existence of two distinct types of α-receptor, the postjunctional receptor generally mediating excitatory effects, known as the α_1-receptor, and the receptors whose activation inhibits the release of NE, referred to as the α_2-receptor.

The most logical location of the release-modulating α_2-receptor at sympathetic neuroeffector junctions and central NE synapses would be on the noradrenergic terminal membrane, and indeed in peripheral tissues it is hard to envisage any other location whereby the signal to depress NE release could be referred back to the sympathetic terminal. In the brain, however, experiments demonstrating α_2-receptor mediation of NE release in slice (20) or synaptosome (21,22) preparations cannot be said to conclusively prove that α_2-receptors with this function reside on NE terminal membranes; it is possible to imagine the α_2-receptor effect resulting from glial–neuronal interactions or short neuronal circuits in the vicinity of NE synapses, involving α_2-receptors not located on NE terminals (13,23). Binding studies suggest that the majority of brain α_2-receptors are not "autoreceptors" (i.e., NE terminal receptors) as commonly defined (see below). Berthelsen and Pettinger (24) proposed the general existence of postjuctional α_2-receptors in several tissues, including kidney and frog melanocyte, and α-receptors on platelets, a nonneural tissue, which promote aggregation (25), are of the α_2-type (26).

3   EFFECTOR SYSTEMS COUPLED TO α-RECEPTORS

Since the initial demonstration that EPI acting at β-receptors stimulates the activity of adenylate cyclase (27), an enormous amount of information has been obtained in many tissues concerning the coupling of β-receptors to the catalytic unit of adenylate cyclase, via a regulatory protein with which the nucleotide guanosine triphosphate (GTP) interacts and which has GTPase activity (3,28–38). The coupling role and metabolism of GTP, the conformational β-receptor changes leading to adenylate cyclase activation, or desensitization, and the link between membrane phospholipid methylation and β-receptor responses, are currently areas of intense investigation (36–42). The kinetics of agonist interactions with β-receptors in frog erythrocyte membranes and other tissues are fit best to a model in which two affinity states exist with respect to hormone or agonist; a low affinity state where hormone is bound to free receptor (HR), and a high affinity state in which hormone, receptor, and guanine nucleotide-binding regulatory protein, are bound together in a "ternary complex" (HRN) (42a). Unfortunately, a corresponding insight into α-receptor mechanisms is much less extensive, especially with regard to the α_1-receptor.

In peripheral tissues, activation of postjunctional α-receptors does not in general stimulate adenylate cyclase activity, and it has been argued that adenosine $3',5'$-cyclic monophosphate (cyclic AMP) is not the second messenger for this receptor, unlike the β-receptor (43). Although activation of α-receptors will stimulate guanosine-$3',5'$-cyclic monophosphate (cyclic GMP) in certain tissues, this appears to be an indirect effect due to α-receptor-mediated changes in intracellular Ca^{2+} storage (44). In fact, there is substantial evidence that the intracellular messenger system linked to postjunctional α-receptors (α_1-receptors) in smooth muscle and liver is a Ca^{2+}-mobilization response (43,45–55), although the steps involved between initial α-receptor activation and intracellular Ca^{2+} mobilization are as yet unknown. Interestingly, in the rat liver, there is evidence that Ca^{2+} depletion may induce a direct coupling of the α_1-receptor to adenylate cyclase (54,56), indicating that Ca^{2+} may play an uncoupling role in the membrane in contradistinction to GTP. A Ca^{2+}-independent response mediated by α_1-receptors in various tissues is the breakdown and subsequent resynthesis of membrane phosphatidylinositol, which may be the initial step involved in the mobilization of Ca^{2+} (43,57).

The biochemical events subsequent to α_2-receptor activation have been recently clarified to some extent in several tissues, at least for receptors not associated with neuronal terminal membranes. Platelet α-receptors, stimulation of which causes aggregation (25,58), are pharmacologically of the α_2-type (26) and are coupled to adenylate cylase in such a way that receptor activation lowers both basal and prostaglandin E_1-stimulated adenylate cyclase activity (59,60). This interaction is referred to below as "inverse coupling." Expression of adenylate cylase α_2-receptor interactions requires the presence of GTP (61), as is the case with cyclase coupling to β-receptors and glucagon receptors in many tissues. Similarly, the α-receptor on a neuroblastoma x glioma hybrid clonal cell line, NG108-15 (108CC15), is inversely coupled to adenylate cyclase and appears to be of the α_2 type (62). Opiate receptors and α_2-receptors on NG108-15 membranes are functionally linked to the same adenylate cyclase pool (62), and with both receptors, GTP and Na^+ are obligatory coupling agents (62.63). Blume et al. (64) have suggested that in inversely coupled receptor–cyclase systems, unlike directly coupled systems (e.g., the β-receptor), *both* GTP and Na^+ are coupling agents by virtue of allosteric interactions at intermediary membrane sites between the receptor and the cyclase catalytic unit. Other inversely coupled α-receptors occur in human and hamster fat cells (65,66), and the former yin–yang hypothesis of Robison et al. [i.e., opposing α- and β-effects on adenylate cyclase (67)] may be valid in the narrower field of α_2-receptor interactions. Fain and colleagues (67a) have recently reviewed the general hypothesis that in all tissues, α_2-receptors are inversely coupled to adenylate cyclase, whereas α_1-receptors are coupled to a membrane phospholipase C - catalyzed phosphatidyl inositol breakdown response which activates calcium gating mechanisms.

4 PHARMACOLOGICAL STRUCTURE–ACTIVITY RELATIONSHIPS AT α-RECEPTORS

The structure–activity relationships for agonists and antagonists have been more thoroughly studied at postjunctional α-receptor (α_1) than at α_2-receptor pharmacological test systems. The β-carbon on the catecholamine side chain is asymmetric, and in all test systems studied, the naturally occurring levorotatory (−)-catecholamine steroisomers, having an absolute *R(D)* configuration, are about two orders of magnitude more potent than the *S(* + *)*-enantiomers (68,69). This chirality is also observed for β-receptors, but there is evidence that the stereochemical demands are higher at β-receptors (70,71). It is important to note that catecholamine interactions at α-receptors are stereoselective, not stereospecific; thus (+)-catecholamines do have some (albeit much weaker) affinity for the α-receptor (68). Therefore, while α-receptor binding studies should show the same degree of specificity as pharmacological test systems, receptor-specific binding is somewhat greater than specific binding narrowly defined by differential stereoisomeric displacement, as opiate receptor binding has been defined (72). In many studies of agonist potencies at different pharmacological postsynaptic α-receptor test systems, the absolute potency of a catecholamine agonist such as (−)-NE can vary over two or three orders of magnitude in different tissue preparations (70,73), although the extent of stereoselectivity is constant (68,69). These pharmacological data suggest a heterogeneity in postsynaptic α-receptors, although differences in diffusional barriers in various tissues is probably also an important factor. In particular, $(-)$-NE has a very high affinity for aortic α-receptors (73), compared to other tissues. One goal of α-receptor binding studies is to determine if the pharmacological heterogeneity is based on true differences in postjunctional α-receptors in different tissues or on the fact that tissues contain different density ratios of postjunctional α_1- and α_2-receptors, at which agonists have different potencies, but which both mediate the same pharmacological (e.g., contractile) response. The assumption then would be that the pharmacological characteristics of the two basic α-receptor subtypes (α_1 and α_2) are constant in different tissues. The limited amount of information presently available suggests that the second alternative is generally correct.

Unlike β-receptor antagonists, which are a more or less structurally homogeneous group of drugs with analogies to the catecholamine structure, the molecular diversity of α-antagonists is quite striking. Major drug classes which are structurally quite different from each other and which bear no strong common resemblance to catecholamines, but which are potent α-receptor antagonists, include the imidazolines, benzodioxanes, haloalkylamines, phenothiazines, and butyrophenones. Because of this diversity, it has been suggested that the α-antagonist locus of action is not identical to the catecholamine active site on the α-receptor (71,74), although the nature of the antagonism by these drugs of the effect of the NE in pharmacological experiments is competitive. The possibility of allosteric-type antagonist interactions, and the related hypotheses of specific, interchangeable receptor states with differential agonist and antagonist affinities, is important in the analysis of α-receptor binding data, since at least one group has suggested that differences in inhibitor affinity constants dependent on the use of a radiolabeled agonist or antagonist ligand could be explained by a two-state α-receptor hypothesis (75,76). Subsequent work from these investigators showed that different α-receptor types were actually being labeled (i.e., α_1- and α_2-receptors; see below). The basis for α-antagonist heterogeneity is still, however, unclear from binding experiments, and more work in future may be directed along the lines of the elegant "subsite" model proposed by De Lean and co-workers (77).

5   THE DEVELOPMENT OF α-RECEPTOR BINDING TECHNIQUES

Early attempts to label peripheral α-receptors with radioactive probes involved the use [^{14}C]- or [^{3}H]haloalkylamines, a class of α-receptor antagonists that irreversibly combine with the receptor. Although the use of these covalent ligands is of great advantage for the isolation of receptor components, early studies were largely unsuccessful because of the lack of haloalkylamine specificity for the α-receptor. EPI-preventable binding of [^{14}C]dibenamine was shown in rabbit aorta (78,79), but binding was not saturable, and receptor-saturating concentrations of phentolamine did not protect against [^{14}C]dibenamine binding. Similar problems occurred with the use of [^{3}H]phenoxybenzamine (80) and [^{3}H]N,N-dimethyl-2-bromo-2-phenylethylamine (81). The estimates of α-receptor number using haloalkylamine ligands in the peripheral tissues examined, about 10^{14} receptors per gram of tissue (80–82), are an order of magnitude higher than estimates using reversibly binding ligands.

Early attempts to label adrenergic receptors with the radiolabeled physiological substrates, [^{3}H]epinephrine and [^{3}H]norepinephrine, were equally unsuccessful (83,84). These studies failed to demonstrate the normal biochemical attributes of ligand–receptor interactions, such as saturability and rapid association and dissociation kinetics, and more significantly the sites were not stereoselective for catecholamines, had an absolute dependence for binding interactions on the presence of the catechol moiety, such as is not found in adrenergic receptor pharmacology, and binding was not inhibited by pharmacologically potent noncatecholamine antagonists. The largely irreversible interactions of [^{3}H]catecholamines in these studies probably represent oxidative coupling to membrane catechol binding proteins not associated with adrenergic receptors (85–87; see Ref. 88 for a review of this literature). Although these investigations were carried out with the purpose of demonstrating β-receptor interactions, improvements in technique ultimately led to the successful use of [^{3}H]catecholamines to label α-receptors (see below) and β_2-receptors (89–92).

Lefkowitz in 1976 took advantage of the fact that dihydrogenation of naturally occurring ergot alkaloids results in compounds with a very high potency as α-antagonists. The dihydrogenated ergots are somewhat more potent and selective as α-blockers than the natural alkaloids (93). Thus catalytic reduction of α-ergocryptine with tritium produced [^{3}H]dihydro-α-ergocryptine ([^{3}H]DHEC), which Lefkowitz showed to specifically label α-receptors in the rabbit uterus (94). This ligand has since been successfully used to label α-receptors in many different tissues (see below).

Independently, Snyder and collaborators obtained tritium-labeled versions of clonidine, an antihypertensive drug with potent agonist properties in α-receptor-mediated vascular contractile responses (95), and a suggested preferential affinity for the NE release-modulating α_2-receptor (95,96) and a substituted benzodioxane analog WB-4101 (2-([2′,6′-dimethoxy]phenoxyethylamino)methylbenzodioxan), which was a highly potent α-antagonist in the rat vas deferens, with a pA$_2$ of 9.8 (17), and almost as potent in the guinea pig taenia caeci (18). NE release experiments subsequently showed that WB-4101 was about 100 times more potent at postjunctional α-receptors than at release-modulating receptors in central and peripheral tissues (19), although these experiments indicated different relative affinities of WB-4101 and phentolamine at release-modulating receptors, implying some tissue heterogeneity for this receptor (19).

It was reasoned a priori that [^{3}H]clonidine ([^{3}H]CLO) would, because of its selective affinity, preferentially label α_2-"autoreceptors" on NE nerve terminals in the brain. However, an initial investigation of [^{3}H]CLO binding in the brains of rats treated with the neurotoxic agent 6-hydroxydopamine (6-OHDA) revealed not the expected loss in sites, but rather a slight increase, perhaps reflecting denervation supersensitivity (75). Central α-receptor binding studies then proceeded on the assumption that both [^{3}H]WB-4101 ([^{3}H]WB) and [^{3}H]CLO labeled postsynaptic α-receptors, as inhibitor specificity studies indicated (see below). Differences in the absolute inhibitory affinities of agonists and antagonists at the two binding sites led to the postulate that the agonist and antagonist ligands bound to two distinct, but interchangeable, conformational states of the same postsynaptic α-receptor (75,76,97), akin to the "two-state" receptor model proposed for central opiate (98) and dopamine (99) receptors. More detailed investigation of affinities at both sites of inhibitors which in pharmacological experiments were selective either for the release-modulating α-receptor, such as yohimbine (100), or for the postjunctional α-receptor, such as the antihypertensive drug prazosin (15,16), led to the conclusion in several laboratories that in the brain [3H]WB selectively labeled α_1-receptors, [3H]CLO selectively labeled α_2-receptors, and [3H]DHEC labeled with approximately equal affinity both α_1- and α_2-receptors (101–104). This general relationship of different α-ligand binding sites to the two fundamental pharmacological classes of α-receptor has been shown to apply in many peripheral tissues as well (105–107). The experimental data indicating that, in the brain and many peripheral tissues, the site labeled by [3H]CLO, which has pharmacological α_2-characteristics, is predominantly postjunctional, is discussed in detail below.

Since the catecholamines (−)-EPI and (−)-NE had dissociation constants in the nanomolar range in inhibiting the α-receptor binding of [3H]CLO (108), it seemed feasible to use the tritiated forms of these drugs to label the [3H]CLO site, provided that precautions were taken to minimize catechol-directed nonspecific binding, and [3H]catecholamine autooxidation and enzyme metabolism. In pargyline-pretreated brain tissue, in the presence of 1.0 mM pyrocatechol, EDTA, and diothiothreitol, 50–75% of the binding of (±)-[3H]EPI or (−)-[3H]NE was to an α-receptor with the same pharmacological characteristics as [3H]CLO binding (109). Subsequently, [3H]EPI and [3H]NE have been used to label α-receptors in a few peripheral tissues (110; see below). (±)-[3H]EPI can also be used to identify β_2-receptors in the brain and lung in the presence of the above reagents plus phentolamine to prevent α-receptor interactions (90,91).

Other α-receptor radioligands have been used recently, but so far to a more limited extent. Patil and co-workers have described the binding interactions of [3H]dihydroazapetine, another potent α-receptor blocking agent, in the rat vas deferens (111). Prazosin is one of the most potent antagonists at postjunctional α_1-receptors (17) and in inhibiting [3H]WB binding (103). [3H]Prazosin has been recently utilized as an α-receptor probe in the brain (112) and other tissues. This radioligand appears to label the same α_1-receptor site as [3H]WB; its advantages are a slightly higher affinity, resulting in a more favorable ratio of specific binding, and, especially in some peripheral tissues, a greater selectivity for the α_1-receptor (19). A derivative of clonidine, p-aminoclonidine, which is an exceptionally potent agonist at vascular α-receptors (113), has also been tritium-labeled. [3H]p-amino-clonidine ([3H]PAC) labels the same α_2-receptor sites as [3H]CLO in the brain (114) and other tissues (see below), but with higher affinity, resulting in increased usefulness in tissues containing a low density of α_2-receptors. The nonselective antagonist [3H]phentolamine also labels α-receptors in platelets (115).

The foregoing ligands (see Figure 1) have all now been labeled to high specific

Agonists and Partial Agonists

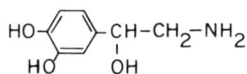

(-)-NOREPINEPHRINE

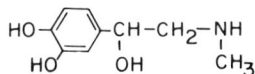

(-)-EPINEPHRINE

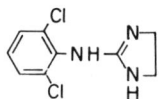

CLONIDINE

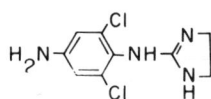

p-AMINOCLONIDINE

Antagonists

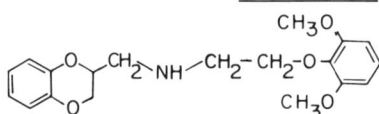

WB-4101

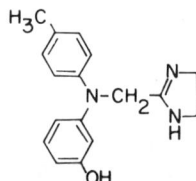

PHENTOLAMINE

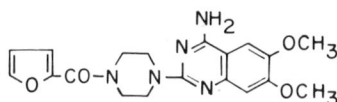

PRAZOSIN

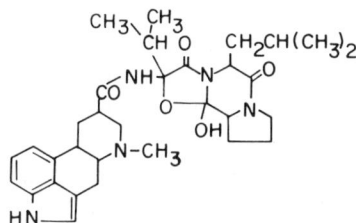

DIHYDROERGOCRYPTINE

Figure 1   Structures of α-adrenergic receptor radioligands.

activity (20–80 Ci/mmol), including the active catecholamine isomer, (−)-[³H]EPI, such that α-receptor binding assays require steadily decreasing amounts of tissue, an important factor in experiments involving cultured cells or small discrete regions from heterogeneous tissue such as brain. The utility of the receptor binding method for examining the mechanisms of receptor–effector interactions is greatly enhanced if both radiolabeled agonist and antagonist ligands are available for a specific receptor. Such is now the case for β-receptor studies (116), but unfortunately not for either $α_1$- or $α_2$-receptors, since there is available no agonist ligand with high selectivity or potency for the $α_1$-receptor, and no antagonist ligand that is $α_2$-receptor-specific. However, the $α_2$-component of the binding of potent, nonspecific α-antagonist ligands such as [³H]DHEC and [³H]phentolamine can in theory be localized by selectively eliminating $α_1$-binding through the inclusion in the assay of a drug such as prazosin, which is extremely $α_1$-selective. A few studies have utilized this technique (see below).

5.1 α-Receptor Binding in the Central Nervous System

5.1.1 [³H]WB Binding

Central α-receptors were initially identified using [³H]CLO and [³H]WB (108). Direct pharmacological evidence for an α-antagonist action of WB-4101, similar to its peripheral properties, has been obtained recently in electrophysiological studies (117). The status of clonidine as an α-receptor agonist or antagonist in the brain is less certain. While clonidine has agonist actions at central NE-release-modulating α_2-receptors (118), it inhibits, with much lower potency, NE-stimulated cyclic AMP accumulation in brain slices (119,120). In isolated cellular systems such as platelets (121) and NG108-15 cells (62), clonidine acts functionally as a partial agonist with low intrinsic activity, and recent data suggest that clonidine has similar partial agonist actions peripherally, both at release-modulating and at postjunctional receptors (95,122–124).

NE-displaceable binding of [³H]WB in rat brain has α-receptor characteristics, with (−)-NE having an inhibitor dissociation constant of 1.0 μM, and with a rank order of catecholamine potency (−)-EPI > (−)-NE > (−)-phenylephrine = (−)-α-methyl-NE ≫ (−)-ISO. Binding is markedly stereoselective for all catecholamines, although somewhat less than in peripheral pharmacological studies (75), with the (−)-isomers of NE and EPI being about 50 times more potent than the corresponding (+)-isomers (Table 1). Imidazoline-type α-agonists are all more potent than (−)-EPI. In the initial series of antagonists studied, WB-4101 itself was the most potent drug, with a K_i value of 0.6 nM, corresponding well to its pharmacological potency in the vas deferens (75). Phentolamine, phenoxybenzamine, and natural and dihydrogenerated ergot alkaloids all had K_i values in the range 1–10 nM (75,125), while yohimbine, tolazoline, dibenamine, and piperoxan were 10- to 100-fold weaker (Table 2). Amine ergot alkaloids (ergonovine and methysergide) were much weaker than peptide ergots, as expected from pharmacological studies (126).

Subsequently, prazosin was found to be at least as potent as WB-4101 in inhibiting [³H]WB binding (101), and in view of the extreme pharmacological specificity of prazosin for α_1-receptors, it appeared that [³H]WB was selectively labeling brain α_1-receptors. This was confirmed by the highly significant correlation obtained in comparing the affinities of over 30 drugs at brain [³H]WB sites, and at the vas deferens contractile response mediated by postjunctional α-receptors (105). Other major drug classes shown to have high affinity (K_i of 1–20 nM) for brain [³H]WB α_1-receptor sites include phenothiazine and butyrophenone neuroleptics (127) and tricyclic antidepressants (128,129). These drugs are pharmacologically potent α_1-receptor antagonists, and the affinity of tricyclics and neuroleptics was correlated with their sedative abilities (127,128). Some tetrahydroquinoline derivatives, condensation adducts of dopamine and aldehydes, have moderate potency at brain [³H]WB sites (130). A direct pharmacological correlation with 3H-WB sites in the brain has recently been reported: namely, activation of neurons in the lateral geniculate nucleus of the rat (117).

TABLE 1 Apparent Dissociation Constants K_i (nM) of Some Agonist and Partial Agonist Inhibitors of Binding to Putative α-Adrenergic Receptor Sites in Rat Brain[a] (76,101,112,114,137,149)

Drug	α_2-Sites			α_1-Sites		$\alpha_1 + \alpha_2$
	[3H]EPI	[3H]CLO	[3H]PAC	[3H]WB	[3H]PRAZ	[3H]DHEC
(−)-Epinephrine	5.2	2.1	2.9	590	600	46
(+)-Epinephrine	92	65	—	28,000	—	2,700
(−)-α-Methylnorepinephrine	11.5	3.4	—	2,800	—	160
(−)-Norepinephrine	20	6.1	10.5	1,000	900	120
(+)-Norepinephrine	2,200	170	363	67,000	43,000	3,000
Dopamine	200	250	—	44,000	—	720
(−)-Phenylephrine	580	115	—	2,600	1,400	1,200
(−)-Isoproterenol	2,800	5,600	2,900	>70,000	—	18,000
p-Aminoclonidine	—	0.63	0.91	130	—	—
Clonidine	3.0	2.0	2.8	430	340	52
Tramazoline	—	2.0	—	110	290	—
Oxymetazoline	3.0	1.2	1.7	24	23	15
Naphazoline	—	1.5	—	110	43	—
Methoxamine	—	940	—	11,000	—	—

[a]Inhibition of the binding of epinephrine (EPI), clonidine (CLO), p-aminoclonidine (PAC), WB-4101 (WB), prazosin (PRAZ), and dihydroergocryptine (DHEC) was determined under equilibrium (sodium- and nucleotide-free) conditions at 25°C, using [3H]ligand concentrations equal to the K_D value or less. The formula of Cheng and Prusoff was applied to obtain K_i values.

TABLE 2 Apparent Dissociation Constants K_i (nM) of Some Antagonist Inhibitors of Binding to Putative α-Adrenergic Receptor Sites in Rat Brain[a]

Drug	α_2-Sites			α_1-Sites		$\alpha_1 + \alpha_2$
	[³H]EPI	[³H]CLO	[³H]PAC	[³H]WB	[³H]PRAZ	[³H]DHEC
Phentolamine	19	2.0	3.1	3.6	—	6.2
Tolazoline	—	80	—	2,100	2,000	—
Phenoxybenzamine[b]	490	20	17.2	4.0	0.9	31
Dibenamine[b]	—	270	—	83	—	260
Ergotamine	6.3	2.4	—	12	—	2.2
Dihydroergotamine	—	3.0	—	3.5	—	1.6
Dihydroergocryptine	7.1	7.0	—	2.4	—	1.7
Ergonovine	1,400	1,400	—	1,800	—	150
Yohimbine	—	47	57	480	1,000	62
Piperoxan	—	36	—	180	360	110
WB-4101	340	108	138	0.6	1.0	29
Prazosin	—	5,000	—	0.49	0.1	39
Indoramin	36,000	26,000	—	5.9	5.0	240

[a]See legend to Table 1.
[b]Noncompetitive inhibition.

In contradistinction to a variety of β-receptor adenylate cyclase-coupled systems where the receptor, as identified by antagonist radioligand bindings recognizes agonist competitors in a heterogeneous manner, suggesting high- and low-agonist affinity components of binding (Hill slope, $n_H < 1$) (3), agonists displace brain [³H]WB α₁-receptor binding in a classical manner with $n_H = 1$ (75).

This, together with the GTP insensitivity of this site (see below), may be construed as very indirect evidence that central α₁-receptor sites are not coupled to adenylate cyclase. However, Davis and collaborators have found a significant correlation between drug potencies at the [³H]WB site in the brain and effects on the α-receptor component of NE-stimulated cyclic AMP accumulation in cortical slices (131). In the absence of direct evidence for α₁-receptor coupling to adenylate cyclase in brain membranes, this correlation cannot, however, be regarded as proof for cyclase coupling.

[³H]WB α₁-receptor sites are fairly uniformly distributed throughout the brain, both in biochemical binding studies in rat and calf (75,132) and in studies examining the autoradiographic localization of [³H]WB (133). In both species, highest levels are in the cerebral cortex, and lowest levels are in the cerebellum. Specific binding is concentrated in gray-matter areas (132). In studies of different rat hippocampal areas, [³H]WB binding was highest in the dentate gyrus, and, unlike β-receptor binding, correlated well with the distribution of noradrenergic innervation (134).

The use of 6-hydroxydopamine (6-OHDA) to destroy central noradrenergic terminals has been a valuable tool in determining in different brain areas the synaptic location of adrenergic receptor sites, and the extent of denervation-induced supersensitivity of postsynaptic receptors. Increases in central β-receptor number after 6-OHDA treatment have been frequently demonstrated (135,136). One report showed no change in [³H]WB binding in rat brain after intracerebroventricular (i.c.v.) 6-OHDA, although there was an increase in α-receptor-mediated cyclic AMP production in cortex slices (136). Other investigators found that with a higher dose of 6-OHDA (i.c.v.) and after a somewhat longer posttreatment period, the number of [³H]WB α₁-receptor sites increased in the cortex by 50% and in the rest of the brain by 23% (137). A specific lesion of the rat dorsal noradrenergic bundle with a small amount of 6-OHDA, which limits damage to central NE systems (138), also led to increases in [³H]WB α₁-receptor number in the cerebral cortex, septum, and thalamus, but not in other brain regions (139). Generally, the "supersensitivity" response of α₁-receptors to this lesion was less extensive and more localized than increases in β-receptor number, suggesting that central α₁-receptor sites may be either more prevalently located on nonneuronal membranes or are less subject to modulation than central β-receptors (139).

The dissociation constants (K_D) for [³H]WB at rat brain α₁-receptors in the experiments which either did or did not demonstrate 6-OHDA-induced supersensitivity differed by an order of magnitude (0.16 and 2.7 nM, respectively) (136,137). Recent evidence suggests that [³H]WB interactions may involve two α-receptor sites with different affinities, as judged by biphasic Scatchard plots (137,140). The lower-affinity component maintains α₁-receptor characteristics and is also postsynaptic (140).

5.1.2 [³H]Clonidine and [³H]Catecholamine Binding

In initial studies (75,108) the NE-displaceable binding of [³H]CLO (1.6 Ci/mmol) was saturable and appeared to have single site characteristics, was rapidly and linearly dissociable, and had the inhibitor potency spectrum of an α-receptor. The rank order of catecholamines is the same as for [³H]WB binding, except that α-methyl-NE is more potent than NE. Imidazolines are generally as or more potent than (−)-EPI (K_i = 2–11 nM) (Table 1). Compared to brain [³H]WB sites, antagonists such as ergot alkaloids and phentolamine are as potent in inhibiting [³H]CLO binding, while yohimbine and tolazoline are more potent, and phenoxybenzamine and WB-4101 are less potent (Table 2). Methoxamine, a selective $α_1$-agonist, is a weak inhibitor of [³H]CLO binding (75).

In general, agonists are much more potent inhibitors of [³H]CLO binding, with K_i values of 6 nM for (−)-EPI and 17 nM for (−)-NE, than of [³H]WB binding (75). The finding that many antagonists were, however, more potent inhibitors of [³H]WB binding led Snyder and collaborators to suggest that [³H]CLO labeled an agonist-preferring conformation, whereas [³H]WB labeled an antagonist preferring conformation, of the same postsynaptic α-receptor in brain, after their initial finding that [³H]CLO binding did not decrease in 6-OHDA-treated animals (75,108). However, different regional distributions of the two binding sites in rat and bovine brain (75,132), together with inconsistencies in this hypothesis with regard to the affinities of some antagonists, culminating in the later discovery that prazosin was 10,000 times more potent in inhibiting [³H]WB than [³H]CLO binding (101), led to the revised suggestion that [³H]CLO labeled in the brain a population of predominantly postsynaptic $α_2$-receptors, whereas [³H]WB labeled $α_1$-receptors (101,105). The much higher potency of yohimbine compared to prazosin at [³H]CLO sites (137), compared to the reverse at [³H]WB sites (75,101), confirms this classification, and [³H]CLO sites in the brain show a significant drug potency correlation with $α_2$-receptor-mediated effects on transmitter release in the rabbit duodenum, cat cardioaccelerator nerves, and brain slices (104,105).

Recent studies with [³H]CLO (27 Ci/mmol) show two components of $α_2$-receptor binding in rat cerebral cortex, differing in affinity for [³H]CLO (K_D values 0.4 nM and 2nM) and other agonists (137). Similar results were obtained by Vetulani et al. (137a). Both equilibrium binding isotherms and dissociation of [³H]CLO were nonlinear, and by selectively dissociating the low-affinity component of specific binding, U'Prichard et al. were able to isolate the high-affinity component (137). The regional variation in [³H]CLO binding is more marked in rat brain than the variation in 3H-WB binding, with highest levels in cortex and lowest in cerebellum (137). The distribution of the high-affinity component of [³H]CLO binding differs from that of the low-affinity component, with the latter being much more uniform and widespread, whereas the high-affinity component is prevalent in forebrain areas, especially cerebral cortex, but not in the corpus striatum. Even in the cortex, however, the binding capacity of the high-affinity component is much less than that of the low-affinity component (137). After i.c.v. 6-OHDA, neither component of [³H]CLO was reduced significantly in any brain region, and the number of high-affinity sites was indeed increased 100% in the cerebral cortex (137). This has led to the speculation that [³H]CLO binds to different types of central α_2-receptor, and that the α_2-receptor species with higher affinity for agonists might be located post-synaptically at central NE synapses, and therefore subject to "denervation super-sensitivity." However, the data could also be interpreted as two states or confor-mations of the α_2-receptor with different agonist affinities, as suggested for the β-receptor (3). Equally important, the absence of a reduction in [³H]CLO binding after NE terminal destruction is not proof for the nonexistence of α_2-receptors on central NE terminal membranes, if one accepts the coexistence of pharmacological identical postsynaptic α_2-receptors. After a more specific lesion of the dorsal no-radrenergic bundle, [³H]CLO binding is increased in rat cortex, and the increase is preferentially associated with the high-affinity component (139). In this study [³H]CLO binding was decreased in lesioned amygdala and septum, suggesting that in these areas there is a greater prevalence of true presynaptic α_2-receptors.

Other laboratories have examined structure–activity relationships at brain [³H]CLO sites in more detail. Many imidazoline components have potencies similar to clonidine, but clonidine metabolites are 1/10 to 1/50 as potent as the parent compound (141). Tanaka and Starke found that among the diastereoisomers of yohimine, yohimbine and rauwolscine (α-yohimbine) were equally potent in in-hibiting [³H]CLO binding, but rauwolscine was much more α_2-selective (i.e., much less potent at the [³H]WB site), agreeing with the pharmacological specificity of rauwolscine (142). Corynanthine, a pharmacological α_1-selective yohimbine iso-mer, was more potent at the [³H]WB site. Some antiarrhythmic drugs are also potent at bovine cortex [³H]CLO sites, with quinidine having a K_i value of 60 nM (142a).

Two interrelated questions are whether the α_2-receptor can indeed exist in low- and high-affinity states, and whether clonidine is an agonist or antagonist at the α_2-receptor binding site. That the central receptor interactions of [³H]CLO are basically agonist in nature is suggested by the high affinities of catecholamines at this site (Table 1) and the sensitivity of [³H]CLO binding to guanine nucleotides (see below). Additional strong evidence is that the sites in brain membranes labeled by the agonists ($\pm$)-[³H]EPI and ($-$)-[³H]NE are pharmacologically and in number iden-tical to [³H]CLO sites, especially with respect to high agonist affinities (109). [³H]Catecholamine α_2-receptor binding was first demonstrated in bovine cortex membranes; specific binding was defined by unlabeled oxymetazoline, a nonca-techolamine (imidazoline) α-agonist. Both ($\pm$)-[³H]EPI and ($-$)-[³H]NE labeled α_2-receptors with high affinity (K_D values at 37°C of 18 and 26 nM, respectively), and equilibrium data indicated one-site interactions (143). However, the dissociation kinetics of both ligands at low concentrations are biphasic at several temperatures, and it was suggested that in the process of brain α_2-receptor labeling with the [³H]agonists, some receptors may be converted to higher-affinity states, akin to the two-step model proposed for β-receptor activation and subsequent desensitization (38,144,145). Recent studies with high specificity activity ($-$)-[³H]EPI (40–50 Ci/mmol) suggest that α_2-binding with this ligand, as with [³H]CLO, is not monophasic at equilibrium (U'Prichard, unpublished data) (Figure 2). A significant difference between the interactions of [³H]catecholamines with brain α_2-receptors and with frog erythrocyte β-receptors (89,144) is that high-affinity interactions of catechol-

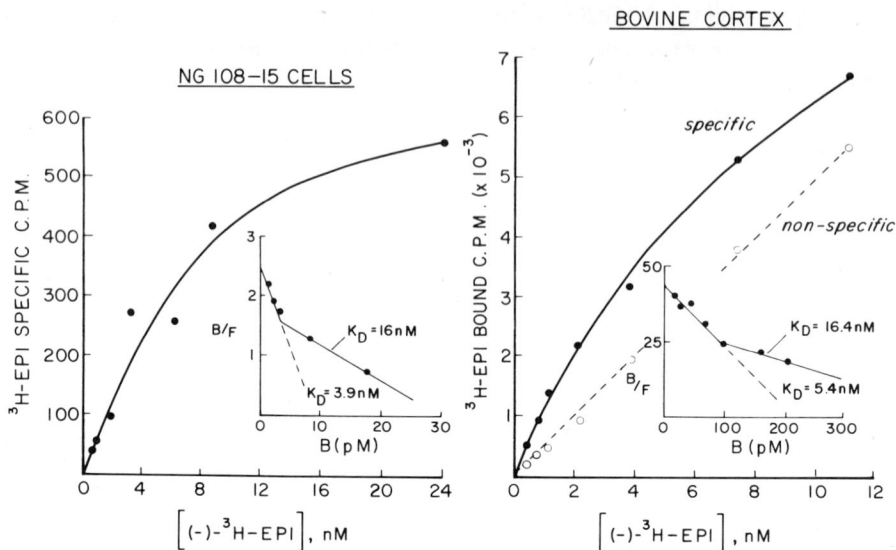

Figure 2 Saturation isotherms of (-)-[³H]epinephrine specific binding to α_2-adrenergic receptors in membranes from neuroblastoma × glioma (NG108-15; 108CC15) hybrid cells, and from gray matter of bovine frontal cerebral cortex. Nonspecific binding was in each case defined using 1.0 μM oxymetazoline. Insets: Scatchard plots of saturation data. Unpublished data.

amines at the brain α_2-receptor are not magnesium-dependent. Reduction of incubation temperatures to 25°C or 0°C increases the affinity of [³H]catecholamines, and other unlabeled agonists, at bovine cortex α_2-receptors two- to threefold (143). A similar agonist-specific increase in affinity with lowered temperature is seen in several β-receptor systems (91,146,147), and a thermodynamic model for adrenergic agonist–receptor interactions has been proposed (148). The regional distribution of [³H]catecholamine α_2-receptor binding in bovine brain is different from that of [³H]WB (132), with highest binding in cortical gray matter and lowest in hindbrain regions.

($\pm$)-[³H]EPI binding to rat cortex α_2-receptors (92,149) is pharmacologically very similar to the bovine cortex [³H]catecholamine site, and also to rat cortex [³H]CLO binding (Table 1). However, the affinity of ($-$)-[³H]EPI and other agonists is somewhat lower in rat than in calf tissue (149), as also seen with the interactions of [³H]dopamine in rat and bovine striatal tissue (150). In 6-OHDA-lesioned rats, ($\pm$)-[³H]EPI α_2-receptor sites were increased in the cortex and the rest of the brain (137), again suggesting that the majority of brain α_2-receptor sites are not localized to NE terminals.

5.1.3 [³H]Dihydroergocryptine Binding

Many ergot alkaloids exhibit high-affinity interactions not only at central and peripheral α_1- and α_2-receptors, but also at brain dopamine and serotonin receptor binding sites, with K_i values in the range 1–10 nM (151,152). Thus the use of radiolabeled ergot alkaloids as specific α-receptor probes is problematic (152), and in initial studies of rat brain [³H]DHEC binding (153), phentolamine-displaceable binding of 10 nM [³H]DHEC appeared to have both α- and serotonin receptor characteristics. Binding to rat brain membranes of another peptide ergot, [³H]dihydroergotamine, appears to be mainly serotonergic (154), and Closse and Hauser, utilizing the [³H]dihydroergotamine binding assay as a marker, obtained evidence for an endogenous ergotlike factor in rat and bovine brain that is not a monoamine (155).

Using a much lower [³H]DHEC concentration (0.3 nM) in rat cortex membranes, Greenberg and Snyder (76,156) obtained specific α-receptor binding that was saturable (K_D of 1.6 nM) and relatively slowly reversible. In other brain regions, especially caudate, [³H]DHEC binding even at this low concentration also labels dopamine receptors, and Seeman's group elegantly demonstrated that, by the use of appropriate concentrations of selective blocking agents, the α- and dopaminergic components of striatal [³H]DHEC binding could be satisfactorily isolated(157,158).

In the rat cortex, however, under the appropriate conditions [³H]DHEC binding is α-receptor specific, with dopamine 15 and 6 times less potent than (−)-EPI and (−)-NE, respectively (Table 1). The affinity constants of agonist and antagonist inhibitors of [³H]DHEC binding are the logarithmic mean of constants at the [³H]CLO and [³H]WB sites (Tables 1 and 2) (76). This evidence, together with the following observations: (a) that drugs selective at either the [³H]CLO or [³H]WB sites had shallow inhibition curves in inhibiting [³H]DHEC (n_H about 0.6; Ref. 76); (b) that in bovine cerebellum, where there are many more [³H]catecholamine than [³H]WB sites (134), [³H]DHEC and [³H]catecholamine inhibitor kinetic constants were identical, and vice versa in bovine pons, where [³H]WB sites are more prevalent (159); (c) that the number of cortical [³H]DHEC sites was equal to the sum of [³H]WB sites and sites labeled with [³H]CLO, [³H]EPI, or [³H]NE (102,105), and that cortical [³H]DHEC binding in the presence of 300 nM indoramin (sufficient to selectively occupy all $α_1$-receptors) assumed the pharmacological and site-number characteristics of [³H]EPI or [³H]CLO binding (159); (d) that inhibition of [³H]DHEC binding by prazosin was markedly biphasic (102), all led to the general conclusion that in the brain, [³H]DHEC labels with equal affinity both $α_1$-receptors (selectively labeled by [³H]WB) and $α_2$-receptors (selectively labeled by [³H]CLO, [³H]EPI, and [³H]NE (101). Similar results suggest that this finding also applies in many peripheral tissues (105). The hypothesis concerning [³H]DHEC interactions in rat cortex were confirmed by Miach et al. (103). More recently, Haga and Haga (160) have suggested that in addition to yohimbine-sensitive ($α_2$-receptor) and prazosin-sensitive ($α_1$-receptor) components of rat brain [³H]DHEC binding, there is a third component also displaceable by phentolamine which has α-receptor characteristics. These results must, however, be treated with caution, since membranes were prepared from the entire brain, and specific binding may include dopamine and serotonin receptor interactions.

Preliminary studies from this laboratory indicate that [³H]DHEC binding to bovine cortex membranes in the presence of 100 nM prazosin has indeed $α_2$-receptor characteristics, and the affinity constants of yohimbine and piperoxan ($α_2$-antagonists) are the same as at the [³H]EPI site (152). However, catecholamine agonists are 6–20 times weaker inhibitors of [³H]DHEC α_2-receptor binding than of [³H]EPI α_2-receptor binding (Table 3), suggesting that central α_2-receptors, like β-receptors in general (3), can exist in both high- and low-agonist affinity states, with the former state selectively labeled by [³H]catecholamines. Interestingly, although [³H]CLO and [³H]catecholamine α_2-receptor sites are very similar, the affinity of clonidine in inhibiting [³H]DHEC α_2-receptor binding is as high as at the [³H]EPI site (Table 3), possibly indicating that at the central α_2-receptor labeled in binding studies, clonidine is a partial agonist rather than a full agonist.

5.1.4   Other Central α-Receptor Ligands

Given the extreme selectivity and high affinity of prazosin for [³H]WB sites (K_i of 0.5 nM; Ref. 101) it is not surprising that [³H]prazosin labeled to high specific activity (33 Ci/mmol) is an effective α_1-receptor probe, with the pharmacological characteristics of phentolamine-displaceable binding in rat brain being very similar to the [³H]WB site (112; Tables 1 and 2). [³H]Prazosin binding may be somewhat more α_1-specific than that of [³H]WB, in line with the pharmacological selectivity of these drugs, since compared to the [³H]WB site, the α_1-antagonists prazosin and phenoxybenzamine are somewhat more potent inhibitors, and the α_2-antagonists piperoxan and yohimbine are somewhat less potent inhibitors of [³H]prazosin binding (Table 2). Recently, the kinetic characteristics of brain [³H]prazosin binding have been described in some detail (161). [³H]p-Aminoclonidine (40–50 Ci/mmol) is two to three times more potent than [³H]CLO, and, like [³H]CLO, the [³H]PAC

TABLE 3   Inhibition of Binding to Bovine Cortex α_2-Receptor Sites Labeled by an Agonist [³H]epinephrine and an Antagonist [³H]dihydroergocryptine: Effects on Drug Affinities of 100 μM GTP and 100 mM NaCl[a]

		[³H]DHEC		
Drug	[³H]EPI	Control	GTP	NaCl
IC₅₀ (nM)				
(−)-Epinephrine	2.5	50	380	1000
(−)-Norepinephrine	24	150	400	5000
Clonidine	3.3	3.6	10	250
Dihydroergocryptine	8.3	4.6	5.6	4.5
Dihydroergotamine	3.2	2.2	1.8	5.7
Yohimbine	60	46		75
Piperoxan	24	20		34

[a]Cortex membranes were incubated with 0.3 nM [³H]DHEC, in the presence of 100 nM prazosin to eliminate α_1-receptor binding, for 60 min at 25°C. IC₅₀ concentrations were determined by probit analysis.

site in rat brain has α_2-receptor characteristics (116; Tables 1 and 2). [³H]PAC has recently been used in autoradiographic studies of brain α_2-receptors (131,162), which show that, unlike central α_1-receptor sites, [³H]PAC grains are highly concentrated in certain brain regions, such as the locus coeruleus, where firing of NE cells is under local α_2-receptor control (163), and the nucleus tractus solitarii, where clonidine may exert its antihypertensive effect (164). The coincidence of the autoradiographic localization of α_2- and opiate receptor sites (162) is intriguing in view of the similarities of adenylate cyclase coupling and allosteric control mechanisms for these receptors (see below), and their similar pharmacological effects upon the release of NE (20,165).

5.2 α-Receptor Binding in the Uterus

The first tissue used to demonstrate specific α-receptor binding of [³H]DHEC was the rabbit uterus (94), where there are fewer problems of ergot specificity than in the brain. Specific binding was saturable (K_D = 10 nM, B_{max} = 170 fmol/mg protein) and appeared to be to a single order of sites. (−)-EPI (K_i = 230 nM) was three times more potent than (−)-NE, and the potency order of agonists at the [³H]DHEC site was the similar to the α-mediated contractile response (166). An exception, however, was clonidine, which was relatively more potent at the binding sites than in eliciting smooth muscle contraction (123,166). Among antagonists, ergot alkaloids, phentolamine, and phenoxybenzamine were potent (K_i = 10–20 nM), while yohimbine was tenfold weaker. In further studies with this preparation, Lefkowitz and co-workers showed that while the interactions at the uterine [³H]DHEC site of ergot alkaloids and phentolamine are competitive and reversible, the haloalkylamine phenoxybenzamine rapidly ($t_{1/2}$ about 1.0 min) inactivates the binding sites in an irreversible manner (167). The uterine [³H]DHEC site also appears to have an essential sulfhydryl group, which phenoxybenzamine may alkylate (167). [³H]DHEC also specifically labels rat myometrial α-receptors (168).

Subsequently, Lefkowitz et al. determined that, as in the brain, prazosin and yohimbine inhibit rabbit uterine [³H]DHEC α-receptor binding in a shallow, biphasic manner. They utilized computer-modeled curve-fitting procedures to determine the number of uterine α_1-sites and α_2-sites with high affinity for prazosin and yohinibine, respectively, and the affinity constants of these inhibitors at each α-receptor population (106,107). α_1-Receptors constituted 45% of the total number of [³H]DHEC sites, at which the K_i values of prozosin and yohimbine were 0.5 and 3000 nM, respectively; at the remaining α_2-type [³H]DHEC sites, the K_i values of prazosin and yohimbine were 7600 and 14 nM (106). These affinity constants correspond well to the values for prazosin and yohimbine as inhibitors in the brain of α_1- and α_2-sites labeled by specific ligands (see Table 2). Lefkowitz and co-workers also showed that the determination of numbers of [³H]DHEC uterine α_1- and α_2-sites by analysis of [³H]DHEC saturation in the presence of 100 nM prazosin to selectively saturate α_1-receptors, as previously described for brain tissue (159), gave similar values to the computerized analysis of prazosin competition of [³H]DHEC binding (169). In this study, however, α_2-receptors constituted 83% of the total [³H]DHEC-labeled α-receptor population, using both methods of calculation. Recently Hoffman et al. (169a) have examined rabbit uterine [³H]DHEC binding in the presence and absence of prazosin and yohimbine in more detail, and have demonstrated by analysis of agonist competition curves that the α_2-receptor component of [³H]DHEC binding exists in high and low affinity states with respect to agonists, whereas the α_1-receptor component exhibits only one agonist affinity state. WB-4101 has high affinity for uterine [³H]DHEC α_2-receptor sites, and prazosin competition curves demonstrate that [³H]WB labels both α_1- and α_2-receptors in the rabbit uterus (169b), providing further evidence that the α_1-selectivity of WB-4101 is quite variable in different tissues (see above).

There has so far been no correlation between the existence of α_1- and α_2-receptors in rabbit uterine binding studies, and possibly selective α_1- or α_2-receptor mediation of uterine responses. In this connection it is noteworthy that Kunos (170), utilizing the irreversible blockade by phenoxybenzamine of both [³H]DHEC binding and contractile response, observed a dissociation between the response and a large part of the phentolamine-displaceable [³H]DHEC binding, indicating that some of the binding sites may not mediate contraction. The interpretation of these data has been questioned (171–173), but more recent studies indicate that uterine contraction is mainly associated with the α_1-receptor component in the tissue (Kunos, unpublished data).

5.3 α-Receptor Binding in the Heart

Williams and Lefkowitz (174) showed that [³H]DHEC bound with high affinity (K_D 2.9 nM) to one order of rat myocardial sites (10 μM phentolamine-displaceable binding). The density of α-receptor sites was similar to that of myocardial β-receptors labeled with [³H]dihydroalprenolol (40–50 fmol/mg protein). (−)-EPI (EC$_{50}$ 200 nM) was 12 and 400 times more potent than (−)-NE and (−)-isoproterenol, respectively, at the [³H]DHEC site. Phentolamine was a potent antagonist (K_i = 20 nM). Using the same tissue preparation, but defining specific [³H]DHEC binding with a higher concentration of phentolamine (50 μM), Guicheney et al. (175) obtained a B_{max} value three times higher than that of Williams and Lefkowitz. The Scatchard plot of the [³H]DHEC binding indicated polyphasic binding (175). By the use of the selective α_1-antagonist ARC 239 to discriminate binding components, these authors determined that ARC 239-sensitive [³H]DHEC binding (to α_1-receptors) was high affinity, classical, and monophasic (K_D = 1.7 nM, B_{max} = 88 fmol/mg protein, n_H = 1.0), whereas residual binding in the presence of ARC 239, which showed high affinity for yohimbine and so presumably was an α_2-receptor, had a sigmoidal concentration–occupancy isotherm and demonstrated "positive homotropic cooperativity," with a Hill slope (n_H) of about 3.0 and B_{max} of 104 fmol/mg protein (175). Yohimbine itself appeared to interact cooperatively with this component of binding. Although the authors referred to the yohimbine-sensitive sites showing positive cooperativity as "presynaptic α_2-receptors," they did not directly demonstrate a sympathetic terminal location for these sites. Another group (176,177) observed that rat ventricular [³H]DHEC binding to α-receptors decreased by 60% after chemical sympathectomy, with a fourfold increase in ligand affinity at the residual sites. [^{3}H]DHEC-specific binding in this study was monophasic in both control and 6-OHDA-treated tissue over a wide range of ligand concentrations, but a pharmacological differentiation of α_1- and α_2-components of binding was not performed.

Both of these reports imply the existence of substantial numbers of rat myocardial α_2-receptors located presynaptically on sympathetic postganglionic nerve terminals; in fact, the α_2-receptor density would be significantly higher than that of cardiac β-receptors (174). In marked contrast, U'Prichard and Snyder (105) could find no evidence of specific [^{3}H]CLO binding to rat heart membranes, although high-affinity [^{3}H]WB binding was observed ($K_D = 0.95$ nM), with affinity constants for ($-$)-EPI ($K_i = 310$ nM), ($-$)-NE ($K_i = 750$ nM), phentolamine ($K_i = 87$ nM), and prazosin ($K_i = 0.92$ nM) (105), similar to values at the [^{3}H]DHEC site described by Williams and Lefkowitz (174). The [^{3}H]WB sites appeared to be postjunctional, since binding was not affected by chemical sympathectomy (105). These findings were confirmed by Langer and co-workers (176,178). Thus, whereas the identification of postjunctional cardiac α_1-receptors in binding studies is not in doubt, there is at time of writing some uncertainty concerning the existence, density, characteristics, and location of α_2-receptors in the heart that are identifiable by receptor labeling techniques. Rabbit heart α-receptor sites have also been characterized using [^{3}H]DHEC (178a), and the number of sites in this tissue was found to be less than the density of β-receptors.

5.4   α-Receptor Binding in Platelets

Platelets present a useful, homogeneous model system with which to examine the biochemical pharmacology of α-receptor function and regulation. Two α-receptor-mediated responses can be determined and correlated with receptor binding studies. One is that stimulation of platelet α-receptors decreases cyclic AMP production in intact cells and depresses the basal or PGE$_1$-stimulated activity of adenylate cyclase in membrane preparations (59,60,179). The other response is platelet aggregation induced by α-agonists (25,58). Both responses are also mediated by a receptor for ADP (180). Phenylephrine and imidazoline compounds, acting at platelet α-receptors, have predominantly antagonist actions for both responses (180a). Several laboratories have investigated the binding of α-receptor radioligands to platelet membranes.

Initial studies with [^{3}H]DHEC indicated that specific binding to human platelet membranes, defined by excess phentolamine or catecholamines, was to a single order of high-affinity sites ($K_i = 35$ nM, $B_{max} = 830$ fmol/mg protein) (121). The full agonists ($-$)-EPI ($K_i = 130$ nM) and ($-$)-NE (250 nM) were equipotent inhibitors of [^{3}H]DHEC binding and of PGE$_1$-stimulated cyclic AMP production in intact cells, while the partial agonist clonidine was almost as potent as ($-$)-EPI in inhibiting binding but 40 times weaker in the cyclic AMP response. Phenoxybenzamine and phentolamine were potent antagonists at the [^{3}H]DHEC binding site ($K_i = 10$–45 nM) and in reversing ($-$)-NE inhibition of cyclic AMP production (121).

Subsequent investigations of [³H]DHEC binding to human platelets compared binding to membranes from lysed cells with binding to intact platelets. Newman et al. (181), in platelet lysates, obtained higher-affinity binding of [³H]DHEC (k_D = 3.1 nM) and a lower number of sites (183 fmol/mg protein), corresponding to 220 receptors per platelet. The affinity constants of catecholamine agonists and phentolamine and ergot alkaloids in inhibiting binding were similar to values reported by Kafka et al. (121); clonidine was a very potent inhibitor of binding (K_i = 17 nM), as were the antagonists yohimbine and dibozane (2 and 4 nM). The potencies of agonists and antagonists in affecting PGE_1-stimulated adenylate cyclase were about tenfold weaker than at the [³H]DHEC binding site (181), and thus tenfold weaker than in affecting cyclic AMP production in intact cells (121). Newman et al. also demonstrated [³H]DHEC binding to intact platelets, with ligand and inhibitor affinity constants and site number corresponding closely to values obtained from cell lysates.

Alexander et al. (182) independently obtained very similar data. In fresh lysates, the K_D of [³H]DHEC was somewhat higher (10.3 nM), while the B_{max} was 100 receptors per cell. In platelet concentrates stored for 3 days at room temperature before membrane preparation, the B_{max} decreased threefold, while the K_D of [³H]DHEC increased a corresponding amount (182). In cell lysates the potencies of agonists and antagonists in inhibiting binding were similar to previous findings, although in intact platelets the catecholamine agonists were weaker inhibitors of [³H]DHEC binding. The catecholamine agonists had a similar range of potencies in three parameters measured: inhibition of [³H]DHEC binding to platelet membranes, inhibition of basal adenylate cyclase activity, and the threshold dose for stimulating platelet aggregation (182). The affinity constants for platelet [³H]DHEC binding and α-receptor-mediated effects on platelet adenylate cyclase activity are shown in Table 4.

The very high inhibitory potency of yohimbine in inhibiting platelet [³H]DHEC binding might indicate that the human platelet α-receptor is of the α_2-type. The aggregation response appears to be α_2-receptor-mediated (26), although the ability of high doses of methoxamine, a selective α-$_1$-agonist, to induce aggregation might also suggest an α_1-component, and, as in the brain and uterus, dihydroergocryptine would be a nonselective blocking agent (183). However, by computer analysis of [³H]DHEC displacement curves in platelets, Hoffman et al. (106) showed that [³H]DHEC labeled exclusively α_2-receptors. In our own laboratory (183a) we have labeled human platelet membranes with (−)-[³H]EPI, a full agonist in the cyclase and aggregating response, and [³H]PAC, whose intrinsic activity at platelet α-receptors is as yet unknown, but which is most likely a partial agonist such as clonidine (121). Both ligands specifically label platelet α-receptors with high affinity. The interactions of these ligands, unlike their binding in the brain, is to some extent magnesium-dependent (with 1.0 mM magnesium, [³H]PAC K_D = 1.9 nM; [³H]EPI K_D = 8.3 nM). As observed by Alexander et al. (182), storage of intact platelets at room temperature for several days radically decreased the number of α-receptors per cell labeled with either [³H]PAC or [³H]EPI, from about 200 to about 20, a more striking loss than observed with [³H]DHEC binding. Table 4

TABLE 4 α-Receptor [³H]Ligand Binding to Human Platelet Membranes: Comparison of Drug Potencies with α-Receptor-Mediated Effects on Platelet Adenylate Cyclase[a]

Drug	[³H]EPI	[³H]PAC	[³H]DHE	Adenylate Cyclase
Agonists				
(−)-Epinephrine	—	2.6	260	1,500
(−)-Norepinephrine	7.7	6.2	850	10,000
(+)-Norepinephrine	380	230	17,000	100,000
(−)-Isoproterenol	400	650	142,000	100,000
(−)-Phenylephrine[b]	—	135	860	—
Clonidine[b]	1.6	1.7	17	—
Antagonists				
Phentolamine	3.2	5.6	6.9	200
Yohimbine	7.7	6.4	2.0	60
Prazosin	10,000	15,000	—	—

[a]Values are K_i or K_D (nM). Platelet membranes were prepared according to Newman et al. (181) and incubated at 25°C with 0.6 nM [³H]p-aminoclonidine (PAC) (114), or with 1–2 nM (−)-[³H]epinephrine in the presence of 1.0 mM pyrocatechol (143). Specific binding was defined as that displaceable by 1.0 μM phentolamine (PAC), or 1.0 μM oxymetazoline (EPI). K_i values were determined using the Cheng–Prusoff equation. [³H]Dihydroergocryptine (DHE) binding and adenylate cyclase data are taken from Refs. 181 and 182.
[b]Partial agonists in inhibiting adenylate cyclase.

shows the displacement characteristics of [³H]PAC and [³H]EPI binding to human platelet membranes. The pharmacological characteristics of the sites are closely similar to the brain α_2-receptor sites labeled by the same ligands (e.g., Tables 1 and 2). No specific platelet binding of [³H]WB or [³H]prazosin was found, indicating, as previously suggested, that platelets have a pure α_2-receptor population. Although antagonists such as yohimbine and phentolamine had similar affinities in inhibiting [³H]EPI, [³H]PAC, or [³H]DHEC binding to platelets, catecholamines and other agonists were much more potent inhibitors of [³H]EPI and [³H]PAC binding than [³H]DHEC binding (Table 4). The data suggest that [³H]EPI and [³H]PAC label a high-agonist affinity conformation of the α_2-receptor, even though PAC may be, like clonidine, only a partial agonist. Thus the platelet α_2-receptor system, without the inconvenience of mixed α_1- and α_2-receptor populations that occurs in the brain, may provide some answers as to whether the α_2-receptor, like the β-receptor (3), exists in two conformational states with respect to agonists. Hoffman et al. (169a) have applied curve-fitting analytical procedures to agonist competition at platelet α_2-receptors labeled by [^{3}H]DHEC, and have shown that, like the β-receptor, the data are best fit to a two-affinity state model of the α_2-receptor. The partial agonist methoxamine induced the formation of fewer high affinity states than the full agonist ($-$)-EPI (169a). These workers have also shown that [^{3}H]EPI at low concentrations selectively labels the high affinity state, which constitutes about 60% of the total α_2-receptor population labeled by [^{3}H]DHEC (183b). Similar data have been obtained using [^{3}H]CLO (183c).

[^{3}H]Yohimbine, a newly developed antagonist radioligand, also labels human platelet α_2-receptors with high affinity (K_D 1.2 nM). As with platelet [^{3}H]DHEC binding, agonists compete at platelet [^{3}H]yohimbine sites in a manner suggesting that [^{3}H]yohimbine labels two affinity states of the receptor (183d). Clonidine and PAC interactions at [^{3}H]yohimbine sites indicate that, compared to catecholamines, fewer high affinity states of the platelet α_2-receptor are being induced. The number of [^{3}H]yohimbine sites is greater than the number of [^{3}H]EPI and [^{3}H]PAC sites (183a). Recently, Steer and co-workers have also utilized [^{3}H]phentolamine to characterize human platelet α_2-receptors (115); the properties of the binding of this ligand resemble those of [^{3}H]DHEC.

The effect of steroids on platelet α_2-receptors has been examined. Cholesterol, which decreases membrane fluidity, increases the ability of ($-$)-EPI to induce human platelet aggregation, but does not alter [^{3}H]DHEC binding characteristics (183e). Estrogen treatment in rabbits reduces the aggregation response to ($-$)-EPI, and also the number of platelet [^{3}H]DHEC sites by 40% (183f).

5.5   α-Receptor Binding in the Liver

Extensive investigations by Exton and collaborators have shown that stimulation of hepatic glycogenolysis in the rat by epinephrine is exceptional in that, as well as involving a hepatocyte β-receptor coupled to adenylate cyclase, the initial activation of phosphorylase a is predominantly mediated via an α-receptor, which, however, does not operate by stimulating adenylate cyclase and cyclic AMP-dependent protein kinase (53–55). Rather, the α-mediated effects on phosphorylase a and other liver metabolic enzymes are achieved by a cylic AMP-independent process involving the release of bound stores of intracellular calcium (55). The precise sequence of events involved in the transduction of the signal at the plasma membrane α-receptor to the intracellular calcium stores is as yet unknown.

Guellaen et al. (184) showed that [^{3}H]DHEC labels a very high number of α-receptor sites in liver, about 1000–1500 fmol/mg protein in plasma membranes prepared according to the procedure of Neville (185). The reversible kinetics of [^{3}H]DHEC binding were temperature-dependent, and the most potent antagonists were ergot alkaloids, including [^{3}H]DHEC itself (K_D or K_i = 2–10 nM) and phentolamine (K_i = 9.5 nM). There were 23 times fewer β-receptor sites labeled by [^{3}H]dihydroalprenolol, but ($-$)-EPI and ($-$)-NE, with similar affinities to each other at the two sites, were about 15 times weaker inhibitors of [^{3}H]DHEC binding than of β-receptor binding (184). Sulfhyldryl groups on liver α-receptors were much more reactive than those on β-receptors, since mercurial compounds rapidly and completely inhibited [³H]DHEC binding at concentrations and times where there was only a small effect on β-receptor binding (186). The presence of α-agonists or antagonists protected the [³H]DHEC binding site against sulfhydryl reagents. Clarke et al. (187) obtained very similar results with [³H]DHEC α-receptor binding to rat liver membranes, with one difference being an increased affinity of the site for (−)-EPI (0.15 μM vs. 2.4 μM in Ref. 184), and a tenfold difference in the affinity of (−)-EPI and (−)-NE. The higher agonist potencies found by Clarke et al. (187) may be due to the incubation temperature of 25°C that was used, compared to 37°C in Guellaen et al. (184), by analogy to the temperature dependency of agonist affinities at brain adrenergic receptor sites (91,143).

Using (±)-[³H]EPI and (−)-[³H]NE as probes for rat liver α-receptors, Exton and co-workers observed two components of specific α-receptor binding by the [³H]catecholamines, a high-affinity site (K_D about 50 nM) comprising about 10% of the total α-receptor sites labeled by [³H]DHEC, and a lower-affinity, larger-capacity site (110). Displacement of [³H]DHEC by unlabeled epinephrine was similarly biphasic. Since the catecholamines have high affinity (K_{50} = 25–50 nM) in two α-receptor-mediated responses, phosphorylase activation and calcium release, the authors concluded that the high-agonist affinity state of the liver α-receptor, selectively labeled by [³H]EPI and constituting 10% of the total liver α-receptor sites, was the physiologically active form of the receptor (110), corresponding to the functional model for the frog erythrocyte β-receptor derived by Lefkowitz and co-workers (144). Subsequent experiments by this group indicated that both [³H]DHEC and (±)-[³H]EPI labeled α₁-receptors in rat liver membranes, since prazosin was more potent than yohimbine at both sites (188). Treatment of liver membranes with trypsin caused reciprocal effects on [³H]EPI and [³H]DHEC binding, and the data were interpreted as a trypsin-induced conversion of [³H]DHEC sites (low affinity states of the α₁-receptor) to [³H]EPI sites (188a). However recent experiments using (−)-[³H]EPI of higher specific activity show that at very low ligand concentrations (1–2 nM), yohimbine is more potent than prazosin, suggesting the existence of an α₂-receptor in rat liver membranes of relatively low density, but with a higher affinity for epinephrine than the α₁-receptor (183b). The existence and α₂-characteristics of [³H]PAC binding to rat liver membranes support this conclusion (U'Prichard and Lefkowitz, unpublished data). Rat liver α₂-receptors have also been characterized using [³H]yohimbine as the radioligand (188b). A 4:1 ratio of α₁-receptors to α₂-receptors was shown either by the use of receptor-selective radioligands, [³H]prazosin and [³H]yohimbine, or by analysis of [³H]DHEC competition curves.

5.6 α-Receptor Binding in Neural Cell Lines

The examination of the properties of α-receptors in neural tissue is hampered by the lack of clearly defined α-receptor-mediated responses at the cellular level. For this reason, it is of considerable significance that a neuroblastoma x glioma hybrid cell line, NG108-15 (108CC15) (189), was recently shown to possess an α-receptor, the stimulation of which, as in platelets, reduces basal and PGE₁-stimulated adenylate cyclase activity (62). These cells, which have "cholinergic" characteristics, also possess opiate (190) and muscarinic (191) receptors which inhibit adenylate cyclase activity, and the three inversely coupled receptors appear to be functionally coupled to the same pool of catalytic cyclase, or regulatory, molecules (62). In our laboratory, we have shown that the NG108-15 α-receptors are, as in platelets, of the α_2-type, since there is no specific binding of [^{3}H]WB-4101 or [^{3}H]prazosin.

[^{3}H]Yohimbine labels NG 108-15 α_2-receptors with high affinity (K_D 7.5 nM), and binds to more sites per cell than [^{3}H]EPI or [^{3}H]PAC (191a). The characteristics of agonist competition suggest, as in platelets, that [^{3}H]yohimbine labels multiple affinity states of the NG 108-15 α_2-receptor.

The α_2-ligands [^{3}H]PAC and (−)-[^{3}H]EPI bind with high affinity to about 20,000 receptors per cell, somewhat less than the number of opiate receptors on these cells (192). Whereas [^{3}H]PAC, which, like clonidine, is a weak partial agonist at the NG108-15 α_2-receptor (62), binds to a single class of sites with a K_D of about 2 nM, the binding of (−)-[^{3}H]EPI is nonlinear (Fig. 2), with a high-affinity component of about 4–5 nM accounting for the same number of sites per cell as labeled by [^{3}H]PAC. The relationship between affinity constants at the sites labeled by these ligands and the adenylate cyclase response is very similar to that in human platelets (Table 5); antagonists are equipotent in inhibiting [^{3}H]PAC binding and in antag-

TABLE 5   Drug Inhibition of [^{3}H]p-Aminoclonidine α-Receptor Binding in Neuroblastoma × Glioma (NG108-15) Cells: Comparison with α-Receptor-Mediated Effects on Adenylate Cyclase[a]

Drug	[^{3}H]PAC: K_i (nM)	Adenylate Cyclase: IC_{50}/K_D (nM)
Agonists		
(−)-Norepinephrine	1.6	400
(+)-Norepinephrine	62.0	30,000
(−)-Epinephrine	1.9	500
(−)-Phenylephrine[b]	30.0	20,000
(−)-Isoproterenol	41.0	60,000
Clonidine[b]	3.3	100
Oxymetazoline[b]	3.6	9,000
Antagonists		
Dihydroergocryptine	2.6	5.0
Phentolamine	9.3	5.0
Yohimbine	11.8	70
WB-4101	100	200

[a][^{3}H]PAC assays were performed on washed cell membranes as in Table 4. Adenylate cyclase data are from Ref. 62.
[b]Partial agonists in inhibiting adenylate cyclase.

onizing catecholamine-induced inhibition of adenylate cyclase, but agonists are in general two orders of magnitude more potent at the α_2-receptor binding site, suggesting that [^{3}H]PAC (and [^{3}H]EPI) label, as in the brain, a high-agonist-affinity conformation of the receptor. One difference between the platelet and NG108-15 receptors is that [^{3}H]PAC and [^{3}H]EPI high-affinity interactions at the NG108-15 α_2-receptor are not dependent on the presence of magnesium. The lack of magnesium-dependence is also seen with brain α_2-receptor agonist interactions, and may indicate a fundamental difference in the properties of neural and nonneural α_2-receptors.

5.7   α-Receptor Binding to Other Tissues

Acinar cells in rat parotid glands respond to adrenergic agonist stimulation by releasing amylase, mediated by β-receptors, and K^+, mediated by α receptors (193,194). The K^+ release response, regulated, in addition to α-receptors, by muscarinic and substance P receptors, appears to be initially dependent on Ca^{2+} influx at sites associated with these receptors (50). In addition, there is evidence for presynaptic α-receptors which control the release of NE from sympathetic nerve endings in rat salivary glands (195). [^{3}H]DHEC labels, with high affinity, α-receptors in parotid membranes and in intact, dispersed acinar cells, with similar potency ratios at the receptor site and in the K^+ release response (196). Using the selective ligands [^{3}H]CLO and [^{3}H]WB, U'Prichard and Snyder showed that the rat submaxillary gland contained equal numbers of α_1- and α_2-receptor sites, and the retention of α_2-receptor sites after peripheral sympathectomy indicated that these sites were postjunctional (105). In a more detailed study, Arnett and Davis also concluded from analysis of inhibition of [^{3}H]DHEC that the α-receptors in rat salivary glands involved in the K^+ release response were of the α_2-type (107,197) and were not diminished by surgical denervation (ganglionectomy), indicating they were postjunctional (197).

Recent reports suggest that the high affinity state of the α_2-receptor in rat salivary glands, labeled by [^{3}H]CLO, is very labile. Surgical denervation induced a very rapid increase in [^{3}H]CLO binding (197a). In normal rat submandibular gland, no [^{3}H]CLO binding was observed, but a significant number of [^{3}H]CLO sites appeared as soon as 12 hr after a single dose of reserpine (197b).

In the rat kidney, epithelial cell α-receptors which modulate the release of renin appeared to be of the α_2-type (24). In a crude membrane fraction obtained from whole kidney, there were significant numbers of [^{3}H]CLO binding sites with the same α_2-characteristics as in brain membranes, and which were unaffected by peripheral sympathectomy (105). In addition, there were three times as many α_1-receptor sites labeled by [^{3}H]WB (105). In view of the extensive vascularization of the kidney, it is likely that the majority of the [^{3}H]WB sites in the kidney represent vascular α_1-receptors. Jarrott and co-workers have more extensively characterized [^{3}H]CLO binding in the guinea pig kidney (198,199) and found that the sites are primarily located in the renal cortex (200).

Contraction of the vas deferens is a classical postjunctional α-receptor response.

Snyder and co-workers observed a significant amount of [^{3}H]-WB binding in the rat vas deferens with α_1-receptor characteristics that correlated, as did the brain [^{3}H]WB site, with pharmacological potencies in the vas deferens over a wide range of agonist and antagonist drugs (105). No [^{3}H]CLO α_2-receptor binding could be easily detected, even though the vas deferens has the highest density of sympathetic innervation, and presumably of prejunctional α-receptors, of all peripheral tissues. A further study from that laboratory investigated the potencies of several benzo-dioxane analogs of WB-4101 at the vas deferens α_1-receptor, and established a strong potency correlation between binding and contraction (201). Another group characterized [^{3}H]DHEC binding in guinea pig vas deferens and achieved the same correlation between binding and contractility affinity constants, except for imida-zolines, which were much more potent inhibitors of [^{3}H]DHEC binding than in influencing contraction (202). This may be another indication of the general phe-nomenon of mixed agonist–antagonist actions at α-receptors of imidazoline drugs.

The α-antagonist azapetine had a similar dissociation constant (100–300 nM) in inhibiting vas deferens [^{3}H]DHEC binding and contractility (202). Patil and co-workers have utilized [^{3}H]dihydroazepetine to label rat vas deferens α-receptors (111). Specific binding defined by 10 μM unlabeled phentolamine was to two classes of sites (K_D = 4 and 370 nM), each of much greater density than in the foregoing vas deferens studies using other radioligands. However, affinity constants for agonists and antagonists at the [^{3}H]dihydroazapetine binding site correlated with the contractility response. The interactions of catecholamine agonists, but not im-idazoline drugs, exhibited positive cooperativity at the [^{3}H]dihydroazapetine site (111, 202a), a finding reminiscent of the nonclassical interactions reported at rat heart [^{3}H]DHEC α-receptor sites (175).

In crude membrane preparations from guinea pig lung, [^{3}H]prazosin was used to label a site with α_1-receptor characteristics, and a density less than 10% that of lung β-receptors (203,204). These sites are most probably vascular α-receptors, although there is some evidence of bronchial smooth muscle α-receptors mediating bronchoconstriction (205). The binding of [^{3}H]CLO to guinea pig ileum α_2-receptors has also been extensively characterized (205a).

Hamster and human adipocytes contain three adrenergic receptors, the β-receptor which is positively coupled to adenylate cyclase, the α_2-receptor which is inversely coupled to adenylate cyclase, and the α_1-receptor, which is linked to phosphatidyl inositol turnover (205b, 205c). The α_2-receptor mediated response in cells from both species is GTP- and Na$^+$-dependent (205d, 205e). Adipocyte α-receptors have been labeled with [^{3}H]DHEC (205f), and [^{3}H]DHEC binding has been dissociated into α_1- and α_2-components, with the latter predominating (205b, 205c). Human adipocyte α_1- and α_2-receptors have also been selectively labeled with [^{3}H]WB and [^{3}H]PAC respectively (205b).

A receptor system of major physiological and therapeutic significance is the postjunctional α-receptor located on vascular smooth muscle, stimulation of which causes vasoconstriction. The identification and characterization in binding studies of vascular α-receptors is of prime importance, but vascular tissue has proven more refractory than other organs to this kind of *in vitro* analysis, the chief problem in small mammals being the low yield of membranes. One published study (206) has characterized the binding of [³H]DHEC to semipurified membranes from dog aorta. Specific binding (10 μM phentolamine blank), which was only 40–55% of total binding, was saturable and reversible. Ergot alkaloids (K_i = 10–20 nM) were significantly more potent than other α-antagonists examined, while clonidine (K_i = 0.3 μM) and (−)-EPI (K_i = 1.0 μM) were the most potent agonists. The relative affinities of drugs at the dog aortic binding site correlated reasonably well with contractile potencies in the rabbit aortic strip preparation (206), although, as with uterine and vas deferens α-receptors, clonidine was more potent in inhibiting binding than in eliciting contraction. α-Adrenergic receptor binding of [³H]DHEC to rat mesenteric arterial smooth muscle membranes has also been characterized (206a). Bovine cerebral microvessel preparations have been reported to contain α-receptor binding sites, labeled with [³H]WB and [³H]PAC (206b), but similar preparations from rat and pig had no observable [³H]WB binding (206c).

Recent studies in our laboratory have demonstrated specific binding of [³H]WB and [³H]prazosin to $α_1$-receptors, and of [³H]CLO and [³H]PAC to $α_2$-receptors, in a purified preparation of smooth muscle membranes from the tunica media of bovine aorta (207). The binding of [³H]PAC is of some interest because p-aminoclonidine has been reported to be exceptionally potent (ED_{50} = 10^{-11} M) in eliciting contraction of rat aortic strips (113). In general, agonists are more potent at the postjunctional α-receptor in the aorta than in other pharmacological α-mediated responses, and the classification of this response as $α_1$- or $α_2$-mediated is not immediately apparent (see above).

Both [³H]WB and [³H]prazosin bound with high affinity to a single class of sites in bovine aorta media membranes (Table 6), with up to 80% specific binding for each ligand (Figure 3). The numbers of sites labeled by these ligands were similar, as were the $α_1$-characteristics of specific binding (Table 7). [³H]CLO and [³H]PAC labeled a site in the same membranes with a lower density (Table 6), and with $α_2$-receptor characteristics in that yohinbine was a considerably more potent displacer than prazosin (Table 7). The interactions of [³H]CLO with the aortic $α_2$-receptor showed significant positive cooperativity (Figure 4), unique among [³H]CLO in-

TABLE 6 Kinetic Constants of α-Receptor Radioligand Binding in Bovine Aorta Tunica Media Membranes[a]

Ligand	K_D (nM)	B_{max} (fmol/mg protein)	n_H
[³H]WB-4101	1.67 ± 0.33	162 ± 6	1.01 ± 0.05
[³H]Prazosin	0.66 ± 0.16	134 ± 5	0.89 ± 0.03
[³H]Clonidine	6.3 ± 1.7	90 ± 35	—

[a]Membranes from a low-speed supernatant fraction were sonicated and washed, and incubated to equilibrium at 25°C with [³H]ligand. Specific binding in each case was defined as that displaced by 10 μM phentolamine. Data are mean ± SE for three experiments.

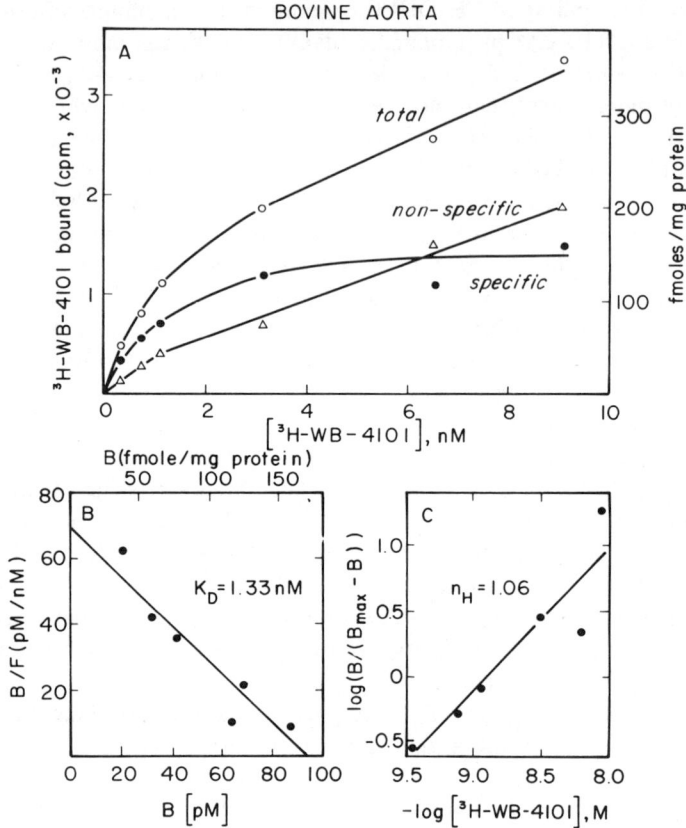

Figure 3   [^{3}H]WB-4101 binding to α_1-receptors in bovine aorta tunica media membranes, as a function of increasing concentrations of [^{3}H]WB-4101. Nonspecific binding was defined using 10 μM phentolamine. *(A)* Saturation isotherm; *(B)* scatchard plot; *(C)* hill plot. Unpublished data.

teractions in various tissues but similar to α_2-receptor interactions reported in the rat heart (175). In general, the α_1- and α_2-receptor sites in the bovine aorta were similar to those in the brain (Figure 5), although the NE isomers were considerably less potent at the aortic than at the brain α_2-receptor. The tissue distribution of aortic binding differed for the two receptor types, with α_1-binding slightly higher in the media than adventitia, but α_2-binding much higher in the adventitia (Table 8). It is not clear from these studies whether the aortic α_2-receptor sites are pre- or postjunctional, but the techniques used in this study are amenable to vascular tissue from other large species where denervation is possible.

6   ALLOSTERIC EFFECTS ON α-RECEPTOR BINDING

Extensive investigation of two hormone receptor systems which are "postively" coupled in plasma membranes to adenylate cyclase (i.e., interaction of the hormone or other agonists with the receptor stimulates adenylate cyclase activity) has shown

TABLE 7   Drug Inhibition of α-Receptor Radioligand Binding in Bovine Aorta Tunica Media Membranes[a]

	[³H]WB-4101	[³H]Prazosin	[³H]Clonidine	[³H]PAC
Agonists				
(−)-Epinephrine	470	3,900	4.3	2.8
(−)-Norepinephrine	3,100	3,500	300	27
(+)-Norepinephrine	72,800	—	3,200	—
(±)-Isoproterenol	45,500	—	1,000	—
Clonidine	1,070	1,320	2.4	13.8
Antagonists				
Prazosin	5.2	1.7	1,200	265
Phentolamine	7.3	23	6.3	0.6
Yohimbine	730	190	20	15.6
Piperoxan	1,080	—	—	—
Propranolol	2,300		>1,000	

[a]Values are K_i (nM). See legend to Table 6.

that the guanine nucleotide guanosine triphosphate (GTP) acts as a "coupling" agent for the receptor and cyclase moieties. At cyclase-coupled glucagon receptors in hepatocytes (208) and β-receptors in frog erythrocytes (209) and other cells (3), GTP is required for full agonist-mediated stimulation of the enzyme. It has been proposed that an integral factor in the stimulation of adenylate cyclase is the metabolism of GTP to guanosine diphosphate (GDP) at a membrane protein site which is linked to both the receptor and the catalytic moiety of the enzyme; the metabolism of GTP is catalyzed by a receptor agonist (30–33) and the regulatory protein to

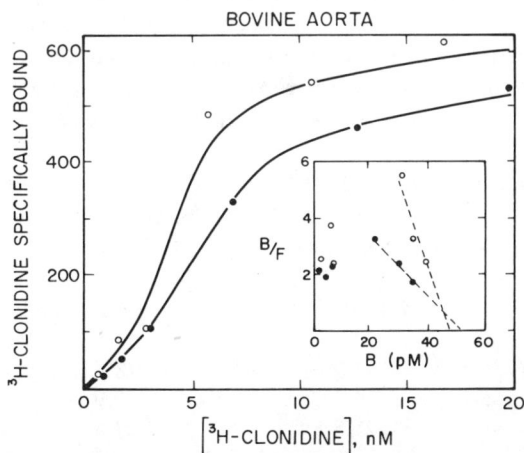

Figure 4   Saturation isotherms of [³H]clonidine specific binding (10 μM phentolamine blank) to α₂-receptors in bovine aorta tunica media membranes. Points are from two different experiments. Inset: Scatchard plots of [³H]clonidine saturation. Unpublished data.

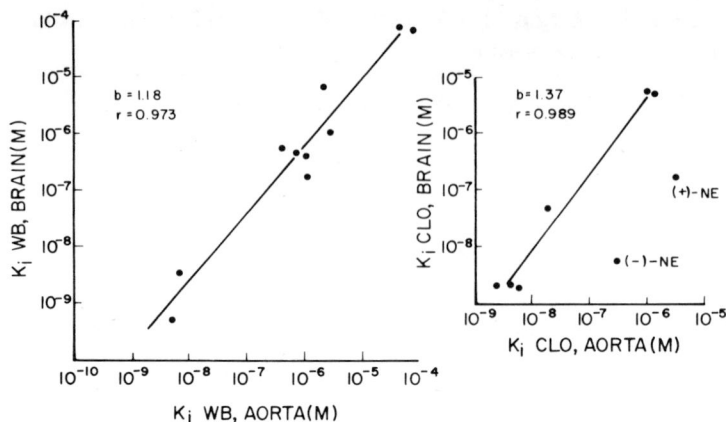

Figure 5 Correlations between potencies of α-receptor agonists and antagonists at rat brain and bovine aorta tunica media α_1-receptor sites labeled by [^{3}H]WB-4101 (WB), and α_2-receptor sites labeled by [^{3}H]clonidine (CLO). IC_{50} values were determined from log probit analysis of drug inhibition curves, and K_i values were derived using the Cheng–Prusoff relationship. Unpublished data.

which GTP binds has recently been isolated (34,35). At the glucagon and β-receptors, GTP has the additional observed effect of decreasing the affinity of agonist ligands specifically by accelerating dissociation (89,144,210,211). It has been suggested that the conversion of the receptor from a high-affinity to a low-affinity state for agonists which is effected by GTP is an integral step in the coupling of the receptor to adenylate cyclase resulting in the stimulation of the enzyme (144). Although two groups has provided evidence that GTP may have independent interactions with both the receptor and cyclase moieties (212–214), and on a theoretical basis therefore a receptor site may be influenced by GTP even if it is not normally linked to the cyclic AMP-generating system, it is currently accepted that the existence of GTP influences on a receptor is indicative of a receptor–cyclase link.

α-Receptors in mammalian brain have been indirectly associated with increased adenylate cyclase activity (119). Soon after the indentification of brain α-receptors with [^{3}H]agonist ligands, it was observed that the guanine nucleotides GTP and GDP, and the phosphohydrolase-resistant analog guanylyl-5′-imidodiphosphate

TABLE 8   **Tissue Distribution (fmole/mg protein) of α-Receptor Radioligand Specific Binding in Bovine Aorta Membranes**[a]

Tissue	[^{3}H]WB-4101	[^{3}H]p-Aminoclonidine
Media	48.9	18.3
Adventitia	45.1	40.9
Intima	25.8	27.4

[a]See legend to Table 6. [^{3}H]ligand concentrations were 1.3 nM (WB) and 1.8 nM (PAC).

[Gpp(NH)p] potently decrease [³H]EPI and [³H]NE specific binding to bovine cortex α-receptors (215). The nucleotides are approximately equipotent (ED_{50} = 1–5 μM) and are several orders of magnitude more effective than adenine nucleotide analogs. The mechanism for decreased binding in the presence of guanine nucleotides appeared to be a decrease in the affinity of all agonist sites caused by accelerated dissociation, and competition experiments showed that the reduction in affinity caused by GTP was agonist-specific (215). These effects on the α-receptor binding of [³H]catecholamines to brain membranes are closely similar to nucleotide effects on [³H]glucagon or [³H]hydroxybenzylisoproterenol to glucagon- and β-receptors, respectively (144,211), suggesting that the brain α-receptor site is associated with the components of an adenylate cyclase system. It was only subsequently appreciated that the brain [³H]catecholamine sites represented α_2-receptors (101). On the other hand, GTP had no effects on agonist affinities at the brain α_1-receptor site labeled by [³H]WB (215), suggesting that this receptor may not be directly cyclase-linked, although a general correlation has been made between this site and α-receptor-mediated cyclic AMP accumulation in intact brain tissue (131). Hornung and co-workers (161) have similarly demonstrated the absence of nucleotide effects on brain α_1-receptor binding of [³H]prazosin. However Yamada et al. have reported that agonist affinities at rat heart [³H]WB α_1-receptor sites are decreased by Gpp(NH)p, and are also influenced by the presence of muscarinic cholinergic agonists in a manner suggesting some kind of allosteric interaction between the two receptor populations (215a). α_1- receptor sites in liver (169a, 188b) and uterine (169a) membranes are not nucleotide-sensitive.

Subsequent to these studies it became apparent that the α_2-receptor in several tissues (e.g., platelets and NG108-15 cells) is coupled to adenylate cyclase, but in an "inverse" manner. Tsai and Lefkowitz observed that GTP and Gpp(NH)p were equally effective (ED_{50} = 4 μM) in decreasing the affinities of agonists, but not antagonists, at [³H]DHEC sites in human platelet membranes (216). Furthermore, in these studies, the extent to which GTP altered the affinity of an agonist was directly correlated to its intrinsic activity in stimulating adenylate cyclase. Computer analysis of agonist competition of platelet [³H]DHEC binding indicated that guanine nucleotides convert the labeled receptors from a mixture of high and low affinity states, to a uniform population of low affinity states (169a). In our laboratory, we have shown that GTP and Gpp(NH)p will directly reduce the high-affinity binding of the α_2-agonist ligands [³H]PAC and (−)-[³H]EPI to human platelet and NG108-15 membranes, with ED_{50}s in the range 0.1–1.0 μM (183a, 191a). The foregoing studies lead to two conclusions: (a) that a receptor like the α_2-receptor, which is "inversely" linked to adenylate cyclase activity, is coupled to the enzyme via a regulatory protein with a GTP site, in a similar manner to "positively" coupled receptors; and (b) that the α_2-receptors labeled in mammalian brain may also be "inversely" coupled to adenylate cyclase. Another "inversely" coupled receptor, the opiate receptor, has also been shown to be very similarly influenced by guanine nucleotides in brain and NG108-15 membranes (217).

The effect of guanine nucleotides on frog erythrocyte β-receptors is dependent on the presence of magnesium to induce a high-agonist-affinity state, which is then influenced by the nucleotide (144). A similar relationship between GTP and Mg^{2+} occurs at the platelet α_2-receptor, in that Mg^{2+} greatly facilitates the ability of GTP to decrease agonist affinities at [^{3}H]DHEC sites (216), and guanine nucleotides decrease specific binding of the agonists [^{3}H]PAC and [^{3}H]EPI much more effectively in the presence of at least 1.0 mM Mg^{2+} (U'Prichard, Mitrius and Kahn, in preparation). However, in bovine and rat brain membranes, Mg^{2+} and other divalent cations, especially Mn^{2+}, antagonize rather than facilitate the reduction of α_2-agonist ligand binding by GTP and GDP (218,219). The significance of this difference between platelet and neural α_2-receptors, and its relation to possible differences in modes of cyclase coupling, is at present unclear, but it is noteworthy that Mg^{2+} and other divalent cations did not affect the ability of the nonhydrolyzable analog Gpp(NH)p to inhibit brain α_2-agonist binding, suggesting that in brain membranes the nucleotide–cation interactions may be associated with GTPase activity (219). Similar cation–nucleotide interactions at brain [^{3}H]clonidine binding sites were observed by Glossmann and Presek (220). More recent work by the same authors (220a) has shown that in the presence of 0.5 mM EDTA, Mg^{2+} increases the affinity of [^{3}H]CLO for rat cortex α_2-receptor sites. This effect is counteracted by Gpp(NH)p, GTPγS, and GTP in the presence of a nucleotide triphosphate regenerating system. The Mg^{2+} effect, and the number of Mg^{2+}-sensitive sites, was temperature-dependent.

In rat cortex membranes, where two α_2-receptor sites with differential affinities for [^{3}H]CLO and other agonists can be ascertained (137), only the high-affinity site is GTP- and divalent cation-sensitive, and Scatchard analysis suggested that GTP decreased the number of high-affinity [^{3}H]CLO sites, with no change in the number of low-affinity sites (221). It was suggested that GTP altered the majority of high-affinity receptors to a state with sufficiently low affinity for [^{3}H]CLO as to be unobservable in the binding assay. An alternative explanation to the partial GTP sensitivity of rat brain [^{3}H]CLO binding is that clonidine is a partial agonist at rat brain α_2-receptors, as it is in platelets and NG108-15 cells. However, binding of [^{3}H]EPI was much less susceptible to GTP in rat cortex than in bovine cortex, suggesting that in rat cortex either fewer α_2-receptors are cyclase-coupled, or that all α_2-receptors are less efficiently coupled (221). Glossmann and Hornung have postulated the existence of four affinity states for rat cortex α_2-receptors, based on [^{3}H]CLO binding data (220a).

Another similarity between opiate- and α_2-receptor sites is the effect on agonist binding of small monovalent cations. Previously, Na^+ was observed to decrease [^{3}H]agonist binding to central opiate receptors (222). Opiate [^{3}H]antagonist binding was elevated by Na^+ (222), probably on account of accelerated dissociation of endogenous opioid peptides. As mentioned above, Blume has suggested that at "inversely" coupled receptors, both Na^+ and GTP regulate the receptor conformation and receptor–cyclase coupling in the same manner, whereas Na^+ influences may be absent from "positively" coupled receptors (63,64,223). However, central β-receptor sites, which are generally insensitive to GTP (91,224) are regulated by Na^+ in a manner exactly analogous to the effects of GTP at peripheral and clonal cell β-receptors (91).

Both Na^+ and Li^+ are potent in reducing the specific binding to brain α_2-receptors of the agonist ligands [³H]EPI, [³H]NE, and [³H]CLO (ED_{50} = 10–20 mM) (225). Monovalent cations with a larger hydrated radius, such as K^+ and Cs^+, are much less effective. At submaximal Na^+ and GTP concentrations, the actions of these allosteric effectors on brain α_2-agonist binding are strictly additive (215), whereas at opiate receptors, Na^+ has a significant permissive effects on the ability of GTP to reduce agonist affinities (226). Divalent cations, especially Mn^{2+}, antagonize Na^+ influences at central α-receptors with the same potency as for their antagonism of GTP effects (218,219), suggesting that regulation of α_2-agonist affinities by nucleotides and monovalent cations is exerted via a common mechanism. It should be noted, however, that in the rat cortex, both the high- and low-affinity components of [³H]CLO α_2-receptor binding were Na^+-sensitive, whereas only the high-affinity component is GTP-sensitive (221). At bovine cortex α_2-receptor sites specifically labeled by the antagonist [³H]DHEC (in the presence of 100 nM prazosin), the potencies of agonists, but not antagonists, were reduced by both Na^+ and GTP, but the "Na^+ shift" was much more extensive (152) (Table 3).

Similar ionic influences have been observed on α_2-receptor binding of [³H]DHEC to rabbit platelet membranes. Na^+ selectively decreased agonist potencies, while Mg^{2+} selectively increased against potencies and counteracted the effects of Na^+ (227).

Agonist affinities at brain α_1-receptors were as equally unaffected by Na^+ as by GTP (215), in the direction of reduced affinity. However Glossmann and Hornung have reported a slight increase in agonist affinities at brain [³H]prazosin α_1-receptor sites, in the pressence of Na^+ (227a, 227b). Exton and co-workers observed that rat hepatocyte binding of [³H]EPI was reduced by GTP (110), even though GTP did not decrease agonist potencies at the liver [³H]DHEC α_1-site (110,184,187). As indicated above, binding of [³H]EPI to rat liver membranes may be complex in that it involves both α_1- and α_2-receptors. In liver membranes, both [³H]yohimbine binding, and the α_2-component of [³H]DHEC binding, but neither [³H]prazosin binding, nor the α_1-component of [³H]DHEC binding, appeared to be GTP-sensitive (169a, 188b).

7 AGONIST-INDUCED REGULATION OF α-RECEPTORS

The general topic of regulation of adrenergic receptor sites by hormones and drugs is covered elsewhere in this book. In this section, agonist-induced alterations in α-receptor binding characteristics are discussed from two perspectives: (a) the comparative ability of α-agonists to downregulate α_2-receptors, which are probably cyclase-coupled, and α_1-receptors, which may not be associated with adenylate cyclase; and (b) the characteristics of α-receptor compared to β-receptor downregulation, and β-agonist-induced changes in α-receptors. The properties of agonist-induced downregulation or "desensitization" of β-receptors in various types of isolated, intact cells have been extensively examined in recent years (2,3), and by comparison very little is known about corresponding agonist-induced changes at α-receptors.

Two well-established α_2-receptor systems, those in platelets and NG108-15 cells, have been examined for downregulation. When intact platelets were incubated with 0.1 mM (−)-EPI, refractoriness of the aggregating response to this drug developed gradually to reach a maximum by 4 hr. In the platelets incubated with (−)-EPI for 4 hr, the number of α-receptor sites labeled by the antagonist [³H]DHEC was decreased by 50% (228). The slow time course and extent of loss of antagonist sites in these experiments are very similar to β-receptor downregulation in intact frog eryrthrocytes (2,145). The reversibility of the loss of receptor sites and the aggregating response was not examined. Interestingly, rapid exposure of platelets to 100 μM (−)-EPI increased [³H]DHEC binding, a phenomenon without known parallel in β-receptor systems (228). There is no ready explanation for this occurrence, although it could conceivably be due to ligand interactions exhibiting positive cooperativity at this site, as in the rat heart (175) and bovine aorta (207). The finding that the entire aggregating response was lost after (−)-EPI incubation, whereas the decrease in [³H]DHEC binding was only 50% (228), suggests that, as in downregulation of the frog erythrocyte β-receptor (144,145) only a fraction (possibly the high-agonist-affinity conformation) of the platelet α_2-receptors are normally coupled to the cellular response and can be "desensitized."

In NG108-15 cells which have, like platelets, "inversely" coupled α_2-receptors, Sabol and Nirenberg observed a gradual acquisition of tolerance to the inhibitory effects of (−)-NE on adenylate cyclase, in cells incubated with (−)-NE, over a period of 10–24 hr (229). Withdrawal of (−)-NE, or addition of an α-antagonist after chronic (−)-NE incubation, resulted in a large increase in basal or PGE_1-stimulated adenylate cyclase activity compared to nonincubated cells. The ED_{50} of (−)-NE for the induction of tolerance was 0.2–0.4 μM (229). Although no NG108-15 binding experiments were reported in these studies, the results suggest a downregulation of NG108-15 α_2-receptors. In the same cells, very similar phenomena occur with tolerance to opiate agonists and associated opiate receptor downregulation (192). One interesting difference, however, is that the development of tolerance to the opiate receptor mediated response in NG 108-15 cells requires the presence of serum lipids in the extracellular medium; this is not the case for α-receptor tolerance (229a).

A much more rapid downregulation occurs in dissociated rat parotid acinar cells after (−)-EPI incubation (230). Reduction in the α-receptor response (K^+ release), and a parallel reduction in the number of [³H]DHEC sites, occurred within a few minutes of (−)-EPI incubation at 38°C. The ED_{50} for (−)-EPI in downregulating the α-receptor was 5–10 μM. "Resensitization" of the response and the reappearance of [³H]DHEC sites was somewhat slower, but was markedly accelerated by increasing external K^+ concentration to cause cell depolarization (230). It is not as yet known whether both α_1- and α_2-receptors, which seem to be present in rat salivary glands (see above), are equally susceptible to downregulation.

In brain slices, downregulation of β-receptors (a decrease in β-receptor sites and β-mediated cyclic AMP accumulation) also occurs much more rapidly than in intact frog erythrocytes, and is much more easily reversible (231–233). Incubation of brain slices with 100 μM clonidine will also rapidly reduce the number of α-receptor sites labeled by [³H]PAC (U'Prichard and Enna, unpublished observations). It is interesting in this connection that in rat vas deferens there is a reported lack of cross-desensitization of postjunctional α_1-receptors by phenylethylamine and imidazoline α-agonists (234). There is indirect evidence that central α_1-receptors can be downregulated by an increase in transmitter concentration, since lesion of the ascending dorsal NE bundle in adult rats, which elevates NE levels in the cerebellum, causes a decrease in α_1-receptor ([^{3}H]WB) as well as β-receptor sites (139). However, chronic immobilization stress, which downregulates brain β-receptors, does not alter the brain α_1-specific binding of [^{3}H]WB, although in the same animals α_1-receptors in the heart and vas deferens appeared to be down-regulated (235).

An intriguing phenomenon is that several *in vivo* and *in vitro* procedures which downregulate β-receptors cause a parallel increase in the number of α_2-receptors in the brain. Thus incubation of rat cortical slices with 100 μM ISO causes a rapid decrease in the number of β-receptor sites, but an even more rapid increase in the number of cortical α_2-receptor sites labeled by [^{3}H]PAC (236). Both events are readily reversible by washing out the isoproterenol, and the increase in α_2-receptor number is prevented by concurrent incubation with the β-antagonist sotalol. Similarly, chronic immobilization stress decreases cortical β-receptor number, whereas acute stress (one 150-min period) does not. However, both acute and chronic stress increase the number of cortical α_2-receptor sites labeled by [^{3}H]CLO (235). Rats treated with a combination of amphetamine and iprindole (which retards the metabolism of amphetamine) for 3 days show a loss of cortical β-receptors, whereas each drug given separately for the same time has no effect on β-receptors. However, 3 days of treatment of the drugs given either separately or together causes an increase in the number of cortical α_2-receptor sites labeled by [^{3}H]PAC (237). The increase is localized to the high-affinity component of [^{3}H]PAC binding, which denervation studies indicate is postsynaptic in the cortex. These data are summarized in Table 9. More recently, we have observed that continuous infusion of ISO into the lateral cerebral ventricle of the rat causes a dose- and time-dependent decrease in the number of [^{3}H]dihydroalprenolol β-receptor sites, and an increase in the number of [^{3}H]PAC sites, in the cortex (237a).

A reasonable explanation for data showing reciprocal modulation of β- and α_2-receptors in the brain is that (*a*) the α_2-receptor population involved is postsynaptic at NE synapses and may occur on the same plasma membranes as cortical postsynaptic β-receptors; (*b*) in rat cortex, α_2-receptors and β-receptors are respectively "inversely" and "positively" coupled to the same pool of adenylate cyclase, possibly via the same pool of guanine nucleotide regulatory sites; and (*c*) the opposite changes in the receptor binding sites are reflected in parallel changes in catecholamine-mediated cyclic AMP accumulation (i.e., changes in the direction of "desensitization"). The regulation of cortical α_2-receptors appears to be dependent on initial β-receptor activation, but not perhaps on β-receptor "desensitization." In the cortex, the postsynaptic α_2-receptor may be allosterically linked to β-receptors, in a manner similar to the benzodiazepine/γ-aminobutyric acid receptor interaction (238). Support for this hypothesis at a functional level is the finding of Skolnick and Daly that clonidine potentiates the increase in cyclic AMP accumulation in cortical slices produced by isoproterenol, an effect that is blocked by phentolamine (119). Another kind of interaction between β- and α_2-receptors has been noted in rat kidney mem-

TABLE 9   Reciprocal Effects on Numbers of Rat Cortex β- and α_2-Adrenergic Receptors *in vitro* and *in vivo*[a]. (fmole/mg protein)

Treatment	β-Receptor ([³H]DHA B_{max})	α_2-Receptor ([³H]PAC or [³H]CLO B_{max})
A *Incubation of cortex slices with isoproterenol*		
Control	550 ± 50	1200 ± 150
100 μM isoproterenol	230 ± 30[b]	1800 ± 200[c]
B *Chronic immobilization stress*		
Control	67 ± 6	58 ± 8
Stressed (14 days)	41 ± 2[c]	78 ± 4[c]
C *Treatment with amphetamine plus iprindole*		
Control	113 ± 7	140 ± 15
Treated (3 days)	81 ± 3[c]	185 ± 15[c]

[a]In saturation experiments, β-receptor binding was determined in each case using [³H]dihydroalprenolol (DHA), and α_2-receptor binding with [³H]p-aminoclonidine (PAC), except in the stress experiments where [³H]clonidine (CLO) was used. In treatment A, slices were incubated for 120 min at 37°C; in treatment B, rats were immobilized for a 150-min period every day and sacrificed immediately after the last period; in treatment C, rats were treated twice daily with amphetamine (10 mg/kg) and once daily with iprindole (25 mg/kg), and sacrificed 16 hr after the last amphetamine treatment.
[b]$P < 0.01$.
[c]$P < 0.05$.

branes, where the presence of clonidine decreases the affinity of ISO at [³H]dihydroalprenolol β-receptor sites (238a). The above α_2-receptor changes may be more complex, since the [³H]imidazolines used in these studies may label not the total brain α_2-receptor population, but rather a component in the high-affinity state (239), in which case increases in [³H]CLO and [³H]PAC binding may reflect a shift in equilibrium in favor of the high-affinity form of the receptor, rather than an increase in total α_2-receptor number.

8   ISOLATION AND PURIFICATION OF α-RECEPTORS

Considerable progress has been made toward purifying β-receptors, especially from frog erythrocytes, where it has been possible to label the receptors in solution with radioligands after membranes are solubilized with digitonin (240). In other β-receptor-containing tissues such as the mouse S49 lymphoma cell, the physicochemical characteristics of the receptor in solution have been examined by taking advantage of the very slow dissociation rate of the β-antagonist ligand [^{125}I]hydroxybenzylpindolol (IHYP). Thus membranes can be prelabeled with IHYP and subsequently solubilized, with significant retention of the label to the solubilized receptor (241). From the foregoing experiments, there is now fairly conclusive evidence that the β-receptor and adenylate cyclase are independent entities with a basically protein structure.

As in other aspects of adrenergic receptor biochemistry and pharmacology, solubilization and purification of α-receptors has proceeded at a slower pace. The major approach taken so far has been to attempt to irreversibly bind a radiolabeled probe to the α-receptor. The covalent link will then allow the identification of ligand-binding material (hopefully to a large extent receptor-specific) through subsequent solubilization and purification steps. The disadvantage of this technique is that after covalent-bond formation the kinetic characteristics of the receptor in solution cannot be adequately studied.

The haloalkylamine compounds have long been recognized as interacting with α-receptors in an irreversible manner, although their specificity for the α-receptor as compared to other monoaminergic receptors is not very great (242). Hanoune and co-workers have recently utilized [^{3}H]phenoxybenzamine to irreversibly label rat liver α-receptors (243). Binding was nondissociable after 2 hours of incubation, and the maximum number of sites was equivalent to the B_{max} obtained with the reversible ligand [^{3}H]DHEC. Protection experiments showed that both α-agonists and α-antagonists could inhibit the covalent binding of [^{3}H]phenoxybenzamine, with a relative potency order that correlated well with the inhibition by these drugs of reversible ligand binding. As expected, however, all the protecting agents were one to two orders of magnitude weaker in antagonizing [^{3}H]phenoxybenzamine binding, compared to [^{3}H]DHEC. Membranes prebound with phenoxybenzamine were solubilized with Lubrol PX, and physicochemical and hydrodynamic constants for the binding site obtained (243). The estimated molecular weight of the liver α-receptor was 96,000 daltons. Although these experiments did indicate that [^{3}H]phenoxybenzamine may covalently label liver α-receptors, the lack of marked α-receptor specificity of the haloalkylamines suggests that they are not ideal irreversible α-receptor probes. For example, the stereoselectivity of catecholamine isomers in inhibiting [^{3}H]phenoxybenzamine binding was much less apparent than in α-receptor responses, or reversible ligand binding studies.

Kunos and colleagues independently obtained covalent binding of [^{3}H]phenoxybenzamine to isolated, intact rat hepatocytes (244). Binding was saturable, and [^{3}H]phenoxybenzamine ($EC_{50} = 10$ nM) showed a similar high affinity in binding and in inhibiting epinephrine-induced phosphorylase activation. Agonists were relatively weak in suppressing binding, but unlike the studies of Hanoune and co-workers with liver membranes (243), [^{3}H]phenoxybenzamine binding in intact cells showed marked stereoselectivity for (−)- and (+)-norepinephrine (244).

Another approach has been to utilize the observation that catecholamines are photoreactive under ultraviolet light and that the photolysis products can apparently activate adrenergic receptors irreversibly in pharmacological experiments (245). Hendley and Heidenreich have examined the interactions of [³H]catecholamines with semipurified rat brain membranes under ultraviolet light, and have demonstrated nondissociable binding, a portion of which can be protected against by noncatecholamine α-receptor drugs, such as phentolamine and oxymetazoline (246). The density of oxymetazoline-inhibitable ("specific") sites is comparable to that found in studies of the reversible α_2-receptor binding of [³H]catecholamines, and the dissociation constants for oxymetazoline and phentolamine inhibition are in the nM range (246). However, SDS slab gel electrophoresis after solubilization showed a multiplicity of protein bands labeled by [³H]EPI, all of which could be inhibited by oxymetazoline (246). Clearly, much more work is necessary to establish the validity of [³H]catecholamines as α-receptor photoaffinity labels; more impressive results may be obtained using an α-receptor-containing tissue which is less heterogeneous than the brain. Liver β- and α_1-receptors have been labeled with [³H]dihydroalprenolol and [³H]WB respectively, after digitonin solubilization (246a). Differences in physicochemical constants indicate that the two binding sites are associated with different macromolecules. These experiments provide a first direct answer to the long-standing question of whether α- and β-adrenergic receptors are separate protein species in plasma membranes.

9 CONCLUSIONS AND FUTURE PROSPECTS

Since the initial identification of α-receptors by radioligand binding methods in 1976, both α_1- and α_2-receptor subtypes have been extensively characterized in brain and peripheral tissues by a variety of selective, high-affinity probes. Sophisticated curve-fitting techniques have allowed the precise quantitation of α-receptor subtypes in various tissues, although this method must be used with caution if cooperative interactions do indeed occur in some α-receptor systems. Studies on α-receptor-effector mechanisms are still hampered to some extent by the lack of selective, high-affinity, α_2-antagonist and α_1-agonist radioligands, labeled to high specific activity. Preliminary experiments, however, in this and other laboratories suggest that [³H]yohimbine (80 Ci/ mmol) is an excellent α_2-antagonist probe (see above). [³H]Yohimbine labels brain α_2-receptors with high affinity (K_D 4nM), and the characteristics of agonist competition at these sites indicate that brain α_2-receptors, like NG 108-15, platelet, rat liver and rabbit uterine α_2-receptors, exist in at least two affinity states (246b). [³H]Yohimbine has recently been used to label solubilized platelet α_2-receptors, and chromatographic analysis shows that agonist-bound receptor is a larger protein species (possibly incorporating the nucleotide regulatory site) than antagonist-bound receptor (Limbird and Smith, personal communication). One outstanding unresolved issue is a conclusive demonstration of the existence of true NE "autoreceptors" (i.e., α_2-receptors located on sympathetic or central noradrenergic terminal membranes). On the other hand, binding studies have suggested that postjunctional or postsynaptic α_2-receptors are more prevalent than previously suspected, and there is strong evidence from both binding and functional studies that these α_2-receptors are in general "inversely" coupled to adenylate cyclase. However, the physiological role of postjunctional α_2-receptors is in most instances very unclear.

It is likely that by correlation of binding and functional studies, the mechanisms of α_2-receptor coupling to adenylate cyclase, and of α_2-receptor regulation, will soon be clarified in some cellular systems at least to the same extent as our present knowledge of β-receptors. Similar studies on α_1-receptor function may be hampered until the biochemical "second messenger" system for this receptor is adequately understood. Solubilization and purification of the α_2-receptor, and reconstitution of the α_2-receptor–effector system, may also proceed at a more rapid rate because of the existence of simple homogeneous cell systems with α_2-receptors. That this field of research is presently in a rapid state of flux is indicated by the perception that α-receptor radioligand binding studies have disclosed many more problematic issues than have been resolved.

ACKNOWLEDGMENTS

This work is supported by U.S. Public Health Service Grant NS-15595 and by a grant from the American Heart Association. The author is indebted to the following people for allowing him access to manuscripts prior to publication: Drs. J. N. Davis, J. H. Exton, G. Kunos, R. J. Lefkowitz, P. B. Molinoff, P. N. Patil, S. Z. Langer, A. H. Mulder, M. J. Rand, K. Starke, D. H. Jenkinson, J. Hanoune, J. W. Daly, and R. J. Summers.

REFERENCES

1 S. H. Snyder, in H. I. Yamamura, S. J. Enna, and M. J. Kuhar, Eds., *Neurotransmitter Receptor Binding*, Raven Press, New York, 1978, pp. 1–11.

2 L. T. Williams and R. J. Lefkowitz, *Receptor Binding Studies in Adrenergic Pharmacology.*, Raven Press, New York, 1978.

3 M. E. Maguire, E. M. Ross, and A. G. Gilman, *Adv. Cyclic Nucleotide Res.*, **8**, 1 (1977).

4 A. Levitzki, D. Atlas, and M. L. Steer, *Proc. Natl. Acad. Sci. U.S.A.*, **71**, 2773 (1974).

5 G. D. Aurbach, S. A. Fedak, C. J. Woodard, J. S. Palmer, D. Hauser, and F. Troxler, *Science*, **186**, 1223 (1974).

6 R. J. Lefkowitz, C. Mukherjee, M. Coverstone, and M. G. Caron, *Biochem. Biophys. Res. Commun.*, **60**, 703 (1974).

7 H. H. Dale, *J. Physiol. (Lond.)*, **34**, 163 (1906).

8 R. P. Ahlquist, *Am. J. Physiol.*, **153**, 586 (1948).

9 A. M. Lands, A. Arnold, J. P. McAuliff, F. P. Luduena, and T. G. Brown, Jr., *Nature (Lond.)*, **214**, 597 (1967).

10 G. L. Brown and J. S. Gillespie, *J. Physiol. (Lond.)*, **138**, 81 (1957).

11 S. Z. Langer, *Biochem. Pharmacol.*, **23**, 1973 (1974).

12 S. Z. Langer, in D. M. Paton, Ed., *The Release of Catecholamines from Adrenergic Neurons*, Pergamon Press, Oxford, 1979, pp. 59–85.

13 K. Starke, in D. M. Paton, Ed., *The Release of Catecholamines from Adrenergic Neurons,* Pergamon Press, Oxford, 1979, pp. 143–183.

14 S. Z. Langer, in S. Z. Langer, K. Starke, and M. L. Dubocovich, Eds., *Presynaptic Receptors,* Pergamon Press, Oxford, 1979, pp. 13–22.

15 D. Cambridge, M. J. Davey, and R. Massingham, *Med. J. Austr. Spec. Suppl.,* **2,** 2 (1977).

16 R. M. Graham, H. F. Oates, L. M. Stoker, and G. S. Stokes, *J. Pharmacol. Exp. Ther.,* **201,** 747 (1977).

17 D. R. Mottram and H. Kapur, *J. Pharm. Pharmacol.,* **27,** 295 (1975).

18 M. Butler and D. H. Jenkinson, *Eur. J. Pharmacol.,* **52,** 303 (1978).

19 M. L. Dubocovich, in S. Z. Langer, K. Starke, and M. L. Dubocovich, Eds., *Presynaptic Receptors,* Pergamon Press, Oxford, 1979, pp. 29–36.

20 H. D. Taube, K. Starke, and E. Borowski, *Naunyn-Schmiedebergs Arch. Pharmacol.,* **299,** 123 (1977).

21 A. H. Mulder, W. B. van den Berg, and J. C. Stoof, *Brain Res.,* **99,** 419 (1975).

22 C. D. J. De Langen, F. Hogenboom, and A. H. Mulder, *Eur. J. Pharmacol.,* **60,** 79 (1979).

23 K. Starke, in *Presynaptic Receptors,* Pergamon Press, Oxford, 1979, pp. 129–136.

24 S. Berthelesen and W. A. Pettinger, *Life Sci.,* **21,** 595 (1977).

25 C. Y. Hsu, D. R. Knapp, and P. V. Halushka, *J. Pharmacol. Exp. Ther.,* **208,** 366 (1979).

26 J. A. Grant and M. C. Scrutton, *Nature (Lond.),* **277,** 659 (1979).

27 F. Murad, Y. M. Chi, T. W. Rall, and E. W. Sutherland, *J. Biol. Chem.,* **237,** 1233 (1962).

28 A. M. Tolkovsky and A. Levitzki, *Biochemistry,* **17,** 3795 (1978).

29 E. Hanski, G. Rimon, and A. Levitzki, *Biochemistry,* **18,** 846 (1979).

30 D. Cassel and Z. Selinger, *Biochim. Biophys. Acta,* **452,** 538 (1976).

31 D. Cassel and Z. Selinger, *Proc. Natl. Acad. Sci. U.S.A.,* **75,** 4155 (1978).

32 D. Cassel, F. Eckstein, M. Lowe, and Z. Selinger, *J. Biol. Chem.,* **254,** 9835 (1979).

33 D. Cassel, H. Lefkowitz, and Z. Selinger, *J. Cyclic Nucleotide Res.,* **3,** 393 (1977).

34 T. Pfeuffer, *J. Biol. Chem.,* **252,** 7224 (1977).

35 E. M. Ross, A. C. Howlett, K. M. Ferguson, and A. G. Gilman, *J. Biol. Chem.,* **253,** 6401 (1978).

36 A. C. Howlett, P. M. Van Arsdale, and A. G. Gilman, *Mol. Pharmacol.,* **14,** 531 (1978).

37 R. J. Lefkowitz, D. Mullikin, and M. G. Caron, *J. Biol. Chem.,* **251,** 4686 (1976).

38 M. R. Wessels, D. Mullikin, and R. J. Lefkowitz, *Mol. Pharmacol.,* **16,** 10 (1979).

39 J. M. Stadel and R. J. Lefkowitz, *Mol. Pharmacol.,* **16,** 709 (1979).

40 H. Arad and A. Levitzki, *Mol. Pharmacol.,* **16,** 745 (1979).

41 G. Rimon, E. Hanski, S. Braun, and A. Levitzki, *Nature (Lond.),* **276,** 394 (1978).

42 F. Hirata, W. J. Strittmatter, and J. Axelrod, *Proc. Natl. Acad. Sci. U.S.A.,* **76,** 368 (1979).

42a A. DeLean, J. M. Stadel, and R. J. Lefkowitz, *J. Biol. Chem.,* **255,** 7108 (1980).

43 L. M. Jones and R. H. Michell, *Biochem. Soc. Trans.,* **6,** 673 (1978).

44 N. D. Goldberg and M. H. Haddox, *Annu. Rev. Biochem.,* **46,** 823 (1977).

45 R. Deth and C. Van Breemen, *J. Membr. Biol.,* **30,** 363 (1977).

46 Z. Selinger, S. Eimerl, and M. Schramm, *Proc. Natl. Acad. Sci. U.S.A.,* **71,** 128 (1974).

47 D. G. Haylett and D. H. Jenkinson, *J. Physiol. (Lond.),* **225,** 721 (1972).

48 D. G. Haylett and D. H. Jenkinson, *J. Physiol. (Lond.),* **225,** 751 (1972).

49 D. H. Jenkenson, D. G. Haylett, K. Koller, and G. Burgess, in E. Szabadi, C. M. Bradshaw, and P. Bevan, Eds., *Recent Advances in the Pharmacology of Adrenoceptors,* Elsevier/North-Holland, Amsterdam, 1978, pp. 23–33.

50 J. W. Putney, Jr., *J. Physiol. (Lond.)*, **268**, 139 (1977).

51 S. Keppens, J. R. Vandenheede, and H. De Wolf, *Biochim. Biophys. Acta*, **496**, 448 (1977).

52 A. D. Cherrington, F. D. Assimacopoulos, S. C. Harper, J. D. Corbin, C. R. Park, and J. H. Exton, *J. Biol. Chem.*, **251**, 5209 (1976).

53 F. D. Assimacopoulos-Jeannet, P. F. Blackmore, and J. H. Exton, *J. Biol. Chem.*, **252**, 2662 (1977).

54 T. M. Chan and J. H. Exton, *J. Biol. Chem.*, **253**, 6393 (1978).

55 P. F. Blackmore, F. T. Brumley, J. L. Marks, and J. H. Exton, *J. Biol. Chem.*, **253**, 4851 (1978).

56 T. M. Chan and J. H. Exton, *J. Biol. Chem.*, **252**, 8645 (1977).

57 R. H. Michell, L. M. Jones, and S. S. Jafferji, *Biochem. Soc. Trans.*, **5**, 77 (1977).

58 W. Barthel and F. Markwardt, *Biochem. Pharmacol.*, **24**, 37 (1974).

59 E. W. Salzman and L. L. Neri, *Nature (Lond.)*, **224**, 609 (1969).

60 G. A. Robison, A. Arnold, and R. C. Hartmann, *Pharmacol. Res. Commun.*, **1**, 325 (1969).

61 K. H. Jakobs, W. Saur, and G. Schultz, *FEBS Lett.*, **85**, 167 (1978).

62 S. L. Sabol and M. Nirenberg, *J. Biol. Chem.*, **254**, 1913 (1979).

63 D. Lichtshtien, G. Boone, and A. Blume, *Life Sci.*, **25**, 985 (1979).

64 A. J. Blume, D. Lichtshtien, and G. Boone, *Proc. Natl. Acad. Sci. U.S.A.*, **76**, 5626 (1979).

65 K. J. Hittelman and R. W. Butcher, *Biochim. Biophys. Acta*, **316**, 403 (1973).

66 R. J. Schimmel, *Biochim. Biophys. Acta*, **428**, 379 (1976).

67 T. W. Burns, P. E. Langley, and G. A. Robison, *Ann. N.Y. Acad. Sci.*, **185**, 115 (1971).

67a J. N. Fain and J. A. Garcia-Sainz, *Life Sci.*, **26**, 1183 (1980).

68 P. N. Patil, D. G. Patel, and R. D. Krell, *J. Pharmacol. Exp. Ther.*, **176**, 622 (1971).

69 P. N. Patil, K. Fudge, and D. Jacobowitz, *Eur. J. Pharmacol.* **19**, 79 (1972).

70 D. J. Triggle, in A. Burger, Ed., *Medicinal Chemistry*, Wiley, New York, 1970, pp. 1235–1295.

71 D. J. Triggle and C. R. Triggle, *Chemical Pharmacology of the Synapse*, Academic Press, New York, 1976, Chap. 3.

72 E. J. Simon, in H. H. Loh and D. H. Ross, Eds., *Neurochemical Mechanisms of Opiates and Endorphins*, Raven Press, New York, 1979, pp. 31–51.

73 E. M. Sheys and R. D. Green, *J. Pharmacol. Exp. Ther.*, **180**, 317 (1972).

74 E. J. Ariens, *Molecular Pharmacology*, Vol. I, Sect. IIA, Academic Press, New York, 1964.

75 D. C. U'Prichard, D. A. Greenberg, and S. H. Snyder, *Mol. Pharmacol.*, **13**, 454 (1977).

76 D. A. Greenberg and S. H. Snyder, *Mol. Pharmacol.*, **14**, 38 (1978).

77 A. De Lean, P. J. Munson, and D. Rodbard, *Mol. Pharmacol.*, **15**, 60 (1979).

78 M. S. Yong, M. R. Parulekar, J. Wright, and G. S. Marks, *Biochem. Pharmacol.*, **15**, 1185 (1966).

79 M. S. Yong and G. S. Marks, *Biochem. Pharmacol.*, **18**, 1609 (1968).

80 J. E. Lewis and J. W. Miller, *J. Pharmacol. Exp. Ther.*, **154**, 46 (1966).

81 M. May, J. F. Moran, H. Kimelberg, and D. J. Triggle, *Mol. Pharmacol.*, **3**, 28 (1967).

82 R. D. Green, G. S. LeFever, E. M. Sheys, and M. Bristow, *J. Pharmacol. Exp. Ther.*, **187**, 524 (1973).

83 R. J. Lefkowitz and G. Haber, *Proc. Natl. Acad. Sci. U.S.A.*, **68**, 1773 (1971).

84 R. J. Lefkowitz, G. W. G. Sharp, and E. Haber, *J. Biol. Chem.*, **248**, 342 (1973).

85 P. Cuatrecasas, G. P. E. Tell, V. Sica, I. Parikh, and K. J. Chang, *Nature (Lond.)*, **247**, 92 (1974).

86 M. E. Maguire, P. H. Goldmann, and A. G. Gilman, *Mol. Pharmacol.*, **10**, 563 (1974).

87 B. B. Wolfe, J. A. Zirolli, and P. B. Molinoff, *Mol. Pharmacol.*, **10**, 582 (1974).

88 H. P. Bar, in D. M. Paton, Ed., *The Mechanism of Neuronal and Extraneuronal Transport of Catecholamines,* Raven Press, New York, 1976, pp. 247–257.

89 R. J. Lefkowitz and L. T. Williams, *Proc. Natl. Acad. Sci. U.S.A.,* **74,** 515 (1977).

90 D. C. U'Prichard and S. H. Snyder, *Nature (Lond.),* **270,** 261 (1977).

91 D. C. U'Prichard, D. B. Bylund, and S. H. Snyder, *J. Biol. Chem.,* **253,** 5090 (1978).

92 D. C. U'Prichard and S. H. Snyder, *J. Supramol. Struct.,* **9,** 189 (1978).

93 B. Berde and E. Sturmer, in B. Berde and H. O. Schild, Eds., *Ergot Alkaloids and Related Compounds,* Springer-Verlag, Berlin, 1978, pp. 1–28.

94 L. T. Williams and R. J. Lefkowitz, *Science,* **192,** 791 (1976).

95 K. Starke, H. Montel, W. Gayk, and R. Merker, *Naunyn-Schmiedebergs Arch. Pharmacol.,* **285,** 133 (1974).

96 K. Starke, T. Endo, and H. D. Taube, *Naunyn-Schmiedebergs Arch. Pharmacol.,* **291,** 55 (1975).

97 S. H. Snyder, D. C. U'Prichard, and D. A. Greenberg, in M. A. Lipton, A. DiMascio, and K. F. Killam, Eds., *Psychopharmacology: A Generation of Progress,* Raven Press, New York, 1978, pp. 361–370.

98 C. B. Pert and S. H. Snyder, *Neurosci. Res. Program Bull.,* **13,** 73 (1975).

99 D. R. Burt, I. Creese, and S. H. Snyder, *Mol. Pharmacol.,* **12,** 800 (1976).

100 K. Starke, E. Borowski, and T. Endo, *Eur. J. Pharmacol.,* **34,** 385 (1975).

101 D. C. U'Prichard, M. E. Charness, D. Robertson, and S. H. Snyder, *Eur. J. Pharmacol.,* **50,** 87 (1978).

102 D. C. U'Prichard, D. A. Greenberg, and S. H. Snyder, in P. Meyer and H. Schmitt, Eds., *Nervous System and Hypertension,* Wiley–Flammarion, New York, 1979, pp. 38–48.

103 P. J. Miach, J. P. Dausse, and P. Meyer, *Nature (Lond.),* **274,** 492 (1978).

104 M. Titeler, J. L. Tedesco, and P. Seeman, *Life Sci.,* **23,** 587 (1978).

105 D. C. U'Prichard and S. H. Snyder, *Life Sci.,* **24,** 79 (1979).

106 B. Hoffman, A. DeLean, C. L. Wood, D. D. Schocken, and R. J. Lefkowitz, *Life Sci.,* **24,** 1739 (1979).

107 C. L. Wood, C. D. Arnett, W. R. Clarke, B. S. Tsai, and R. J. Lefkowitz, *Biochem. Pharmacol.,* **28,** 1277 (1979).

108 D. A. Greenberg, D. C. U'Prichard, and S. H. Snyder, *Life Sci.,* **19,** 69 (1976).

109 D. C. U'Prichard and S. H. Snyder, *Life Sci.,* **20,** 527 (1977).

110 M. El-Refai, P. F. Blackmore, and J. H. Exton, *J. Biol. Chem.,* **254,** 4375 (1979).

111 R. R. Ruffolo, Jr., J. W. Fowble, D. D. Miller, and P. N. Patil, *Proc. Natl. Acad. Sci. U.S.A.,* **73,** 2730 (1976).

112 P. Greengrass and R. Bremner, *Eur. J. Pharmacol.,* **55,** 323 (1979).

113 B. Rouot and G. Leclerc, *Eur. J. Med. Chem.,* **13,** 521 (1978).

114 B. R. Rouot and S. H. Snyder, *Life Sci.,* **25,** 769 (1979).

115 M. L. Steer, J. Khorana, and B. Galgoci, *Mol. Pharmacol.,* **16,** 719 (1979).

116 R. J. Lefkowitz and M. Hamp, *Nature (Lond.),* **268,** 453 (1977).

117 M. A. Rogawski and G. K. Aghajanian, *Brain Res.,* **182,** 345 (1980).

118 K. Starke and H. Montel, *Neuropharmacology,* **12,** 1073 (1973).

119 P. Skolnick and J. W. Daly, *Mol. Pharmacol.,* **11,** 545 (1975).

120 P. Skolnick and J. W. Daly, *Eur. J. Pharmacol.,* **39,** 11 (1976).

121 M. S. Kafka, J. F. Tallman, and C. C. Smith, *Life Sci.,* **21,** 1429 (1977).

122 C. Medgett, M. W. McCulloch, and M. J. Rand, *Naunyn-Schmiedebergs Arch. Pharmacol.,* **304,** 215 (1978).

123 R. J. Ress, F. D. Field, O. E. Lockley, and M. J. Fregly, *Pharmacology,* **18,** 149 (1979).

124 M. E. M. Tolbert, A. C. White, K. Aspry, J. Cutts, and J. N. Fain, *J. Biol. Chem.*, **255**, 1938 (1980).

125 M. Goldstein, J. Y. Lew, F. Hata, and A. Lieberman, *Gerontology*, **24**, 76 (1978).

126 M. Nickerson and N. K. Hollenberg, in W. S. Root and F. G. Hofmann, Eds., *Physiological Pharmacology*, Vol. 4, *The Nervous System, Part D, Autonomic Nervous System Drugs*, Academic Press, New York, 1967, pp. 243–305.

127 S. J. Peroutka, D. C. U'Prichard, D. A. Greenberg, and S. H. Snyder, *Neuropharmacology*, **16**, 549 (1977).

128 D. C. U'Prichard, D. A. Greenberg, P. Sheehan, and S. H. Snyder, *Science*, **199**, 197 (1978).

129 A. Maggi, D. C. U'Prichard, and S. J. Enna, *Eur. J. Pharmacol.*, **59**, 297 (1979).

130 Y. Nimitkitpaisan and P. Skolnick, *Life Sci.*, **23**, 375 (1978).

131 J. N. Davis, C. D. Arnett, E. Hoyler, L. P. Stalvey, J. W. Daly, and P. Skolnick, *Brain Res.*, **159**, 125 (1978).

132 D. C. U'Prichard, D. A. Greenberg, P. Sheehan, and S. H. Snyder, *Brain Res.*, **138**, 151 (1977).

133 W. S. Young and M. J. Kuhar, *Eur. J. Pharmacol.*, **59**, 317 (1979).

134 K. A. Crutcher and J. N. Davis, *Brain Res.*, **182**, 107 (1980).

135 J. R. Sporn, T. K. Harden, B. B. Wolfe, and P. B. Molinoff, *Science*, **194**, 624 (1976).

136 P. Skolnick, L. P. Stalvey, J. W. Daly, E. Hoyler, and J. N. Davis, *Eur. J. Pharmacol.*, **47**, 201 (1978).

137 D. C. U'Prichard, W. D. Bechtel, B. Rouot, and S. H. Snyder, *Mol. Pharmacol.*, **16**, 47 (1979).

137a J. Vetulani, M. Nielsen, A. Pilc, and K. Golembiowska-Nikitin, *Eur. J. Pharmacol.*, **58**, 95 (1979).

138 S. T. Mason and S. D. Iverson, *J. Comp. Physiol. Psychol.*, **91**, 165 (1977).

139 D. C. U'Prichard, T. D. Reisine, S. T. Mason, H. C. Fibiger and H. I. Yamamura, *Brain Res.*, **187**, 143 (1980).

140 M. Rehavi, B. Yavetz, O. Ramot, and M. Sokolovsky, *Life Sci.*, **26**, 615 (1980).

141 B. Jarrott, W. J. Louis, and R. J. Summers, *Biochem. Pharmacol.*, **28**, 141 (1979).

142 T. Tanaka and K. Starke, *Eur. J. Pharmacol.*, **63**, 191 (1980).

142a F. R. Ciofalo, *Eur. J. Pharmacol.*, **65**, 309 (1980).

143 D. C. U'Prichard and S. H. Snyder, *J. Biol. Chem.*, **252**, 6450 (1977).

144 L. T. Williams and R. J. Lefkowitz, *J. Biol. Chem.*, **252**, 7207 (1977).

145 M. R. Wessels, D. Mullikin, and R. J. Lefkowitz, *J. Biol. Chem.*, **253**, 3371 (1978).

146 L. J. Pike and R. J. Lefkowitz, *Mol. Pharmacol.*, **14**, 370 (1978).

147 P. A. Insel and M. Sanda, *J. Biol. Chem.*, **254**, 6554 (1979).

148 G. A. Weiland, K. P. Minneman, and P. B. Molinoff, *Nature (Lond.)*, **281**, 114 (1979).

149 D. C. U'Prichard and S. H. Snyder, *Eur. J. Pharmacol.*, **51**, 145 (1978).

150 I. Creese, K. Stewart, and S. H. Snyder, *Eur. J. Pharmacol.*, **60**, 55 (1979).

151 D. C. U'Prichard and S. H. Snyder, in I. Hanin and E. Usdin, Eds., *Animal Models in Psychiatry and Neurology*, Pergamon, Oxford, 1977, pp. 477–495.

152 D. C. U'Prichard, in M. Goldstein, Eds., *Ergot Compounds and Brain Function: Neuroendocrine and Neuropyschiatric Aspects*, Raven Press, New York, 1979, pp. 103–115.

153 J. N. Davis, W. J. Strittmatter, E. Hoyler, and R. J. Lefkowitz, *Brain Res.*, **132**, 327 (1977).

154 A. Closse and D. Hauser, *Life Sci.*, **19**, 1851 (1976).

155 A. Closse and D. Hauser, *Brain Res.*, **147**, 401 (1978).

156 D. A. Greenberg and S. H. Snyder, *Life Sci.*, **20**, 927 (1977).

157 M. Titeler, P. Weinreich, D. Sinclair, and P. Seeman, *Proc. Natl. Acad. Sci. U.S.A.*, **75**, 1153 (1978).

158 M. Titeler and P. Seeman, *Proc. Natl. Acad. Sci. U.S.A.*, **75**, 2249 (1978).

159 S. J. Peroutka, D. A. Greenberg, D. C. U'Prichard, and S. H. Snyder, *Mol. Pharmacol.*, **14**, 403 (1978).

160 T. Haga and K. Haga, *Life Sci.*, **26**, 211 (1980).

161 R. Hornung, P. Presek, and H. Glossman, *Naunyn-Schmiedebergs Arch. Pharmacol.*, **308**, 223 (1979).

162 W. S. Young and M. J. Kuhar, *Proc. Natl. Acad. Sci. U.S.A.*, **77**, 1696 (1980).

163 J. M. Cedarbaum and G. K. Aghajanian, *Eur. J. Pharmacol.*, **44**, 375 (1977).

164 D. J. Reis, T. H. Joh, M. A. Nathan, B. Renaud, D. W. Snyder, and W. T. Tallman, in P. Meyer and H. Schmitt, Eds., *Nervous System and Hypertension*, Wiley–Flammarion, New York, 1979, pp. 147–164.

165 H. Montel, K. Starke, and F. Weber. *Naunyn-Schmiedebergs Arch. Pharmacol.*, **283**, 357 (1974).

166 L. T. Williams, D. Mullikin, and R. J. Lefkowitz, *J. Biol. Chem.*, **251**, 6915 (1976).

167 L. T. Williams and R. J. Lefkowitz, *Mol. Pharmacol.*, **13**, 304 (1977).

168 J. F. Krall, H. Mori, M. L. Tuck, S. L. LeShon, and S. G. Korenman, *Life Sci.*, **23**, 1073 (1978).

169 B. B. Hoffman and R. J. Lefkowitz, *Biochem. Pharmacol.*, **29**, 452 (1980).

169a B. B. Hoffman, D. Mullikin-Kilpatrick, and R. J. Lefkowitz, *J. Biol. Chem.*, **255**, 4645 (1980).

169b B. B. Hoffman and R. J. Lefkowitz, *Biochem. Pharmacol.*, **29**, 1537 (1980).

170 G. Kunos, B. Hoffman, Y. N. Kwok, W. H. Kan, and L. Mucci, *Nature (Lond.)*, **278**, 254 (1979).

171 J. M. Roberts, A. Goldfien, and P. A. Insel, *Nature (Lond.)*, **283**, 108 (1980).

172 A. DeLean and R. J. Lefkowitz, *Nature (Lond.)*, **283**, 109 (1980).

173 G. Kunos, Y. N. Kwok, W. H. Kan, and L. Mucci, *Nature (Lond.)*, **283**, 110 (1980).

174 R. S. Williams and R. J. Lefkowitz, *Circ. Res.*, **43**, 721 (1978).

175 P. Guicheney, R. P. Garay, C. Levy-Marchal, and P. Meyer, *Proc. Natl. Acad. Sci. U.S.A.*, **75**, 6285 (1978).

176 M. S. Briley, S. Z. Langer, and D. F. Story, *Br. J. Pharmacol.*, **66**, 90P (1979).

177 D. F. Story, M. S. Briley, and S. Z. Langer, *Eur. J. Pharmacol.*, **57**, 423 (1979).

178 R. Raisman, M. Briley, and S. Z. Langer, *Naunyn-Schmiedebergs Arch. Pharmacol.*, **307**, 223 (1979).

178a H. J. Schumann and O. E. Brodde, *Naunyn-Schmiedebergs Arch. Pharmacol.*, **308**, 191 (1979).

179 K. H. Jakobs, W. Saur, and G. Schultz, *Mol. Pharmacol.*, **14**, 1073 (1978).

180 D. M. F. Cooper and M. Rodbell, *Nature (Lond.)*, **282**, 517 (1979).

180a P. Lasch and K. H. Jakobs, *Naunyn-Schmiedebergs Arch. Pharmacol.*, **306**, 119 (1979).

181 K. D. Newman, L. T. Williams, N. H. Bishopric, and R. J. Lefkowitz, *J. Clin. Invest.*, **61**, 395 (1978).

182 R. W. Alexander, B. Cooper, and R. I. Handin, *J. Clin. Invest.*, **61**, 1136 (1978).

183 M. C. Scrutton and J. A. Grant, *Nature (Lond.)*, **280**, 700 (1979).

183a J. C. Mitrius, D. J. Kahn, and D. C. U'Prichard, *Neuroscience Soc. Abstr.*, **6**, 852 (1980).

183b B. B. Hoffman, T. Michel, D. M. Kilpatrick, R. J. Lefkowitz, M. E. M. Tolbert, H. Gilman, and J. N. Fain, *Proc. Natl. Acad. Sci. U.S.A.*, **77**, 4569 (1980).

183c S. J. Shattil, M. McDonough, J. Turnbull, and P. A. Insel, *Mol. Pharmacol.*, **19**, 179 (1981).

183d M. Daiguji, H. Y. Meltzer, and D. C. U'Prichard, *Life Sci.*, in press (1981).

183e P. A. Insel, P. Nirenberg, J. Turnbull, and S. J. Shattil, *Biochemistry*, **17**, 5269 (1978).

183f J. M. Roberts, R. D. Golfien, A. M. Tsuchiya, A. Goldfien, and P. A. Insel, *Endocrinology*, **104**, 722 (1979).

184 G. Guellaen, M. Yates-Aggerback, G. Vauquelin, D. Strosberg, and J. Hanoune, *J. Biol. Chem.*, **253**, 1114 (1978).

185 D. M. Neville, *Biochim. Biophys. Acta*, **154**, 540 (1968).

186 G. Guellaen, M. Aggerback, and J. Hanoune, in E. Szabadi, C. M. Bradshaw, and P. Bevan, Eds, *Recent Advances in the Pharmacology of Adrenoceptors*, Elsevier/North-Holland, Amsterdam, 1978, pp. 343–344.

187 W. R. Clarke, L. R. Jones, and R. J. Lefkowitz, *J. Biol. Chem.*, **253**, 5975 (1978).

188 M. F. El-Refai and J. H. Exton, *Eur. J. Pharmacol.*, **62**, 201 (1980).

188a M. F. El-Refai and J. H. Exton, *J. Biol. Chem.*, **255**, 5853 (1980).

188b B. B. Hoffman, D. F. Dukes, and R. J. Lefkowitz, *Life Sci.*, **28**, 265 (1981).

189 B. Hamprecht, *Int. Rev. Cytol.*, **49**, 99 (1977).

190 W. A. Klee and M. Nirenberg, *Proc. Natl. Acad. Sci. U.S.A.*, **71**, 3474 (1974).

191 H. Matsuzawa and M. Nirenberg, *Proc. Natl. Acad. Sci. U.S.A.*, **72**, 3472 (1975).

191a D. J. Kahn, J. C. Mitrius, and D. C. U'Prichard, *Neuroscience Soc. Abstr.*, **6**, 852 (1980).

192 W. A. Klee, in K. Blum, Ed., *Alcohol and Opiates*, Academic Press, New York, pp. 299–308.

193 S. Batzri and Z. Selinger, *J. Biol. Chem.*, **248**, 356 (1973).

194 S. Batzri, Z. Selinger, M. Schramm, and M. R. Robinovitch, *J. Biol. Chem.*, **248**, 361 (1973).

195 E. J. Filinger, S. Z. Langer, C. J. Perec, and F. J. E. Stefano, *Naunyn-Schmiedebergs Arch. Pharmacol.*, **304**, 21 (1978).

196 W. J. Strittmatter, J. N. Davis, and R. J. Lefkowitz, *J. Biol. Chem.*, **252**, 5472 (1977).

197 C. D. Arnett and J. N. Davis, *J. Pharmacol. Exp. Ther.*, **211**, 394 (1979).

197a C. Pimoule, M. S. Briley, and S. Z. Langer, *Eur. J. Pharmacol.*, **63**, 85 (1980).

197b D. B. Bylund and J. R. Martinez, *Nature (Lond.)*, **285**, 229 (1980).

198 R. J. Summers, B. Jarrott, and W. J. Louis, *Proc. Austr. Physiol. Pharmacol. Soc.*, **9**, 90P (1978).

199 B. Jarrott, W. J. Louis, and R. J. Summers, *Br. J. Pharmacol.*, **65**, 663 (1979).

200 B. Jarrott, R. J. Summers, A. J. Culvenor and W. J. Louis, *Circ. Res.*, **46**, I15 (1980).

201 H. Kapur, B. Rouot, and S. H. Snyder, *Eur. J. Pharmacol.*, **57**, 317 (1979).

202 M. I. Holck, B. H. Marks, and C. A. Wilberding, *Mol. Pharmacol.*, **16**, 77 (1979).

202a R. R. Ruffolo, B. S. Turowski, and P. N. Patil, *J. Pharm. Pharmacol.*, 30, 498 (1978).

203 P. J. Barnes, C. T. Dollery, C. A. Hamilton, and J. S. Karlinger, *Br. J. Pharmacol.*, **68**, 138P (1980).

204 P. J. Barnes, J. S. Karliner, C. A. Hamilton, and C. T. Dollery, *Life Sci.*, **25**, 1207 (1980).

205 B. J. Everitt and R. D. Cairncross, *J. Pharm. Pharmacol.*, **21**, 97 (1969).

205a T. Tanaka and K. Starke, *Naunyn-Schmiedebergs Arch. Pharmacol.*, **309**, 207 (1979).

205b T. W. Burns, P. E. Langley, B. E. Terry, D. B. Bylund, B. B. Hoffman, M. D. Tharp, R. J. Lefkowitz, J. A. Garcia-Sainz, and J. N. Fain, *J. Clin. Invest.*, in press (1981).

205c J. A. Garcia-Sainz, B. B. Hoffman, S. Y. Li, R. J. Lefkowitz, and J. N. Fain, *Life Sci.*, **27**, 953 (1980).

205d K. Aktories, G. Schultz, and K. H. Jakobs, *Naunyn-Schmiedebergs Arch. Pharmacol.*, **312**, 167 (1980).

205e H. Kather, J. Pries, V. Schrader, and B. Simon, *Eur. J. Clin. Invest.*, **10**, 345 (1980).

205f R. Pecquery, L. Malagrida, and Y. Guidicelli, *FEBS Letters*, **98**, 241 (1979).

206 B. S. Tsai and R. J. Lefkowitz, *J. Pharmacol. Exp. Ther.*, **204**, 606 (1978).

206a W. S. Colucci, M. A. Gimbrone, and R. W. Alexander, *Hypertension,* **2**, 149 (1980).

206b   S. J. Peroutka, M. A. Moskowitz, J. F. Reinhard, and S. H. Snyder, *Science,* **208,** 610 (1980).

206c   S. I. Harik, V. K. Sharma, J. R. Wetherbee, R. H. Warren, and S. P. Banerjee, *Eur. J. Pharmacol.,* **61,** 207 (1980).

207   D. C. U'Prichard and C. Rosendorff, in preparation.

208   M. Rodbell, L. Birnbaumer, S. L. Pohl, and H. M. J. Krans, *J. Biol. Chem.,* **246,** 1877 (1971).

209   R. J. Lefkowitz, *J. Biol. Chem.,* **249,** 6119 (1974).

210   M. Rodbell, H. M. J. Krans, S. L. Pohl, and L. Birnbaumer, *J. Biol. Chem.,* **246,** 1872 (1971).

211   M. C. Lin, S. Nicosia, P. M. Lad, and M. Rodbell, *J. Biol. Chem.,* **252,** 2790 (1977).

212   P. M. Lad, A. F. Welton, and M. Rodbell, *J. Biol. Chem.,* **252,** 5942 (1977).

213   A. F. Welton, P. M. Lad, A. C. Newby, H. Yamamura, S. Nicosia, and M. Rodbell, *J. Biol. Chem.,* **252,** 5947 (1977).

214   R. Iyengar and L. Birnbaumer, *Proc. Natl. Acad. Sci. U.S.A.,* **76,** 3189 (1979).

215   D. C. U'Prichard and S. H. Snyder, *J. Biol. Chem.,* **253,** 3444 (1978).

215a   S. Yamada, H. I. Yamamura, and W. R. Roeske, *Eur. J. Pharmacol.,* **63,** 239 (1980).

216   B. S. Tsai and R. J. Lefkowitz, *Mol. Pharmacol.,* **16,** 61 (1979).

217   A. J. Blume, *Life Sci.,* **22,** 1845 (1978).

218   D. C. U'Prichard and S. H. Snyder, in E. Szabadi, C. M. Bradshaw, and P. Bevan, Eds, *Recent Advances in the Pharmacology of Adrenoceptors,* Elsevier/North Holland, Amsterdam, 1978, pp. 153–162.

219   D. C. U'Prichard and S. H. Snyder, *J. Neurochem.,* **34,** 385 (1980).

220   H. Glossman and P. Presek, *Naunyn-Schmiedebergs Arch. Pharmacol.,* **306,** 67 (1979).

220a   H. Glossmann and R. Hornung, *Naunyn-Schmiedebergs Arch. Pharmacol.,* **314,** 101 (1980).

221   B. R. Rouot, D. C. U'Prichard, and S. H. Snyder, *J. Neurochem.,* **34,** 374 (1980).

222   C. B. Pert and S. H. Snyder, *Mol. Pharmacol.,* **10,** 868 (1974).

223   D. Lichtshtein, G. Boone, and A. J. Blume, *J. Cyclic Nucleotide Res.,* **5,** 367 (1979).

224   L. R. Hegstrand, K. P. Minneman, and P. B. Molinoff, *J. Pharmacol. Exp. Ther.,* **210,** 215 (1979).

225   D. A. Greenberg, D. C. U'Prichard, P. Sheehan, and S. H. Snyder, *Brain Res.,* **140,** 378 (1978).

226   S. R. Childers and S. H. Snyder, *Life Sci.,* **23,** 759 (1978).

227   B. S. Tsai and R. J. Lefkowitz, *Mol. Pharmacol.,* **14,** 540 (1978).

227a   H. Glossmann and R. Hornung, *Eur. J. Pharmacol.,* **61,** 407 (1980).

227b   H. Glossmann and R. Hornung, *Naunyn-Schmiedebergs Arch. Pharmacol.,* **312,** 105 (1980).

228   B. Cooper, R. I. Handin, L. H. Young, and R. W. Alexander, *Nature (Lond.),* **274,** 703 (1978).

229   S. L. Sabol and M. Nirenberg, *J. Biol. Chem.,* **254,** 1921 (1979).

229a   D. Wilkening and M. Nirenberg, *J. Neurochem.,* **34,** 321 (1980).

230   W. J. Strittmatter, J. N. Davis, and R. J. Lefkowitz, *J. Biol. Chem.,* **252,** 5478 (1977).

231   M. D. Dibner and P. B. Molinoff, *J. Pharmacol. Exp. Ther.,* **210,** 433 (1979).

232   H. R. Wagner and J. N. Davis, *Proc. Natl. Acad. Sci. U.S.A.,* **76,** 2057 (1979).

233   D. C. U'Prichard and S. J. Enna, *Eur. J. Pharmacol.,* **59,** 297 (1979).

234   R. R. Ruffolo, B. S. Turowski, and P. N. Patil, *J. Pharm. Pharmacol.,* **29,** 378 (1977).

235   D. C. U'Prichard and R. Kvetnasky, in E. Usdin, R. Kvetnansky, and I. J. Kopin, Eds, *Catecholamines and Stress,* Elsevier/North-Holland, Amsterdam, pp. 299–308.

236   A. Maggi, D. C. U'Prichard, and S. J. Enna, *Science,* **207,** 645 (1980).

237   T. D. Reisine, D. C. U'Prichard, N. L. Wiech, R. C. Ursillo, and H. I. Yamamura, *Brain Res.,* **188,** 587 (1980).

237a   C. H. Wang and D. C. U'Prichard, *Neuroscience Abstr.,* **6,** 1 (1980).

238 J. F. Tallman, S. M. Paul, P. Skolnick, and D. W. Gallager, *Science, 207,* 274 (1980).

238a E. A. Woodcock and C. I. Johnston, *Nature (Lond.), 386,* 159 (1980).

239 D. C. U'Prichard, J. Mitrius, D. Kahn, and M. Daiguji, in H. J. Yamamura, R. W. Olsen and E. Usdin, Eds., *Psychopharmacology and Biochemistry of Neurotransmitter Receptors,* Elsevier/ North-Holland, Amsterdam, 1980, pp. 247–259.

240 L. E. Limbird and R. J. Lefkowitz, *J. Biol. Chem., 252,* 799 (1977).

241 T. Haga, K. Haga, and A. G. Gilman, *J. Biol. Chem., 252,* 5776 (1977).

242 K. G. Walton, P. Liepmann, and R. J. Baldessarini, *Eur. J. Pharmacol., 52,* 231 (1978).

243 G. Guellaen, M. Aggerbeck, and J. Hanoune, *J. Biol. Chem., 254,* 761 (1979).

244 W. H. Kan, C. Farsang, H. G. Preiksaitis, and G. Kunos, *Biochem. Biophys. Res. Commun., 91,* 303 (1979).

245 I. Takayanagi, M. Yoshioka, K. Takagu, and Z. Tamura, *Eur. J. Pharamacol., 35,* 121 (1976).

246 K. A. Heidenreich and E. Hendley, in E. Usdin, I. J. Kopin, and J. Barchas, Eds., *Catecholamines: Basic and Clinical Frontiers,* Pergamon, Oxford, 1979, pp. 367–369.

246a C. L. Wood, M. G. Caron, and R. J. Lefkowitz, *Biochem. Biophys. Res. Commun., 88,* 1 (1979).

246b D. C. U'Prichard, M. Daiguji, and D. J. Kahn, *Neuroscience Abstr., 6,* 852 (1980).

CHAPTER SIX

RECENT DEVELOPMENTS IN STRUCTURE–ACTIVITY RELATIONSHIPS AMONG INHIBITORS OF THE ADRENERGIC α-RECEPTOR

Carlo Melchiorre

Institute of Organic and Pharmaceutical Chemistry, University of Camerino (MC), Italy

Bernard Belleau

Department of Chemistry, McGill University, Montreal, Quebec, Canada

1 INTRODUCTION

To achieve receptor specificity with drugs or affinity reagents remains a formidable challenge in molecular pharmacology and medicine. The eventual understanding of neurotransmitter receptor structure and function requires accessibility to receptor-specific reagents and drugs. Future drug design depends in turn on accessibility of basic knowledge. Although numerous drugs affecting neurotransmitter receptors are presently widely used in medicine, the vast majority of them cause unwanted side effects mainly because they lack receptor specificity. For the purpose of neurotransmitter receptor localization and characterization, specific reagents possessing high affinity for the receptor protein must be discovered and developed. The discovery a few years ago of the anticholinergic neurotoxins from cobra venom (1,2) constitutes a major contribution to a basic understanding of the nicotinic receptor structure and function. These neurotoxins are specific for this receptor and possess such an affinity that their binding onto the protein may be said to be virtually irreversible (1,2). Rapid progress in our understanding of the nicotonic receptor structure and function has been made thanks to the discovery of the neurotoxins (3). Unfortunately, comparatively little progress has been made as regards other neurotransmitter receptors primarily because no other neurotoxins are known that allow similar specific affinity labeling of the proteins concerned. Of prime relevance to human pharmacology are the receptors for catecholamines because of their involvement in the regulation of the cardiovascular system, the central nervous system, metabolic pathways, and platelet aggregation among other vital functions (Table 1).

TABLE 1   Some Adrenergic Responses

Effector System	α-Response	β-Response
Smooth muscle		
Uterus (rabbit)	Contraction	Relaxation
Bronchial		Relaxation
Bladder (detrusor)		Relaxation
Ciliary muscle (lens)		Relaxation
Pyloric sphincter	Contraction	Relaxation
Bladder (trigone and sphincter)	Contraction	
Iris (radial muscle)	Contraction	
Intestine	Decreased motility	Decreased motility
Arterial	Contraction	Relaxation
Platelets	Aggregation	Inhibition of aggregation
Cardiac muscle		
Contractility	Increase	Increase
Heart rate		Increase
Adipose tissue		lipolysis
Salivary glands	$K^+ + H_2O$ secretion	Amylase secretion
Lymphocytes		Inhibition of cytolysis

Many years ago, structure–activity relationships (SAR) studies led Ahlquist (4) to recognize the existence of two distinct types of catecholamine receptors: α-adrenergic and β-adrenergic receptors. While the α-type is concerned with sympathetic neurotransmission mediated by noradrenaline (NA), the β-type appears to be generally involved in the regulation of metabolic responses where adrenaline (A), a circulating hormone, plays a key role. However, both types of receptors respond well to either one of the two natural agonists and insofar as we are aware, these receptor subtypes are the only ones that may be said to possess two natural, endogenous agonists. Recently, many pharmacological observations indicate that the α-class of receptors consists of at least two subtypes, α_1 and α_2 (5,6). The α_1-type is referred to as the classical postsynaptic α-receptor, which initiates the response of effector tissues, while the α_2-type constitutes the presynaptic α-receptor, which regulates transmitter release. However, this classification should not be applied too rigorously because α_1- and α_2-receptors may also occur in regions other than the post- and presynaptic areas (6). More recently, it has been proposed that the classification into α_1- and α_2-receptors should be based on pharmacological differences between the relative affinities of agonists and antagonists without taking into account either location or function of the adrenergic α-receptor (7). This topic will be discussed in detail in the next volume of this series.

In this chapter we describe and correlate recent developments in the gross chemistry of the adrenergic α-receptor as judged from structure–activity relationship studies. Various other aspects of these developments have already been reviewed and hence are not covered exhaustively here (8–12).

2   α-ADRENORECEPTOR BLOCKING DRUGS

Of the various effectors acting on the α-adrenoreceptors, only antagonistic agents are dealt with here, since the SAR of the α-agonists have been reviewed recently (11) and no new significant developments have been reported.

Two broad categories of adrenoreceptor inhibitors are known: *(a)* those that inhibit by forming covalent bonds with some component of the receptor apparatus (irreversible blockade), and *(b)* those that bind reversibly and thus prevent access of agonists to the receptor (competitive or reversible inhibition).

Structure–activity relationship studies can be very useful in promoting a better understanding of the mechanism of drug–receptor interactions and, of course, in the design of new drugs. However, much caution is needed in drawing firm conclusions about the binding-site structural chemistry because of the several factors to be taken into account. The molecular pharmacology of a series of compounds may be usefully analyzed if the following basic requirement is fulfilled: the biological response that they produce should be mediated through selective interaction with a unique receptor system. Even if this requirement is satisfied, SAR studies are often complicated by the possibility of interaction with more than one binding site on the receptor. SAR studies of compounds acting on the α-adrenoreceptor are also difficult to interpret because of the relatively large number of chemical structures that are unrelated to one another and which all possess some antagonistic properties. A partial list of α-antagonists commonly used as pharmacological tools is given in Table 2.

With few but important exceptions, no significant improvements in the selectivity or affinity of α-blocking drugs have been achieved in the recent past. The following discussion will therefore not cover all the classes listed in Table 2, which in any event have already been reviewed (8–12).

2.1 Benzodioxanes

Benzodioxanes form a class of competitive α-antagonists that incorporate 2-aminomethylbenzodioxane as a basic feature. Some of them have been used clinically as antihypertensive agents, but they were eventually discarded because of their lack of specificity, as reflected in the several adverse effects that they produce (such as sedation, tachycardia, and muscle relaxation) (13–15). Many benzodioxanes were reported by Fourneau and Bovet nearly 50 years ago (16), but little progress has been made in improving the specificity of piperoxane (**1**) and prosympal (**2**), which

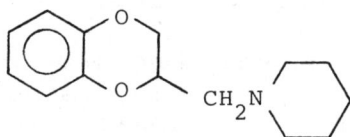

1

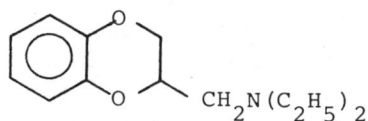

2

are the prototypes of this class (16). Recently, however, Mottram and Kapur (17) and Kapur et al. (18,19) have described the α-blocking activity of a series of benzodioxane derivatives previously synthesized several years ago by others (20,21). They found that WB-4101 (**3**) displays high postsynaptic α-antagonism

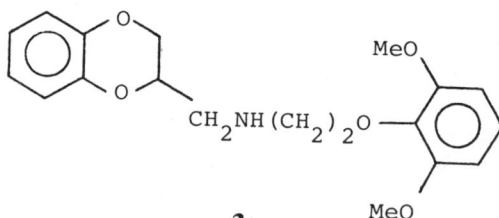

3

against NA with a pA$_2$ value of 9.8. On the basis of SAR studies with related benzodioxanes (Table 3), these authors suggested that the high activity of **3** would depend both on the presence of a benzodioxane nucleus and of a 2,6-dimethoxyphenyl substituent at a specific distance from the basic nitrogen (18). The hypothesis was advanced that WB-4101 may bind on the same site where catecholamines interact, one part of the molecule projecting over an accessory area (17,18), in agreement with the general theoretical model proposed earlier by Ariens and Simonis

**TABLE 2   Some Classes and Representative
Prototypes of α-Adrenergic Blocking Compounds**

Reversible Blockade	*Irreversible Blockade*
Imidazolines	1,2-Dihydroquinoline *N*-carbamates
Phentolamine	N-ethoxycarbonyl-2-ethoxy-1,2-
Tolazoline	dihydroquinoline
Benzodioxanes	β-Haloalkylamines
Prosympal	Dibenamine
Piperoxan	Phenoxybenzamine
Dibozane	Tetramine disulfides
WB4101	BHC
Ergot alkaloids	AOC
Ergotamine	
Ergocryptine	
Dihydroergotamine	
Dihydroergocryptine	
Others	
Yohimbine	
Chlorpromazine	
Haloperidol	

(22) for the binding of agonists and antagonists on the α-receptor. However, Avner and Triggle (23) have shown that benzodioxane antagonists related to **2** but carrying an alkylating *N*(2-chloroethyl) substituent bind similarly to the classical β-haloalkylamines, which interact with two kinetically distinguishable sites. Although the alkylating part of the molecule may change the binding process, it is nevertheless likely that benzodioxanes also interact with the receptor at least with two sites, one related perhaps to the NA recognition site and another possibly concerned with calcium binding (24,25).

The importance of the ether oxygens on binding in the benzodioxane series was studied briefly. Thus substitution of the oxygen in position 4 with less-polar groups such as sulfur (26) or methylene (19) led to a dramatic decrease in potency. However, the presence of sulfur or methylene produced different effects, the sulfur analog acting as a partial agonist (26) and the methylene analog of WB-4101 behaving only as a weak antagonist (19) (Table 4). Furthermore, when the chroman moiety was changed to a 2,3-dihydrobenzofuran, a further decrease in activity was observed. Kapur et al. (19) explained this drop in activity in terms of differences in the relative rigidity of 2,3-dihydropyran and 2,3-dihydrofuran rings which would alter the optimal distance required for a good fit on the receptor between the aromatic ring and the nitrogen atom. In the light of these results, especially with the sulfur analog, it is obvious that the role of the ether oxygen at position 4 cannot be undervaluated. It seems that the alterations in the optimal distance between the phenyl and the nitrogen are too small to play a major role. Rather, it appears that the 4-oxygen plays a key role by a way of an electronic effect and/or engages in

TABLE 3 Biological Activity of 1,4-Benzodioxane Derivatives on α-Receptor Preparations (17–19,29)

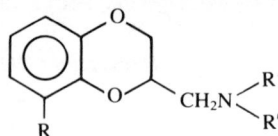

Compound	R	R'	R''	pA$_2$ Rat Vas Deferens	Rabbit Aorta	S/R Ratio
S-Prosympal					5.60	2.2
	H	Et	Et			
R-Prosympal					5.26	
S-Piperoxan					7.12	18.2
	H	-(CH$_2$)$_5$-				
R-Piperoxan					5.86	
S,S-Dibozane					7.66	19.1
	H					
R,R-Dibozane					6.38	
meso-Dibozane					7.53	
WB-4101	H	H	(CH$_2$)$_2$O-2',6' (MeO)$_2$C$_6$H$_3$	9.80[a]		
WB-4085	H	H	(CH$_2$)$_2$O-2'-MeOC$_6$H$_4$	6.62		
WB-4107	H	H	(CH$_2$)$_2$O-3'-MeOC$_6$H$_4$	5.87		
WB-4105	H	H	(CH$_2$)$_2$O-4'-MeOC$_6$H$_4$	4.35		
WB-4082	H	H	(CH$_2$)$_2$OC$_6$H$_5$	4.88		
WB-4110	H	H	(CH$_2$)$_2$O-2',6'-Me$_2$C$_6$H$_3$	5.39		
WB-4108	H	H	(CH$_2$)$_2$O-6'-MeC$_6$H$_4$	4.92		
WB-4111	H	H	(CH$_2$)$_3$O-2',6' (MeO)$_2$C$_6$H$_3$	4.40		
WB-4093	H	H	(CH$_2$)$_3$OC$_6$H$_5$	<3.00		
WB-4099	H	H	(CH$_2$)$_4$OC$_6$H$_5$	4.30		
WB-4109	H	H	(CH$_2$)$_2$O(CH$_2$)$_2$OCH$_3$	5.30		
WB-4371	H	H	(CH$_2$)$_3$SO$_2$C(Et$_2$)CH$_3$	5.10		
WB-4116	H	CH$_3$	(CH$_2$)$_2$O-2',6' (MeO)$_2$C$_6$H$_3$	5.49		
WB-4267	Me	H	(CH$_2$)$_2$O-2',6' (MeO)$_2$C$_6$H$_3$	7.50		

[a]Butler and Jenkinson (27) reported a pA$_2$ value of 8.9.

hydrogen-bond formation. It would be interesting to evaluate the effects of substitution of the oxygen at position 2 with other groups. N-methylation of WB-4101 resulted in a compound with a significantly lower potency, thus emphasizing the importance of a secondary amino group for optimal activity (Table 3) (19).

WB-4101 has been widely used recently in binding studies (see Section 3). However, few pharmacological studies have been reported on this compound and this in spite of its unusually high pA$_2$ value (17,27,28). Its receptor selectivity remains to be ascertained. The use of any reagent of unknown receptor selectivity

TABLE 4 Biological Activity of Compounds Related to Benzodioxans on Rat Vas Deferens (19,26)

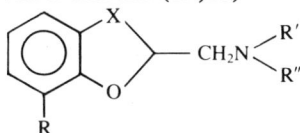

R	R'	R''	X	pA_2	pD_2	Intrinsic Activity
H	H	$(CH_2)_2O\text{-}2',6'(MeO)_2C_6H_3$	$(CH_2)_2$	7.35		
H	H	$(CH_2)_2O\text{-}2'\text{-}MeOC_6H_4$	$(CH_2)_2$	6.77		
H	H	$(CH_2)_2OC_6H_5$	$(CH_2)_2$	5.39		
Me	H	$(CH_2)_2O\text{-}2',6'(MeO)_2C_6H_3$	$(CH_2)_2$	7.19		
H	H	$(CH_2)_2O\text{-}2',6'(MeO)_2C_6H_3$	CH_2	6.26		
H	H	$(CH_2)_2O\text{-}2'\text{-}MeOC_6H_4$	CH_2	6.02		
H	H	$(CH_2)_2OC_6H_5$	CH_2	4.17		
Me	H	$(CH_2)_2O\text{-}2',6'(MeO)_2C_6H_3$	CH_2	6.32		
H	Me	Me	SCH_2		3.48	0.47
H	Et	Et	SCH_2		3.99	0.91
H	\-\-$(CH_2)_5$\-\-		SCH_2		4.02	0.85

precludes definite conclusions about its mechanism of action. The effects of optical isomerism on biological activity can yield critical information on binding-site selectivity and associated receptor events. Surprisingly, it is only recently that the optical isomers of some benzodioxanes have been synthesized and compared at the α-receptor level (Table 3) (29). It was found that the S-enantiomer is twice as potent as the R-enantiomer for prosympal and 18–19 times for piperoxan and dibozane (4). The *meso* isomer of 4 was equiactive with the S,S-isomer. It follows that a

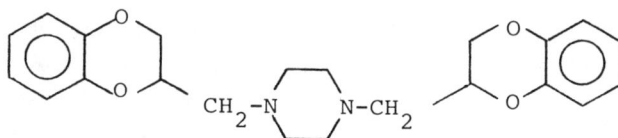

4

single S-chiral center is needed for activity in the dibozane molecule. From a comparison of molecular models of the various conformers, one conformation of the S-benzodioxanes appears similar to that of R-adrenaline as far as spatial relationships among the amine, oxygen, and aromatic ring are concerned. On the basis of this similarity, the hypothesis that S-benzodioxanes may bind to the same site occupied by R-adrenaline has been suggested (29). In this connection, it is noteworthy that the S-enantiomer of WB-4101 has been found to be 40–50 times more potent than the R-enantiomer (137).

Piperoxan and WB-4101 behave differently as far as selectivity toward α_1- and α_2-receptors is concerned. For instance, WB-4101 was reported to be a selective α_1-receptor antagonist (27,30), while piperoxan displayed selectivity toward the α_2-receptor of the rat (31–36), the guinea pig ileum (37), and the rat vas deferens (38). However, on the rabbit pulmonary artery (39) it showed no selectivity. It can be seen that meaningful SAR studies are difficult to achieve even when compounds of the same structural class are used. The assumption that structural similarities favor binding on identical sites appears invalid when drugs such as piperoxan and WB-4101 are compared.

2.2   Imidazolines

Imidazolines substituted at position 2 are a class of compounds with a complex and wide range of pharmacological activities. They incorporate a 4,5-dihydroimidazole and a phenyl ring separated by a one carbon or one nitrogen unit as a basic feature. This class of compounds may act either as direct agonists or competitive antagonists at the α-receptor level, depending on the nature of the 2-substituent. However, there is no clear-cut delimitation between agonistic and antagonistic actions. In fact, some imidazolines are agonists in a tissue and antagonists in another, which causes major difficulties in the interpretation of SAR. Many compounds of this class, which act either as antagonists (40–43) or agonists (44), have been used clinically. Imidazolines with agonistic activity have been reviewed (45–52) and hence will not be given much attention in this section.

Hartmann and Isler (53) first studied a series of 2-substituted imidazolines on blood vessels. They reported that optimum vasodilator and vasodepressor activities for a series of 2-alkylimidazolines were associated with a six- to eight-carbon chain length (**5**). Substitution of the alkyl group for a benzyl group (**6a**, tolazoline) retained antagonistic activity, whereas exchange for an α-naphthylmethyl substituent (**6b**, naphazoline) or the introduction of hydroxy or methoxy groups on the aromatic ring reversed the activity from depressor to pressor (54). Moreover, substitution of the benzene for a quinuclidine ring resulted in a loss of pharmacological activity (55). Opening of the imidazoline ring of tolazoline also resulted in inactive compounds (56). These latter findings were interpreted in terms of structural differences between the amidinium and imidazolinium ions (56). Furthermore, the introduction of a substituted 2-amino group onto the imidazoline ring resulted in potent, directly acting α-agonists such as clonidine (**6c**). This class of drugs has been extensively reviewed. Nevertheless, it is worthwhile pointing out that this class of compounds exhibits a wide variety of pharmacological actions mediated through interactions with the central or peripheral α-receptors. In spite of the fact that substituted 2-aminoimidazolines usually act as pure α-agonists they can behave as partial agonists in some tissues such as rat aorta (48,57) and as competitive antagonists in others, such as rat parotid cells (58).

The most active α-antagonist of this class is phentolamine (**7**) (54), which differs markedly from tolazoline in its chemical structure. It has been widely investigated and is uniformly more active than tolazoline (59–62). It was reported that imida-

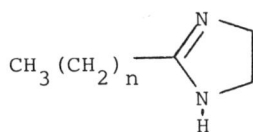

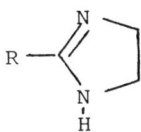

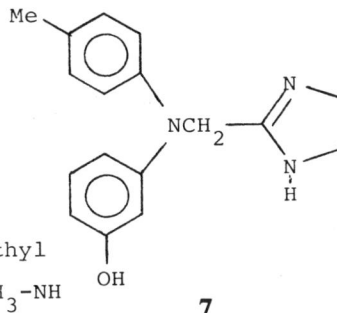

5; n = 5-7   **6a**; R=benzyl

6b; R=α-naphthylmethyl

6c; R=2',6'-Cl$_2$-C$_6$H$_3$-NH   **7**

zolines, either agonists or antagonists, act at a common site on rabbit aorta (48). More recently, Ruffolo et al. (63,64) gave evidence that agonists of the imidazoline class such as clonidine may possibly bind on a site different from that for the agonists of the phenethylamine class. In spite of these extensive studies with the imidazolines, only few results are available as regards stereochemical requirements. It has been reported that the optical forms of 4-methylnaphazoline did not show any selectivity toward the α-receptors (65). Interestingly, substitution of a hydrogen in position 4 by a methyl group gave the expected two optical isomers, which now behaved as α-antagonists on rabbit aorta, thus suggesting that both classes of agonists and antagonists may act at a common site.

Both tolazoline and phentolamine are not receptor-specific, since they display several other effects in addition to α-receptor effects (48,66–73). Furthermore, both compounds do not show significant selectivity for the α$_1$- and α$_2$-receptors either *in vivo* (31,33,74–79) or *in vitro* (37,38,80,81). However, on the cat nictitating membrane, tolazoline failed to block the α$_2$-receptor (82), while on the rabbit pulmonary artery (39) and in the rat (36) it blocked this receptor preferentially. On the other hand, phentolamine showed higher affinity for the α$_1$- than the α$_2$-receptors of the rabbit pulmonary artery (39) and also in the rat (35,36), but the reverse applied to the rat submaxillary gland (83). In conclusion, these drugs appear to be nonselective toward α$_1$- and α$_2$-receptors.

2.3 β-Haloalkylamines

In the field of α-adrenoreceptor antagonists, the β-haloalkylamines are probably the best known pharmacological tools. Since the discovery of dibenamine (84), the prototype of this class of irreversible α-blockers, an enormous number of related compounds have been synthesized in attempts to improve both selectivity and activity. Unfortunately, the β-haloalkylamines are not receptor-specific and the considerable number of reported efforts at improving selectivity have not been too rewarding. As a result, the active-site chemistry and topography of the α-receptor interaction with β-haloalkylamines has remained quite obscure. The two best known prototypes are dibenamine (**8**) and phenoxybenzamine (**9**), and their mechanism of action, site of interaction, and selectivity have been reviewed extensively

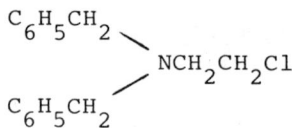

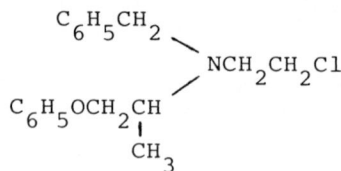

8   9

(8–11,85–87). Several analogs are known to be at least as active as phenoxybenzamine. An interesting characteristic of these agents is the biphasic response that
they induce in α-adrenergic preparations. An initial short phase of competitive
antagonism is followed by a long-lasting irreversible blockade of the α-receptor.
To account for this biphasic nature of the response, Nickerson (88) first suggested
the sequence of reactions shown in Figure 1. The β-haloalkylamine would cyclize
in the biophase to form the pharmacologically active aziridinium ion, which initially
would interact with the receptor in a reversible manner and would subsequently
alkylate a nucleophile on the receptor to form a covalent bond.

   This mechanism has been widely accepted on the basis of a considerable amount
of evidence (11). Additional support for the formation of a reactive aziridinium ion
intermediate in the mechanism of receptor inactivation was recently offered by
Henkel et al. (89), who synthesized the aziridinium ions corresponding to dibenamine and phenoxybenzamine and determined their rates of formation and decom-
position under physiological conditions as well as their *in vitro* activities. As expected, the aziridinium salts were equipotent with the precursor amines as regards
blockade of the contractile response to NA of rat vas deferens (Figure 2). It has
been also demonstrated that the precursor amines have little or no receptor alkylating
ability, strongly suggesting the formation of an aziridinium ion intermediate as the
only active species at the receptor level. Moreover, the difference in potency between phenoxybenzamine and dibenamine may be accounted for primarily by differences in rate of transport and receptor affinity rather than by differences in
alkylating abilities (89). However, for other types of β-haloalkylamines, the aziridinium ion as such may not account completely for α-receptor inactivation. In one
specific case, Belleau and Triggle (90) obtained evidence indicating that the alkylating species may have significant carbonium-ion character. This possibility is
especially attractive for the case of *N,N*-dimethyl-β-chlorophenethylamine (Figure
3). It is logical to expect protein active sites to display chiral selectivity (as is the

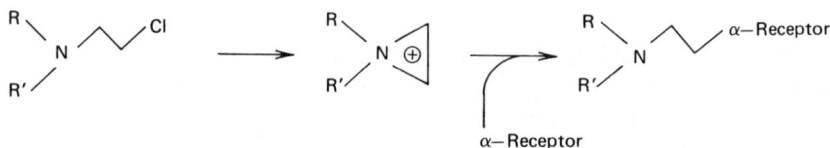

Figure 1   Mechanism of action of β-haloalkylamines.

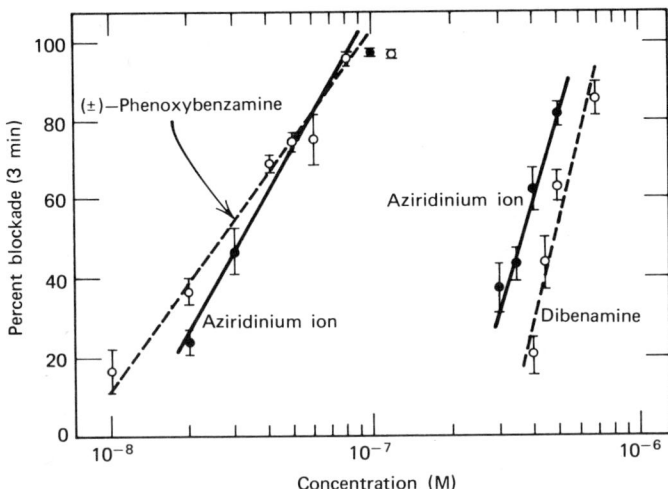

Figure 2 Blockade of the response to 5×10^{-5} M NA on rat vas deferens after a 3-min exposure to the parent amines and their derived aziridinium salts. Reprinted with permission from J. G. Henkel et al., *J. Med. Chem., 19, 6 (1976). Copyright by the American Chemical Society.*

case for NA) and it would be expected that the two enantiomers of **10** should not be equiactive if the alkylation of the receptor involves direct attack of the aziridinium ion by a nucleophile. The observation that the two enantiomers of **10** were equiactive may signify that a species with considerable carbocationic character may instead attack a nucleophile (as in Figure 3).

However, enantiomeric selectivity of the more common type has also been examined (85,91–93). For instance, the enantiomers of phenoxybenzamine had a potency ratio of 14.5 (at 50% blockade, rat vas deferens), the R-enantiomer being the more active. This relatively good stereoselectivity may be indicative of multiple point interaction with the receptor, and it has been argued that the difference in the enantiomer potencies reflects a difference in affinity rather than different alkylating

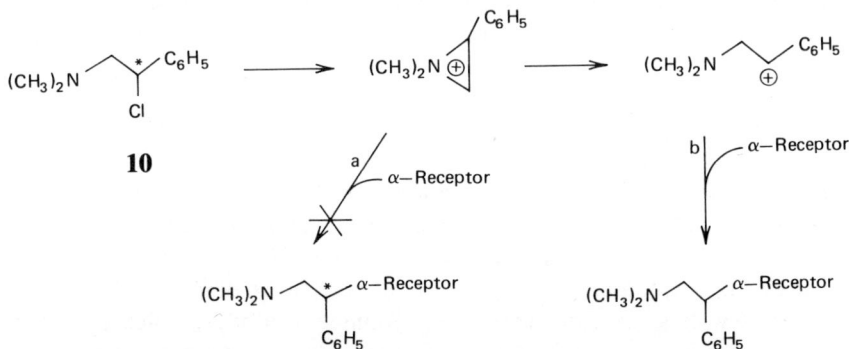

Figure 3 Possible mechanism of alkylation by N,N-dimethyl-β-chlorophenethylamine (**10**). Inactivation by way of (a) aziridinium ion and (b) carbonium ion more or less open.

capacities. The contrasting behavior of the enantiomers of **10** was explained on the basis that their chiral center would be located in a part of the molecule that is insensitive to steric effects at the receptor level (91). A definitive explanation for the absence of chiral selectivity toward R- and S-**10** is difficult at this time. The effect of chirality on the 2-chloroethyl substituent, on the benzyl carbon of dibenamine, and on desmethylphenoxybenzamine produced contrasting effects on blocking activity. Thus the enantiomers of the N-2-chloropropyl analogs had a potency ratio higher than 1.0, the R-isomer being the more active of the pair. On the other hand, with the α-methylbenzyl enantiomers, it was the S-isomer that was more active. In addition, α-methylation of the β-chloroethyl part had a detrimental effect on blocking potency as was the case for related compounds (94), and these results agree with the expectation that approach of an aziridinium ion to a counter anionic site of the receptor should be hindered when the cation is bulky. Enantiomeric N-β-haloalkylphenethylamines carrying a benzylic hydroxyl as in NA differed marginally in α-blocking potency, which of course is in sharp contrast to the high stereoselectivity of the receptor toward the chiral benzylic carbon of NA and A. For instance, S-N(2-chloroethyl)-N-methyl-2-hydroxy-2-phenethylamine was only about six times more active than the R-enantiomer at 50% blockade of the rat vas deferens response to phenylephrine (92). These results indicate that the β-hydroxyl of this molecule does not interact like the equivalent group of catecholamines. The observed partial stereoselectivity of some chiral β-haloalkylamines may not allow a unique interpretation because the enantiomeric stereoselectivity may reflect only differences in receptor affinity as suggested (91). Several other factors should be taken into account, such as possible differences in alkylating ability and the strength of inhibition of uptake sites. Moreover, the presence of at least two kinetically distinguishable sites which characterize the reaction of β-haloalkylamines with α-receptors (24,25) further complicates the interpretation of observed differences in enantiomeric selectivity.

Phenoxybenzamine is an effective antagonist of both α_1- and α_2-receptors (31,34–36,39,80,95–97) but displays a selectivity toward the α_1-receptor of the rat (35,36), the rabbit pulmonary artery (39), the rat vas deferens (80), and the perfused cat spleen (95). Apart from its general lack of receptor specificity, phenoxybenzamine can block both the α_1- and α_2-receptors, with only few exceptions (36, 95,135).

2.4   Tetramine Disulfides

What emerges from the preceding discussion is that in spite of considerable efforts and well-documented SAR studies among several classes of agonists and antagonists, little progress has been achieved as regards the structural and functional chemistry of the adrenergic α-receptor. In this section we describe SAR studies of an entirely new class of α-blockers, the polyamine disulfides, which provide new insights into adrenergic pharmacology and the molecular structure of the receptor binding sites.

Some 10 years ago, Demaree et al. (98) and Herman et al. (99) reported that some sulfur-containing amines showed a persistent, albeit modest, α-blocking activity either *in vitro* or *in vivo*. These authors found that the diaminothiol-S-phosphate (**11**; $n = 5$, R = PO_3H_2), the diaminothiol (**11**; $n = 5$, R = H), and the corresponding disulfide [**15a**, N,N′-bis(5-amino-n-pentyl)cystamine (APC)] (Figure 4) were competitive α-antagonists. The slow onset of action of the diaminothiol-S-phosphate was attributed to the necessary action of a phosphatase as a step leading to the liberation of the corresponding thiol, which would then suffer oxidation to the disulfide, the species subsequently established to be responsible for α-blockade. That the disulfide is indeed the active species was demonstrated by Lippert and Belleau (100), who reinvestigated the mechanism of the α-receptor inhibition by

$$H_2N(CH_2)_nNH(CH_2)_2S-R$$

(**11**; $n=3-6$)

12; R=2′-MeO-benzyl, $n=6$
13; R=H, $n=8$

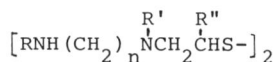

(**15**; $n=5-12$)

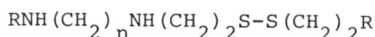

(**14**; $n=6,8$)

15a (APC); R=R′=R″=H, $n=5$

15b (BHC); R=2′-MeO-benzyl, R′=R″=H, $n=6$

15c (AOC); R=R′=R″=H, $n=8$

15d; R=R′=R″=H, $n=7$

15e; R=3′,4′-(HO)$_2$-benzyl, R′=R″=H, $n=6$

15f; R=2′-MeO-benzyl, R′=Me, R″=H, $n=6$

15g; R=R″=H, R′=Me, $n=8$

15h; R=α-Me-benzyl, R′=R″=H, $n=6$

15i; R=3′,4′-(HO)$_2$-benzyl, R′=R″=H, $n=8$

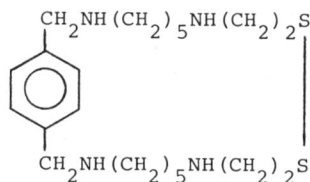

16

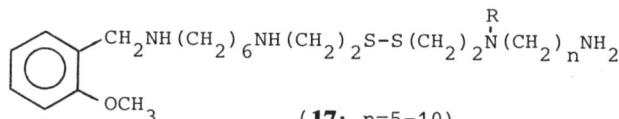

(**17**; $n=5-10$)

17a; R=H, $n=8$

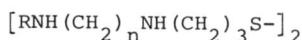

18a; R=2′-MeO-benzyl, $n=5$

18b; R=H, $n=7$

Figure 4 General structures of polyaminedisulfides.

TABLE 5 Relative α-Blocking Activities of Tetramine Disulfides 15 (Figure 4) on Rat Vas Deferens (101, 102, 140) and Rabbit Aorta (139).

| | Structure 15 | | | | α-Blockade (% ± 5) | | | | | |
| | | | | | Vas Deferens[b] | | | Aorta[c] | | |
	n	R'	R''	R	10^{-5}	2×10^{-5}	5×10^{-5} M	5×10^{-5}	10^{-5}	3×10^{-6} M
(APC)	5	H	H	H		10	34			
	6	H	H	H		27		42		
(15d)	7	H	H	H		63		86		
(AOC)	8	H	H	H		100		68		
(15g)	8	Me	H	H		90				
	9	H	H	H		92		69		
	10	H	H	H		56		50		
	12	H	H	H		24				
	5	H	H	Benzyl			87			
	6	H	H	Benzyl		90				
	7	H	H	Benzyl		58				
(BHC)	5	H	H	2'-MeO-benzyl		59	81			
(15f)	6	H	H	2'-MeO-benzyl	96	100				
	6	Me	H	2'-MeO-benzyl		48				
	7	H	H	2'-MeO-benzyl		71				
	8	H	H	2'-MeO-benzyl		41				11
	6	H	H	2'-EtO-benzyl		81				77
	6	H	H	2'-Allyloxy-benzyl		71				17

6	H	H	3'-MeO-benzyl		72	
6	H	H	4'-MeO-benzyl		81	82
5	H	H	2'-Me-benzyl			
6	H	H	2'-Me-benzyl		69	
7	H	H	2'-Me-benzyl		56	
8	H	H	2'-Me-benzyl		49	
5	H	H	2'-HO-benzyl			57
6	H	H	2'-HO-benzyl	36		
6	H	H	3'-HO-benzyl	24		
6	H	H	4'-HO-benzyl	14		
5	H	H	2'-Cl-benzyl			44
6	H	H	2'Cl-benzyl		44	
7	H	H	2'-Cl-benzyl		36	
5	H	H	Guanyl			14
5	H	H	3'-Pyridylmethyl			21
5	H	H	1'-Naphtylmethyl			38
5	H	Me	H (l,l)			23
5	H	Me	H (d,d)			28
5	H	Me	H (meso)			25
5	H	H	3',4'-(MeO)₂-benzyl			32
6	H	H	3',4'-(MeO)₂-benzyl		24	
6	H	H	2',3'-(MeO)₂-benzyl		26	
6	H	H	2',4'-(MeO)₂-benzyl		26	
6	H	H	2',5'-(MeO)₂-benzyl		76	
6	H	H	3',5'-(MeO)₂-benzyl		41	
6	H	H	2',6'-(MeO)₂-benzyl	77	100	
6	H	H	2',6'-Cl₂-benzyl		40	
7	H	H	2',6'-Cl₂-benzyl		39	

TABLE 5 (cont'd.)

				α-Blockade (% ± 5)					
Structure 15				Vas Deferens[b]			Aorta[c]		
n	R'	R''	R	10^{-5}	2×10^{-5}	5×10^{-5} M	5×10^{-5}	10^{-5}	3×10^{-6} M
6	H	H	2',3'-(HO)$_2$-benzyl	44					
6	H	H	2',4'-(HO)$_2$-benzyl	60					
6	H	H	2',5'-(HO)$_2$-benzyl	28					
6	H	H	2',6'-(HO)$_2$-benzyl	41					
6	H	H	3',5'-(HO)$_2$-benzyl	23					
(15e) 6	H	H	3',4'-(HO)$_2$-benzyl	70	98			85	
5	H	H	3',4'-(HO)$_2$-benzyl	0	5			13	
7	H	H	3',4'-(HO)$_2$-benzyl	20	50			76	
(15i) 8	H	H	3',4'-(HO)$_2$-benzyl	83	98			97	
9	H	H	3',4'-(HO)$_2$-benzyl	46	72			46	
(meso) 6	H	H	α-Me-benzyl		87				
(R,R) 6	H	H	α-Me-benzyl		84				
(S,S) 6	H	H	α-Me-benzyl		75				

[a]Responses to NA after 30-min incubations followed by washing for 30 min.
[b]Percent decrease of maximum response to 10^{-4} M NA.
[c]Percent decrease of maximum response to 10^{-5} M NA.

these sulfur-containing amines. They found that when the thiol (11; $n = 5$, R = H) was tested on aortic tissue in the absence of oxygen, little blocking activity occurred. In addition, they found that the α-blockade was not reversible as previously inferred (98,99) and that it was time-dependent and truly of the nonequilibrium type.

2.4.1 Chain Length and Substituent Effects

It had been reported that optimum activity was associated with a five-carbon chain as in APC (99). However, these conclusions were based on observations with diaminothiol-S-phosphate salts (11; $n = 3–6$, R = PO_3H_2), which are less active than APC. The finding that the active species is the disulfide prompted further investigations of the effects of chain length as well as substituent effects on α-blocking activity. A wide variety of compounds related to APC were synthesized and tested (Figure 4; Tables 5–7) (101–106). First, it was noted that substitution of the terminal nitrogens of APC with benzyl-type substituents was not detrimental to activity, whereas changing the terminal amine functions for guanidyl groups caused a drop in activity, suggesting that the terminal nitrogens may not interact with anionic phosphate groups near the binding sites (Table 5) (101). Our attention was next focused on both chain-length and substituent effects (102). This work led to the discovery of two distinct classes of α-blockers: in the first class, optimum activity was obtained with a six-carbon chain carrying a benzyl-type substituent on the terminal nitrogens {the prototype being N,N'-bis[6-(2'-methoxybenzyl)amino-n-h-exyl]cystamine (BHC), **15b; Benextramine***} and in the second class, optimum activity was sharply associated with an unsubstituted eight-carbon chain [the prototype being N,N'-bis(8-amino-n-octyl)cystamine (AOC), **15c**] (Table 5). The blocking activity of the two distinct prototypes, Benextramine and AOC, surpassed that of APC by two orders of magnitude. Next, the question of whether the common cystamine part of Benextramine and AOC is necessary for optimum activity was examined. The dramatic drop in activity observed with the analogs **18a** and **18b,** where a three-carbon chain separates the sulfurs from the inner nitrogens, strongly indicates that a two-carbon chain length (i.e., cystamine moiety) is essential for optimum activity (138). In order to verify whether four basic nitrogens are necessary for optimum activity, some unsymmetrical analogs of Benextramine and AOC incorporating one or two fewer nitrogens (**14**) were evaluated (103). The results assembled in Table 6 clearly show that removal of one or two nitrogens from BHC or AOC resulted in sharp drops in activity, indicating that four charges symmetrically disposed are necessary for optimum activity. Moreover, the observed drop in activity caused by folding of the molecule as in the macrocycle (**16**) further suggests that the receptor counteranionic charges may prefer a linear arrangement of cationic nitrogens (101). The divergent effects of chain length and N-benzyl substitution indicates that the two classes of tetramine disulfides interact with the α-receptor by two topographically distinct mechanisms. The possibility that BHC and AOC may

---

*Available from Aldrich Chemical Company under the generic name "Benextramine."

TABLE 6   Relative α-Blocking Activities of Polyamine Disulfides 14 (Figure 4) at 2×10^5 M on Rat Vas Deferens[a](103)

	Structure 14		
n	*R*	*R¹*	*α-Blockade (% ± 5)[b]*
6	2'-MeO-benzyl	Me	30
6	2'-MeO-benzyl	NH_2	52
8	H	Me	20
8	H	NH_2	56

[a]Responses to NA were measured after 30-min incubations followed by washing for 30 min. Under these conditions, BHC and AOC produce 100% blockade.
[b]Percent decrease of maximum response to 10^{-4} M NA.

occupy a single set of sites which can adapt conformationally to either one of the two structures (Figure 5) found little support in the observation that the effect of *N*-methylation of the common cystamine segments on activity had a significant effect only on one (Benextramine) of the two prototypes (Table 5). In fact, the activity of **15g** was quite close to that of AOC itself, while that of **15f** was much lower than the parent BHC (Figure 6). Furthermore, the *N,N'*-dimethyl analog of **15f** showed very little activity, suggesting that four secondary amine functions may be necessary for optimum activity in the BHC class of antagonists. These findings strongly suggest that BHC and AOC bind to two different sets of sites, and this raises the question of whether they are connected or topographically independent. A preliminary answer to this question was obtained by comparing the α-blocking activities of analogous but unsymmetrical disulfides (**17,** Table 7). The observation

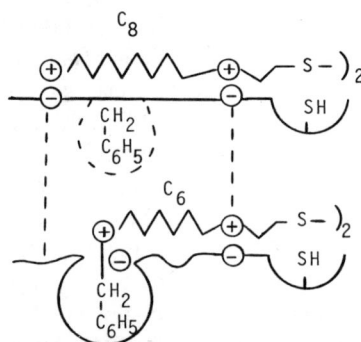

Figure 5   Hypothetical conformational change in the α-receptor induced by terminal *N*-benzyl substitution of tetramine disulfides. Reprinted with permission from C. Melchiorre et al., in *Recent Advances in Receptor Chemistry,* F. Gualtieri et al., Eds., Elsevier/North-Holland Biomedical Press, 1979, p. 207.

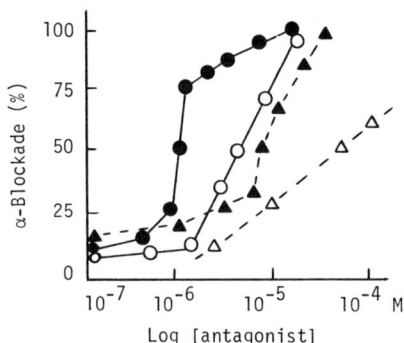

Figure 6 Percent decrease of maximum α-response (to NA) of vas deferens vs. increasing concentrations of BHC (●———●), AOC (○———○), **15f** (△---△), and **15g** (▲---▲) after 30-min incubations followed by 30-min washing. Reprinted with permission from C. Melchiorre et al., in *Recent Advances in Receptor Chemistry*, F. Gualtieri et al., Eds., 1979, Elsevier/North-Holland Biomedical Press, 207.

that, among the various homologs, **17a** turned out to be as active as BHC and AOC is consistent with predictions based on a topographical model where the two prototypes (BHC and AOC) would occupy two different sets of anionic sites while sharing the same target thiol. The effects of inner nitrogen methylation of **17a** could be readily accounted for on the basis of this "cross" model (Figure 7) (103,104). The question of whether the model can apply not only to rat vas deferens but to other tissues, such as rabbit aorta, was examined next. The data assembled in Table 5 clearly show that all the compounds investigated displayed affinity for the aortic α-receptor. The two classes of compounds revealed again the existence for this receptor of a remarkable topographical dualism, as shown in Figure 8. However, the detailed SAR were significantly different from the results obtained with rat vas deferens. Whereas the series carrying a 2'-methoxybenzyl group on the terminal nitrogens followed the same pattern found for the vas deferens α-receptors [optimum activity being associated with a six-carbon chain (BHC)], the unsubstituted series

TABLE 7 Relative α-Blocking Activities of Unsymmetrical Tetramine Disulfides 17 (Figure 4) at 10^{-5} M on Rat Vas Deferens[a](103)

Structure 17: n	α-Blockade (% ± 5)[b]
5	47
6	75
7	77
8 (17a)	90
9	70
10	54

[a]Responses to NA were measured after 30-min incubations followed by washing for 30 min.
[b]Percent decrease of maximum response to 10^{-4} M NA.

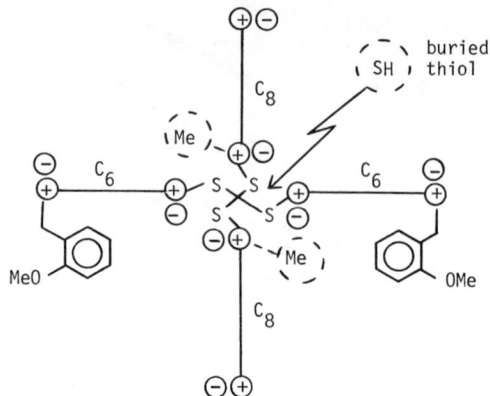

Figure 7 Schematic representation of the topographical arrangement of the α-receptor anionic sites interacting with BHC, AOC, and hybrid **17a.** For AOC, the cystamine segment can be *N*-methylated without loss of activity. A common receptor thiol is shown at the crossover point. Reprinted with permission from C. Melchiorre et al., in *Recent Advances in Receptor Chemistry,* F. Gualtieri et al., Eds., 1979, Elsevier/North-Holland Biomedical Press, 207.

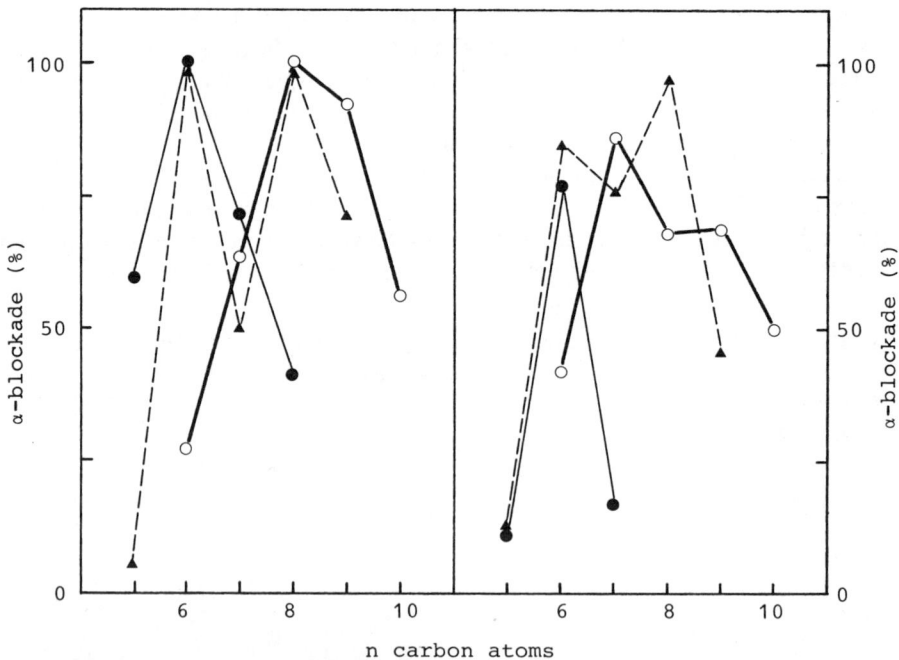

Figure 8 Relative α-blocking activities of BHC homologs (●———●), AOC homologs (○———○), and **15e** homologs (▲---▲) on rat vas deferens (left) and rabbit aorta (right). Data from Table 5.

differed in that optimum activity was now associated sharply with a seven- instead of an eight-carbon chain as in **15d**. Again, these results are consistent with the existence of two sets of sites. However, the detailed structural requirements for optimum activity on rabbit aorta and rat vas deferens are not identical (139).

2.4.2 *Nature of the Inhibition Reaction*

Blockade of the α-receptor by tetramine disulfides is time-dependent and the blocking effects of BHC against NA on rabbit aorta is persistent and remains unchanged after 18 hr (102). This is clearly consistent with covalent-bond formation with a protein thiol through a disulfide–thiol interchange reaction. Concrete evidence supporting this conclusion is based on the observation that the all-carbon analogs of BHC and AOC (**12** and **13**; Figure 4) are ineffective as α-blockers under the same conditions. Lippert and Belleau (100) reported that prior exposure of rabbit aorta to common thiol oxidizing reagents (such as ferricyanide and Ellman's reagent) does not affect the α-receptor, thus suggesting that the target thiol for the tetramine disulfide may not be exposed on the surface. Moreover, α-blockade by APC was shown to be stable to high concentrations of mercaptoethanol (100), suggesting that the target receptor thiol may reside under the surface or at the interface between protomers or other protein constituents (102). Accordingly, one may conclude that the multiple cation–anion interactions in the initial addition complexes with the receptor surface would promote accessibility of a thiol group for the interchange reaction, presumably by way of a conformational change as schematically shown in Figure 9. In the covalent reaction product (Figure 9), only one-half of the initial cationic charges is retained, which suggests that the product conformation probably differs from that which is stabilized by four charges in the addition complex. The intriguing possibility emerges that an α-receptor transition involving an "inside–outside" thiol conformation may play a functional role in receptor-associated

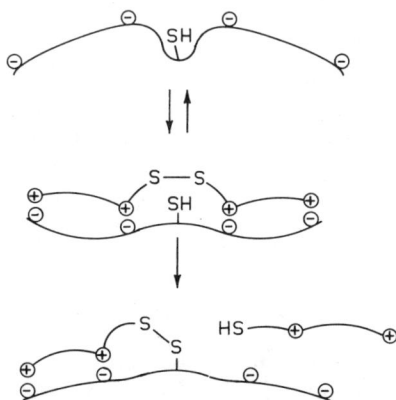

Figure 9 Schematic representation of the interaction of tetramine disulfides with the α-receptor surface and a buried thiol function. The latter is shown to become susceptible to covalent-bond formation in the initial addition complex.

events (102). Recently, Lukas et al. (107) suggested that disulfide and thiol groups may be involved in receptor function in the central nicotinic acetylcholine receptor.

It has been reported by several workers (108–113) that a thiol function may be involved in the makeup of the α-receptor and may be the target site for alkylating α-blockers and organomercuric reagents. As pointed out before (108,114) the reaction product of alkylating agents of the β-haloalkylamine family would be a sulfide whose chemical stability under physiological conditions would not allow spontaneous recovery over a few hours of the blocked receptor. However, spontaneous regeneration is consistent with a hydrolytic mechanism possibly involving an ester function at or near a receptor site. On the other hand, organomercuric reagents interact selectively with protein thiols and can inhibit noradrenergic responses. On that basis, it was suggested that a thiol function would be involved in α-receptor chemistry (115). In disagreement with this hypothesis, Mottram (116) reported that α-blockade by methylmercuric iodide is not receptor-specific, acetylcholine and histamine responses being blocked equally well. Moreover, it was argued that the effects of the reagent would be the result of interference with the contractile process so that thiol groups would not be involved in the makeup of the α-receptor. It is evident that the behavior of the α-receptor toward the tetramine disulfides does not support this conclusion. Clearly, the possibility was overlooked that at least one receptor thiol sterically hindered from alkylating agents and organomercuric reagents is nevertheless involved in the active site chemistry of the α-receptor.

The possibility was investigated that the thiol function involved in the reaction with tetramine disulfides may be sensitive to chiral effects since protein thiols are of necessity in chiral environments. Accordingly, the optical isomers of the α,α'-dimethyl analogs of APC (Figure 4: **15**, $n = 5$, R = R' = H, R" = CH$_3$; D,D-, L,L-, and meso-forms) were synthesized and tested (Table 5) (101). They turned out to be equiactive but less active than APC itself, suggesting that the environment of the receptor thiol is not sensitive to such chirality effects. However, a definite conclusion is not possible because the much poorer fit of APC compared to Benextramine may not allow chirality effects to manifest themselves. Access to the meso-isomer provided an opportunity to see whether one of the mirror image halves of the molecule is preferentially ejected in the interchange reaction. The results obtained with ^{35}S-meso-isomers (**15**; $n = 5$, R = R' = H, R" = CH$_3$; Figure 4) suggest that both halves of the meso-isomer are equally effective as leaving groups (101).

2.4.3 *Sites of Action of Tetramine Disulfides*

The most striking results of the SAR studies with tetramine disulfides as regards substituent effects concern the effects of inner *N*-methylation of both BHC and AOC, the high activity of unsymmetrical disulfide **17a** and the effects of *o*-methoxy functions on the terminal benzyls of BHC. These results, discussed earlier in this section, were interpreted on the basis of a "cross" model, which probably applies to both rat vas deferens and rabbit aorta. However, the role of the *o*-methoxy functions appears very special. The presence of this functionality appears uniquely responsible for the unusual steepness of the curve relating α-blockade to BHC concentration (Figure 6). This may indicate that the *o*-methoxy groups are involved in the stabilization of a special receptor conformation displaying some kind of cooperativity. We are then faced with the intriguing question of whether the *o*-methoxybenzyl groups interact with the natural agonists binding sites. To this end, we prepared and tested all of the possible dihydroxybenzyl analogs of BHC (Figure 4; **15**, $n = 5$–9, $R' = R'' = H$, R = dihydroxybenzyl) (Table 5). From the data assembled in Table 5, it is evident that the terminal N-3′,4′-dihydroxybenzyl substituents uniquely confer optimum activity to those compounds incorporating six- and eight-carbon chains, but the activity is somewhat reduced relative to BHC both on vas deferens and rabbit aorta (139). The specificity of these two chain lengths in promoting optimum activity is illustrated in Figure 8. The sharp peaks at six- and eight-carbon chains clearly indicate that **15e** and **15i** interact with the receptor by two distinct topographical mechanisms. This remarkable dualism finds a parallel in the BHC and AOC interaction mechanisms already discussed. The following possibilities offer themselves which may account for these different topographies: *(a)* the two antagonists (**15e** and **15i**) bind with a single set of sites that adapts conformationally to either one of the two structures, *(b)* they react with two distinct sets of sites but involving the same anionic sites occupied by BHC and AOC, or *(c)* one antagonist acts at a site occupied by BHC or AOC while the other would bind on a different set of sites which may or may not involve the same target thiol. The observation that substitution of the 3′,4′-dihydroxybenzyl groups of **15i** for 2′-methoxybenzyl groups (Figure 4: **15**, $n = 8$, $R' = R'' = H$, R = 2′-MeO-benzyl) caused a large drop in activity (in contrast to **15e** and BHC) does not support hypothesis *(a)*. Hypothesis *(b)* appears to provide the simplest explanation of the topographical dualism underlying the properties of **15e** and **15i.** This model implies that the two antagonists would occupy the same anionic sites that are occupied by the cationic nitrogens of Benextramine and AOC respectively (Figure 7). It is unlikely that the 2′-methoxybenzyl and the 3′,4′-dihydroxybenzyl groups interact with the same binding sites because of the large difference in their respective polarities. This difficulty can be circumvented by assuming that the two distinct substituents occupy two different areas at the periphery of the terminal nitrogens of BHC and **15e**. The observed large difference in activity between **15i** and its 2′-methoxybenzyl analog suggests that a binding site for a benzyl-type substituent near the ends of the eight-carbon chain is unavailable. The chain-length effects were parallel for BHC and **15e** on both vas deferens and rabbit aorta (Figure 8), as one would expect if *both* compounds bind to the same counteranionic sites through their four nitrogens. On the other hand, the chain-length effects were significantly different for AOC and **15i** (Figure 8), thus suggesting that they may not bind on the same set of anionic sites. Moreover, BHC and the two dihydroxybenzyl analogs were more active on rabbit aorta than on vas deferens, while the reverse is true of AOC, thus emphasizing topographical differences for the latter and **15i**. It can be seen that these findings do not agree with hypothesis *(b)*. On that basis, we suggest that **15i** and AOC may interact with two topographically different sites, while **15e** and BHC would occupy the same anionic sites (Figure 10). We are thus faced with the question of whether the two sets of sites occupied by **15e** or BHC and **15i** cross each other over the same target thiol (as would be the case for BHC and AOC) or are spatially unconnected and clustered about two different thiols which may or may not be on different subunits of the receptor. We cannot distinguish between these models at the present time, although we are intuitively inclined to favor the "two sets–two subunits" hypothesis because the other two models demand that an exceptionally large number of independent and *specific* binding sites be supplied by a single subunit. It is known that the nicotinic receptor is made of different subunits and since this is likely to be true for other receptors, one may be justified in preferring the "two sets–two subunits" model as a working hypothesis. Work is in progress aimed at testing this hypothesis.

Of considerable significance was the observation that NA at the low bath concentration of 30 μM afforded complete protection of the α-receptor against BHC at 3 μM (a concentration sufficient to block completely the receptor). In contrast, NA at 300 μM afforded only 56% protection against phenoxybenzamine at 0.15 μM (117). Although this unprecedented capacity of an agonist to protect completely against an α-blocker may be explained by invoking an allosteric mechanism, it may well be that the foregoing results with the 3′,4′-dihydroxybenzyl analog **15e** reflect the operation of a topographical mechanism where the terminal nitrogens of BHC and **15e** would occupy the same sites normally occupied by the nitrogen of NA. Whereas the catechol rings of **15e** may bind to the catecholamine binding sites, the 2′-methoxybenzyl groups of BHC would interact with another peripheral site. It is known that the β-hydroxyl of catecholamines shows high stereoselectivity toward the α-receptor. It was of interest to evaluate the effects of chiral centers on the 2′-methoxybenzyl groups of BHC. To this end, the optical isomers of the α-methylbenzyl analog of BHC (Figure 4: **15h,** $n = 6$, R′ = R″ = H, R = α-Me-benzyl;

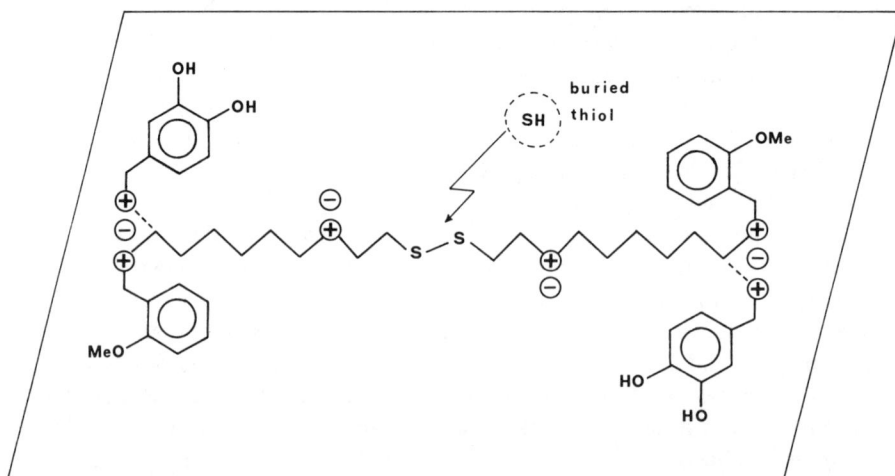

Figure 10 Schematic representation of the topographical arrangement of the α-receptor anionic sites interacting with BHC and the 3′,4′-dihydroxybenzyl analog **15e**.

R,R-, S,S-, and meso-forms) were synthesized and tested (Table 5) (140). They turned out to be uniformly slightly less active than BHC itself and exhibited a low degree of stereoselectivity, the R/S ratio being not higher than 3 (at 50% blockade). The R,R- and *meso*-enantiomers were equiactive at higher concentrations, while the latter was more active at lower concentrations. On the other hand, the S,S-enantiomer was slightly less active. These observations establish that the peripheral binding sites for the benzylic groups of BHC are not significantly sensitive to chiral effects. The model shown in Figure 10 may have considerable topographical and mechanistic relevance as it indicates that the anionic receptor sites and the thiol would be tightly linked to the agonist binding sites so that the "inside–outside" thiol transition would play a role in receptor-linked events in view of the very favorable redox potential of thiols.

2.4.4   *Selectivity of Tetramine Disulfides*

As one of the most active members of this series, BHC (Benextramine) was selected to evaluate the receptor selectivity of this new class of nonequilibrium antagonists. At a bath concentration of 10 μM, BHC (a concentration well above that required for complete α-blockade) left intact tissue responses elicited by 5-hydroxytryptamine (5-HT), histamine, and isoprenaline in potassium-contracted aortic tissue (102).

It is well established that all known adrenergic α-antagonists will block the 5-HT receptor and can also inhibit the muscarinic receptor in some measure (111). As regards the latter receptor, BHC was found to inhibit the acetylcholine- or carbachol-induced responses of the guinea pig atrium (105). However, the antagonism had a *fast onset* and a relatively *fast offset,* indicating a reversible mode of inhibition. Moreover, the possibility that the muscarinic receptor may be inhibited by way of a disulfide–thiol interchange reaction was ruled out by the observation that the all-carbon analog **12** of BHC produced similar effects. Accordingly, it would seem that inhibition of the muscarinic receptor is the result of addition complex formation through ionic forces.

Recently, the potent 3′,4′-dihydroxy analog **15e** of BHC was found to be a much poorer antagonist of the muscarinic receptor. Hence the catechol rings of **15e** markedly increase selectivity toward the α-receptor (141).

It is interesting to note that this unprecedented *covalent* specificity for the α-receptor finds a parallel in the anticholinergic neurotoxins (1,2), which suggests that BHC or **15e** may find similar applications. Work along this line is in progress. Other well-known classes of irreversible α-blockers are not receptor specific.

3   REMARKS ON AFFINITY LABELING OF THE ADRENERGIC α-RECEPTOR

Molecular characterization of the α-receptor and the acquisition of precise knowledge about adrenergic pharmacology necessitates accessibility to selective irreversible (nonequilibrium in the kinetic sense) antagonists. Alkylating agents such as phenoxybenzamine will cause efficacious blockade by way of covalent-bond formation. This suggests that such affinity reagents may be suited for α-receptor characterization, but unfortunately, they are not receptor-specific and their attempted use to selectively label the α-receptor was not successful (117–122). More recently, however, Kan et al. (123) reported that [³H]phenoxybenzamine at low concentrations may be a suitable affinity reagent for the α-receptor of the rat liver cells. The high tritium content of the drug allowed the observation of saturable sites at suitably low concentrations. Effective competitive inhibitors such as phentolamine cannot be used for chemical tagging of the α-receptor protein because the reversible nature of the binding allows dissociation from its binding sites upon attempted isolation of the complex. In any event, this lipophilic blocker also lacks specificity for the α-receptor. The same limitations apply to other known classes of α-blockers. Considerable efforts have been made very recently with the aim of characterizing the α-receptor with [³H]ligands that are either agonists or competitive antagonists. Since this subject is reviewed by U'Prichard elsewhere in this book, we limit ourselves to comments that reflect the point of view of the medicinal chemist. In spite of the virtual impossibility of isolating a [³H]ligand–receptor complex that is dissociable and of the lack of receptor specificity of most affinity reagents, meaningful results are claimed by several authors as regards regional distribution of α-receptors in different organs and the mechanistic aspects of drug–receptor interactions. These topics have been recently reviewed (12).

Results that might help in the elucidation of the mechanism of action of both agonists and antagonists at the active-site level of mammalian brain were obtained through labeling experiments with the agonist [³H]clonidine, the competitive antagonist [³H]WB-4101, and the mixed agonist–antagonist [³H]dihydroergocryptine (30,124–130). The medicinal chemist may find these results confusing, and it appears that the source of any confusion may have to do with the basic lack of receptor specificity of the drugs adopted in such studies. Some aspects of this problem have been discussed recently by Furchgott (131). All the [³H]ligands such as clonidine, WB-4101, and dihydroergocryptine used in such binding studies are known not to be receptor-specific. As an example, dihydroergocryptine binds to the α-receptors as well as to the serotonin and dopamine receptors (132–134). Moreover, it fails to discriminate significantly between α_1- and α_2-receptors, and this may complicate the interpretation of binding results (135,136). In addition, the detailed pharmacology of WB-4101 is very sparse, and its selectivity was assumed on the basis that ease of its displacement from binding sites followed the expected order of potency of α-agonists and antagonists (134). It should be noted that the *specific* binding of these ligands is low relative to overall binding, being less than 60%, 50%, and 30% of the total for WB-4101, clonidine, and dihydroergocryptine, respectively (124,125). The nonspecific binding is always given as the difference in ligand uptake in the absence and presence of a quantity of unlabeled ligand which is assumed to interact selectively with all the available α-receptor sites. This assumption that the unlabeled ligand occupies *only* the α-receptors may be correct only when very low ligand concentrations are used. However, at high concentration (such as 10^{-4} M for NA) it is likely that several other cell constituents will act as binding sites, which can make interpretation of the results quite difficult. Although

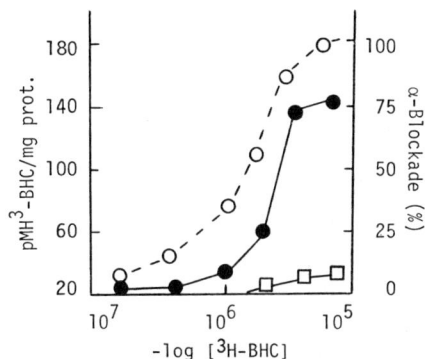

Figure 11 Uptake of [³H]BHC by aortic tissue (left axis) vs. α-blockade (right axis). Plasma membrane fraction (●———●); homogenate (□———□); α-blockade (○---○). Reprinted with permission from C. Melchiorre et al., in *Recent Advances in Receptor Chemistry*, F. Gualtieri et al., Eds., 1979, Elsevier/North-Holland Biomedical Press, 207.

the information obtained remains inherently useful, rigorous deductions in terms of chemistry may be allowed only when receptor specificity of a ligand can be clearly demonstrated. The recently reported tetramine disulfide class of α-blockers (Section 2.4) display unprecedented specificity for the α-receptor and their potential usefulness as receptor markers is enhanced by the covalent nature of their interaction with a genuine protein functionality. Preliminary studies have shown that covalent uptake of [³H]BHC involves only proteins of the plasma membrane of aortic tissue and receptor inhibition was perfectly parallel to uptake (Figure 11) (104).

ACKNOWLEDGMENTS

We are grateful to the Medical Research Council of Canada for the financial support of our work and to Professor B. Benfey for numerous helpful discussions and advice.

REFERENCES

1 R. Miledy, P. Molinoff, and L. T. Potter, *Nature (Lond.)*, **229**, 554 (1971).

2 J. P. Changeux, J. C. Meunier, and M. Huchet, *Mol. Pharmacol.*, **7**, 538 (1971).

3 J. P. Changeux, M. Kasai, and C. Y. Lee, *Proc. Natl. Acad. Sci. U.S.A.*, **67**, 1241 (1970).

4 R. P. Ahlquist, *Am. J. Physiol.*, **153**, 586 (1948).

5 S. Z. Langer, *Biochem. Pharmacol.*, **23**, 1793 (1974).

6 S. Berthelsen and W. A. Pettinger, *Life Sci.*, **21**, 595 (1977).

7 K. Starke and S. Z. Langer, in S. Z. Langer, K. Starke, and M. L. Dubocovich, Eds., *Presynaptic Receptors*, Pergamon Press, Oxford, 1979, p. 1.

8 J. D. P. Graham, *Prog. Med. Chem.*, **2**, 132 (1962).

9 M. S. K. Ghouri and T. J. Haley, *J. Pharm. Sci.*, **58**, 511 (1969).

10 D. J. Triggle, *Neurotransmitter–Receptor Interactions*, Academic Press, New York, 1971, Chap. 4.

11 D. J. Triggle and C. R. Triggle, *Chemical Pharmacology of the Synapse*, Academic Press, New York, 1976, Chap. 3.

12 L. T. Williams and R. J. Lefkowitz, *Receptor Binding Studies in Adrenergic Pharmacology*, Raven Press, New York, 1978.

13 A. B. Demson, S. Bardhanabaedyna, and H. D. Green, *Circ. Res.*, **2**, 537 (1954).

14 W. Rosenblatt, T. M. Haymond, S. Bellet, and G. Koelle, *Am. J. Med. Sci.*, **227**, 179 (1954).

15 M. Nickerson, *Pharmacol. Rev.*, **9**, 246 (1957).

16 E. Fourneau and D. Bovet, *Arch. Int. Pharmacodyn. Ther.*, **46**, 178 (1933).

17 D. R. Mottram and H. Kapur, *J. Pharm. Pharmacol.*, **27**, 295 (1975).

18 H. Kapur, D. R. Mottram, and P. N. Green, *J. Pharm. Pharmacol.*, **30**, 259 (1978).

19 H. Kapur, P. N. Green, and D. R. Mottram, *J. Pharm. Pharmacol.*, **31**, 188 (1979).

20 H. Fenton, P. N. Green, M. Shapero, and C. Wilson, *Nature (Lond.)*, **206**, 725 (1965).

21 P. N. Green, M. Shapero, and C. Wilson, *J. Med. Chem.*, **12**, 326 (1969).

22 E. J. Ariens and A. M. Simonis, *Acta Physiol. Pharmacol.*, **15**, 78 (1969).

23 B. P. Avner and D. J. Triggle, *J. Med. Chem.*, **17**, 197 (1974).

24 J. F. Moran, V. C. Swamy, and D. J. Triggle, *Life Sci.*, **9**, 1303 (1970).

25 V. C. Swamy and D. J. Triggle, *Eur. J. Pharmacol.*, **19**, 67 (1972).

26 C. S. Tsai, U. S. Shah, H. B. Bhargava, R. G. Zaylskie, and W. H. Shelver, *J. Pharm. Sci.*, **61**, 228 (1972).

27 M. Butler and D. J. Jenkinson, *Eur. J. Pharmacol.*, **52**, 303 (1978).

28 H. Kapur and D. R. Mottram, *J. Pharm. Pharmacol.*, **31**, 337 (1979).

29 W. L. Nelson, J. E. Wennerstrom, D. C. Dyer, and M. Engel, *J. Med. Chem.*, **20**, 880 (1977).

30 R. Raisman, M. Briley, and S. Z. Langer, *Naunyn-Schmiedebergs Arch. Pharmacol.*, **307**, 223 (1979).

31 G. M. Drew, *Eur. J. Pharmacol.*, **36**, 313 (1976).

32 J. M. Cedarbaum and G. K. Aghajanian, *Eur. J. Pharmacol.*, **44**, 375 (1977).

33 G. M. Drew, A. J. Gower, and A. S. Marriott, *Br. J. Pharmacol.*, **61**, 468P (1977).

34 R. D. Robson, M. J. Antonaccio, J. K. Saelens, and J. Liebman, *Eur. J. Pharmacol.*, **47**, 431 (1978).

35 K. B. J. Franklin and L. J. Herberg, *Eur. J. Pharmacol.*, **43**, 33 (1977).

36 A. Delini-Stula, P. Baumann, and O. Büch, *Naunyn-Schmiedebergs Arch. Pharmacol.*, **307**, 115 (1979).

37 G. M. Drew, *Br. J. Pharmacol.*, **64**, 293 (1978).

38 G. M. Drew, *Eur. J. Pharmacol.*, **42**, 123 (1977).

39 E. Borowski, K. Starke, H. Ehrl, and T. Endo, *Neuroscience*, **2**, 285 (1977).

40 J. H. Moyer and C. Caplovitz, *Am. Heart J.*, **45**, 602 (1953).

41 S. H. Taylor, G. R. Sutherland, G. J. MacKenzie, H. P. Staunton, and K. W. Donald, *Clin. Sci.*, **28**, 265 (1965).

42 M. Thomas, H. Campbell, and G. Heard, *Br. J. Surg.*, **55**, 588 (1968).

43 M. Nickerson and B. Collier, in L. S. Goodman and A. Gilman, Eds., *The Pharmacological Basis of Therapeutics*, 5th ed., Macmillan, New York, 1975, p. 533.

44 I. R. Innes and M. Nickerson, in Ref. 43, p. 477.

45 M. Mujic and J. M. van Rossum, *Arch. Int. Pharmacodyn. Ther.*, **155**, 432 (1965).

46 H. Struyker Boudier, G. Smeets, G. Brouwer, and J. van Rossum, *Life Sci.*, **15**, 887 (1974).

47 H. Struyker Boudier, J. de Boer, G. Smeets, E. J. Lien, and J. van Rossum, *Life Sci.*, **17**, 377 (1975).

48 J. Sanders, D. D. Miller, and P. N. Patil, *J. Pharmacol. Exp. Ther.*, **195**, 362 (1975).

49 T. Jen, H. van Hoeven, W. Groves, R. A. McLean, and B. Loev, *J. Med. Chem.*, **18**, 90 (1975).

50 B. Rouot, G. Leclerc, and C.-G. Wermuth, *J. Med. Chem.*, **19**, 1049 (1976).

51 P. B. M. W. M. Timmermans and P. A. van Zwieten, *Eur. J. Pharmacol.*, **45**, 229 (1977).

52 I. C. Medgett and M. W. McCulloch, *Arch. Int. Pharmacodyn. Ther.*, **240**, 158 (1979).

53 M. Hartmann and H. Isler, *Naunyn-Schmiedebergs Arch. Exp. Pathol. Pharmakol.*, **192**, 141 (1939).

54 E. Urech, A. Marxer, and K. Miescher, *Helv. Chim. Acta*, **33**, 1386 (1950).

55 T. K. Trubitsyna and M. D. Mashkovskii, *Farmakol. Toksikol. (Mosc.)*, **37**, 553 (1974); *Chem. Abstr.*, **82**, 149262n (1975).

56 L. Villa, V. Ferri, and E. Grana, *Farm. Ed. Sci.*, **22**, 491 (1967).

57 R. R. Ruffolo, E. L. Rosing, and J. E. Waddell, *J. Pharmacol. Exp. Ther.*, **209**, 429 (1979).

58 J. N. Davis and W. Maury, *J. Pharmacol. Exp. Ther.*, **207**, 425 (1978).

59 G. Roberts, A. W. Richardson, and H. D. Green, *J. Pharmacol. Exp. Ther.*, **105**, 466 (1952).

60 R. F. Furchgott, in H. Blaschko and E. Muscholl, Eds., *Catecholamines*, Handbook of Experimental Pharmacology, Vol. 33, Springer-Verlag, New York, 1972, p. 283.

61 K. Starke, H. Montel, W. Gayk, and R. Merker, *Naunyn-Schmiedebergs Arch. Pharmacol.*, **285**, 133 (1974).

62 M. G. Collis and B. J. Alps, *J. Pharm. Pharmacol.*, **25** 621 (1973).

63 R. R. Ruffolo, J. W. Fowble, D. D. Miller, and P. N. Patil, *Proc. Natl. Acad. Sci. U.S.A.*, **73**, 2730 (1976).

64 R. R. Ruffolo, B. S. Turowski, and P. N. Patil, *J. Pharm. Pharmacol.*, **29**, 378 (1977).

65 D. D. Miller, F.-L. Hsu, R. R. Ruffolo, and P. N. Patil, *J. Med. Chem.*, **19**, 1382 (1976).

66 A. Hoszowska-Owczarek, N. Djordjevic, S. Z. Langer, and A. F. De Schaepdryver, *Arch. Int. Pharmacodyn. Ther.*, **168**, 485 (1967).

67 N. D. Edge, *Br. J. Pharmacol.*, **38**, 386 (1970).

68 A. Walz and P. A. van Zwieten, *Eur. J. Pharmacol.*, **10**, 369 (1970).

69 M. L. Cohn, M. Cohn, and F. H. Taylor, *Arch. Int. Pharmacodyn. Ther.*, **217**, 80 (1975).

70 G. Simon and M. Winter, *Biochem. Pharmacol.*, **25**, 881 (1976).

71 P. Skolnick, J. W. Daly, and D. S. Segal, *Eur. J. Pharmacol.*, **47**, 451 (1978).

72 S. Kunitada and M. Ui, *Eur. J. Pharmacol.*, **49**, 169 (1978).

73 B. J. Williams, W. H. Griffith, and C. M. Albrecht, *Eur. J. Pharmacol.*, **49**, 7 (1978).

74 E. Malta, G. A. McPherson, and C. Raper, *Br. J. Pharmacol.*, **65**, 249 (1979).

75 J. C. Doxey and R. E. Easingwood, *Br. J. Pharmacol.*, **63**, 401P (1978).

76 A. G. Roach, F. Lefèvre, and I. Cavero, *Clin. Exp. Hypertension* **1**, 87 (1978).

77 I. Cavero, F. Lefèvre, and A. G. Roach, *Br. J. Pharmacol.*, **61**, 469P (1977).

78 J. W. Constantine, R. A. Weeks, and W. K. McShane, *Eur. J. Pharmacol.*, **50**, 51 (1978).

79 D. R. Algate and J. F. Waterfall, *J. Pharm. Pharmacol.*, **30**, 651 (1978).

80 J. C. Doxey, C. F. C. Smith, and J. M. Walker, *Br. J. Pharmacol.*, **60**, 91 (1977).

81 K. F. Rhodes and J. F. Waterfall, *J. Pharm. Pharmacol.*, **30**, 516 (1978).

82 S. Arbilla and S. Z. Langer, *Br. J. Pharmacol.*, **64**, 259 (1978).

83 E. J. Filinger, S. Z. Langer, C. J. Perec, and F. J. E. Stefano, *Naunyn-Schmiedebergs Arch. Pharmacol.*, **304**, 21 (1978).

84 M. Nickerson and L. S. Goodman, *Fed. Proc.*, **2**, 109 (1945).

85   G. E. Ullyot and J. F. Kerwin, in F. F. Blicke and C. M. Suter, Eds., *Medicinal Chemistry*, Vol. 2, Wiley, New York, 1956, p. 234.

86   B. Belleau, *Can. J. Biochem. Physiol.*, **36,** 731 (1958).

87   B. Belleau, *Ann. N.Y. Acad. Sci.*, **139,** 580 (1967).

88   M. Nickerson, *Pharmacol. Rev.*, **1,** 27 (1949).

89   J. G. Henkel, P. S. Portoghese, J. W. Miller, and P. Lewis, *J. Med. Chem.*, **19,** 6 (1976).

90   B. Belleau and D. J. Triggle, *J. Med. Pharm. Chem.*, **5,** 636 (1962).

91   P. S. Portoghese, T. N. Riley, and J. W. Miller, *J. Med. Chem.*, **14,** 561 (1971).

92   S. McLean, V. C. Swamy, D. Tomei, and D. J. Triggle, *J. Med. Chem.*, **16,** 54 (1973).

93   T. N. Riley and F. W. Crawford, *J. Pharm. Sci.*, **65,** 544 (1976).

94   J. D. P. Graham and G. W. James, *J. Med. Pharm. Chem.*, **3,** 489 (1961).

95   M. L. Dubocovich and S. Z. Langer, *J. Physiol. (Lond.)*, **237,** 505 (1974).

96   D. Cambridge, M. J. Davey, and R. Massingham, *Br. J. Pharmacol.*, **59,** 514P (1977).

97   F. L. Atkins and G. L. Nicolosi, *Biochem. Pharmacol.*, **28,** 1233 (1979).

98   G. E. Demaree, R. E. Brockenton, M. H. Heiffer, and W. E. Rothe, *J. Pharm. Sci.*, **60,** 1743 (1971).

99   E. H. Herman, M. H. Heiffer, G. E. Demaree, and J. A. Vick, *Arch. Int. Pharmacodyn. Ther.*, **193,** 102 (1971).

100 B. Lippert and B. Belleau, in E. Usdin and S. H. Snyder, Eds., *Frontiers in Catecholamine Research,* Pergamon Press, New York, 1973, p. 369.

101 Y. Ueda, C. Melchiorre, B. Lippert, B. Belleau, S. Chona, and D. J. Triggle, *Farm. Ed. Sci.*, **33,** 479 (1978).

102 C. Melchiorre, M. S. Yong, B. G. Benfey, and B. Belleau, *J. Med. Chem.*, **21,** 1126 (1978).

103 C. Melchiorre, D. Giardinà, L. Brasili, and B. Belleau, *Farm. Ed. Sci.*, **33,** 999 (1978).

104 C. Melchiorre, M. S. Yong, B. Benfey, L. Brasili, G. Bolger, and B. Belleau, in F. Gualtieri, M. Giannella, and C. Melchiorre, Eds., *Recent Advances in Receptor Chemistry*, Elsevier/North-Holland, 1979, p. 207.

105 B. G. Benfey, M. S. Yong, B. Belleau, and C. Melchiorre, *Can. J. Physiol. Pharmacol.*, **57,** 41 (1979).

106 B. G. Benfey, M. S. Yong, B. Belleau, and C. Melchiorre, *Can. J. Physiol. Pharmacol.*, **57,** 510 (1979).

107 R. J. Lukas, H. Morimoto, and E. L. Bennett, *Biochemistry*, **18,** 2384 (1979).

108 S. C. Harvey and M. Nickerson, *J. Pharmacol. Exp. Ther.*, **112,** 274 (1954).

109 J. M. Goldman and M. E. Hadley, *J. Pharmacol. Exp. Ther.*, **182,** 93 (1972).

110 J. P. Huidobro-Toro and A. Carpi, *Arch. Int. Pharmacodyn. Ther.*, **222,** 180 (1976).

111 K. N. Salman, H. S. Chai, D. D. Miller, and P. N. Patil, *Eur. J. Pharmacol.*, **36,** 41 (1976).

112 L. T. Williams and R. J. Lefkowitz, *Mol. Pharmacol.*, **13,** 304 (1977).

113 R. R. Ruffolo, B. S. Turowski, and P. N. Patil, *J. Pharm. Pharmacol.*, **30,** 498 (1978).

114 B. Belleau, in J. R. Vane, G. E. W. Wolstenhome, and M. O'Connor, Eds., *Adrenergic Mechanisms,* Ciba Foundation Symposium, Churchill, London, 1960, p. 223.

115 A. D'Iorio and J. C. Lague, *Can. J. Biochem.*, **41,** 121 (1963).

116 D. R. Mottram, *Biochem. Pharmacol.*, **25,** 2104 (1976).

117 M. S. Yong and M. Nickerson, *J. Pharmacol. Exp. Ther.*, **186,** 100 (1973).

118 J. E. Lewis and J. W. Miller, *J. Pharmacol. Exp. Ther.*, **154,** 46 (1966).

119 M. May, J. F. Moran, H. Kimelberg, and D. J. Triggle, *Mol. Pharmacol.*, **3,** 28 (1967).

120 M. S. Yong and G. S. Marks, *Biochem. Pharmacol.*, **18,** 1619 (1969).

121 J. D. P. Graham, C. Ivens, J. D. Lever, R. McQuiston, and T. L. Spriggs, *Br. J. Pharmacol.,* **41,** 278 (1971).

122 R. D. Green, G. S. Lefever, E. M. Sheys, and M. Bristow, *J. Pharmacol. Exp. Ther.,* **187,** 524 (1973).

123 W. H. Kan, C. Farsang, H. G. Preiksaitis, and G. Kunos, *Biochem. Biophys. Res. Commun.,* **91,** 303 (1979).

124 D. A. Greenberg, D. C. U'Prichard, and S. H. Snyder, *Life Sci.,* **19,** 69 (1976).

125 D. A. Greenberg and S. H. Snyder, *Life Sci.,* **20,** 927 (1977).

126 D. C. U'Prichard, D. A. Greenberg, and S. H. Snyder, *Mol. Pharmacol.,* **13,** 454 (1977).

127 D. C. U'Prichard, D. A. Greenberg, P. Sheehan, and S. H. Snyder, *Brain Res.,* **138,** 151 (1977).

128 S. J. Peroutka, D. A. Greenberg, D. C. U'Prichard, and S. H. Snyder, *Mol. Pharmacol.,* **14,** 403 (1978).

129 D. A. Greenberg and S. H. Snyder, *Mol. Pharmacol.,* **14,** 38 (1978).

130 D. C. U'Prichard and S. H. Snyder, *Life Sci.,* **24,** 79 (1979).

131 R. F. Furchgott, *Fed. Proc.,* **37,** 115 (1978).

132 H. Corrodi, K. Fuxe, T. Hokfelt, P. Lidbrink, and U. Ungerstedt, *J. Pharm. Pharmacol.,* **25,** 409 (1973).

133 H. Corrodi, L. O. Farnebo, K. Fuxe, and B. Hamberger, *Eur. J. Pharmacol.,* **30,** 172 (1975).

134 M. Titeler and P. Seeman, *Proc. Natl. Acad. Sci. U.S.A.,* **75,** 2249 (1978).

135 G. Kunos, B. Hoffman, Y. N. Kwok, W. H. Kan, and L. Mucci, *Nature (Lond.),* **278,** 254 (1979).

136 M. C. Scrutton and J. A. Grant, *Nature (Lond.),* **280,** 700 (1979).

137 W. L. Nelson, M. L. Powell, and D. C. Dyer, *J. Med. Chem.,* **22,** 1125 (1979).

138 D. Giardinà, L. Brasili, C. Melchiorre, B. Belleau, and B. G. Benfey, *Eur. J. Med. Chem.,* submitted.

139 C. Melchiorre, M. Giannella, L. Brasili, B. G. Benfey, and B. Belleau, *Eur. J. Med. Chem.,* in press.

140 L. Brasili, M. Giannella, C. Melchiorre, B. Belleau, and B. G. Benfey, *Eur. J. Med. Chem.,* in press.

141 B. G. Benfey, B. Belleau, L. Brasili, and C. Melchiorre, *Fed. Proc. Fed. Amer. Soc. Exptl. Biol.,* **39,** 1106 (1980).

CHAPTER SEVEN

β-ADRENOCEPTORS, ADENYLATE CYCLASE, AND THE ADRENERGIC CONTROL OF CARDIAC CONTRACTILITY

J. Craig Venter

Department of Pharmacology and Therapeutics, School of Medicine, State University of New York at Buffalo, Buffalo, New York

1   INTRODUCTION

Stimulation of cardiac β-adrenergic receptors by catecholamines can produce a substantial variety of metabolic and mechanical responses (1,2). Since the studies of Murad et al. (3) in 1962 and Robison et al. (4) in 1965, there has been much discussion concerning the underlying mechanisms of the β-adrenergic receptor activation process and the potential role for adenosine $3',5'$-cyclic monophosphate (cyclic AMP) in mediating the cardiac effects of catecholamines. The extensive literature concerning cyclic nucleotides and cardiac function has been reviewed numerous times in the past few years (2,5–13); the most recent reviews by Tsien et al. in 1977 (2) and Drummond and Severson in 1979 (8) cover much of the literature through late 1977. While the exact role(s) for cyclic AMP in cardiac inotropic responses remain undefined (2), recent progress in the characterization of β-adrenergic receptors, adenylate cyclase, and cardiac adrenergic responses provides us with a more complete picture of the relationship between β-receptors and the augmentation of cardiac contractility by catecholamines.

2   MOLECULAR CHARACTERIZATION OF CARDIAC β-ADRENERGIC RECEPTORS

Adrenergic receptors, originally classified as α and β in 1948 by Ahlquist (14) have since been further defined. In 1967, Lands et al. proposed that subclasses of β-adrenergic receptors exist in tissues (15). As a result and according to currently accepted classifications, β-receptors in cardiac muscle are commonly referred to as β_1-adrenergic receptors, while the majority of β-receptors found in blood vessels, lung, smooth muscle, and liver are classified as β_2. The principal "physiological" difference between peripheral β_1- and β_2-receptors rests in the variation in potencies of epinephrine and norepinephrine at the two receptors. Epinephrine and norepinephrine have been demonstrated to have essentially equal potency at β_1-receptors, while norepinephrine has attenuated activity at β_2-adrenergic receptors.

While several methods have been used in the subclassification of β-receptors, the majority of experiments have been functional, testing either agonists or antagonists for biological effects. This type of study has not been entirely successful, as wide variations have been found in the apparent potency of agonists and antagonists between species and among various organs of single species (16,17). These variations make it difficult to draw conclusions concerning the underlying molecular nature of the β-receptors.

Recent studies from this laboratory have presented evidence that cardiac β-adrenergic receptors are unique molecular entities, distinct from β-receptors found in the liver and lung (18–21). In characterizing a wide variety of detergents for their ability to solubilize canine cardiac and hepatic β-receptors, substantial differences in the detergent specificity as well as the stabilities of the solubilized β-receptors were noted (19,20). Subsequent studies have demonstrated that the Stokes radii and apparent molecular weight of heart, liver, and lung β-receptors differ substantially (21). The cardiac β-receptor with a Stokes radius of 4.2 nm and a calculated molecular weight of 65,000 is smaller than liver and lung β-receptors which have Stokes radii of 5.8 nm and calculated molecular weights of 90,000 (Table 1) (21). In addition to its unique size and shape, the cardiac β-receptor is extremely susceptible to inactivation by sulfhydryl reagents such as dithiothreitol (DTT), which essentially abolish receptor specific binding (21) (Table 2). The liver β-receptor is not susceptible to inactivation by 1 mM DTT (21). Immunological evidence also supports the concept of the heart β-receptor being a unique molecular entity. Antibodies raised against partially purified [³H]propranolol binding sites from dog heart block [³H]propranolol binding to heart but not liver membranes (22). Conversely, autoantibodies to β-adrenergic receptors from the serum of asthma and allergic rhinitis patients which block ligand binding to lung β_2-receptors have no effect on ligand binding to cardiac β_1-receptors (23). The molecular properties of β-adrenergic receptors from a number of cells and tissues are summarized in Table 2. The molecular basis of β-receptor subclassification supports and extends

TABLE 1   Molecular Parameters of Cardiac, Lung, and Liver β-Adrenergic Receptors[a](21)

Parameter	Heart	Lung	Liver
Stokes radius,	4.2 ± 0.01	5.8 ± 0.02	5.8 ± 0.02
a (nm)	n = 9	n = 4	n = 6
Sedimentation coefficient,	3.69 ± 0.06	3.78 ± 0.13	3.67 ± 0.09
$S_{20,w}$	n = 10	n = 8	n = 13
	(3.4–3.9)	(3.3–4.2)	(3.1–4.2)
Partial specific volume (g/ml)	0.73	0.73	0.73
Molecular weight (MW)	65,000	91,000	90,000
Frictional ratio, f/f_o	1.6	2.0	2.0

[a]Hydrodynamic properties determined from Sepharose 6B gel permeation chromatography (Stokes radius), and sucrose density gradient centrifugation (sedimentation coefficients).

TABLE 2 Molecular Subclassification of β-Adrenergic Receptors^a(21)

Species	Cell or Tissue	IHYP Binding Inhibition by 1 mM DTT (%)	Stokes Radius (nm)	β-Receptor Subtype		Literature
				Molecular Parameters	Literature	
Dog	Liver	0	5.8	β_2 (100%)	β_2	117
Rat	Liver	0	—	β_2 (100%)	β_2 (100%)	118
Cat	Liver	0	—	β_2 (100%)	—	—
Frog	Erythrocytes	0	5.8	β_2	β_2	119
Mouse	Lymphoma (S49 cells)	—	6.4	β_2	β_2	66
Dog	Lung	16 ± 6.4	5.8	β_1 (20%) β_2 (80%)	β_2	117
Human	Lung (VA$_2$ cells)	17 ± 6.0	5.8	β_1 (21%) β_2 (79%)	—	—
Rat	Lung	32 ± 1.1	—	β_1 (40%) β_2 (60%)	β_1 (15%) β_2 (85%)	118
Rabbit	Lung	41 ± 2.0	—	β_1 (52%) β_2 (48%)	β_1 (60%) β_2 (40%)	120
Cat	Lung	43 ± 8.0	—	β_1 (54%) β_2 (46%)	—	—
Dog	Adipocytes	41 ± 1.7	—	β_1 (51%) β_2 (49%)	β_2	117
Rat	Glioma (C6 cells)	44 ± 2.3	—	β_1 (55%) β_2 (45%)	β_1	62
Turkey	Erythrocytes	35 ± 4.6	4.2	β_1 (44%) β_2 (56%)	β_1 ?	121
Rat	Heart	65 ± 6.1	4.8	β_1 (81%) β_2 (19%)	β_1 (83%) β_2 (17%)	118
Dog	Heart	80 ± 1.5	4.2	β_1 (100%)	β_1	117

^{a}β-receptor classification based upon [^{125}I]iodohydroxybenzylpindolol ([^{125}I]IHYP) binding data in the presence and absence of 1 mM DTT. The 80% inhibition of binding to dog heart β-receptors by DTT is assumed to represent the presence of only β_1-receptors. The lack of a DTT effect on liver β-receptors is assumed to represent the presence of only β_2-receptors.

the information available from pharmacological and direct ligand binding classification studies. While there are at least two distinct molecular forms of the β-receptor (Tables 1 and 2), there are increasing data suggesting the existence of more than two types of β-receptor, with the possibility that cardiac β-receptors are distinct from other forms of β_1-receptors (21).

β-Adrenergic receptors from two sources have been purified by this laboratory using immunoaffinity columns (24–26). Calf lung β_2-receptors were labeled with radioactive affinity reagent and purified by gel exclusion chromatography, preparative isoelectric focusing, and immunoaffinity chromatography. (24,25). The β_2-receptor has a molecular weight of 114,000 on SDS gels, with the subunit containing the adrenergic ligand binding site having a molecular weight of 59,000 daltons (Figure 1) (24).

In contrast, a nonmammalian β_1-receptor from turkey erythrocytes was purified using isolelectric focusing and immunoaffinity chromatography with monoclonal antibodies developed against the turkey β-receptor (26). The β-receptor has an apparent molecular weight of 70,000 with a possible subunit at 31,000 daltons

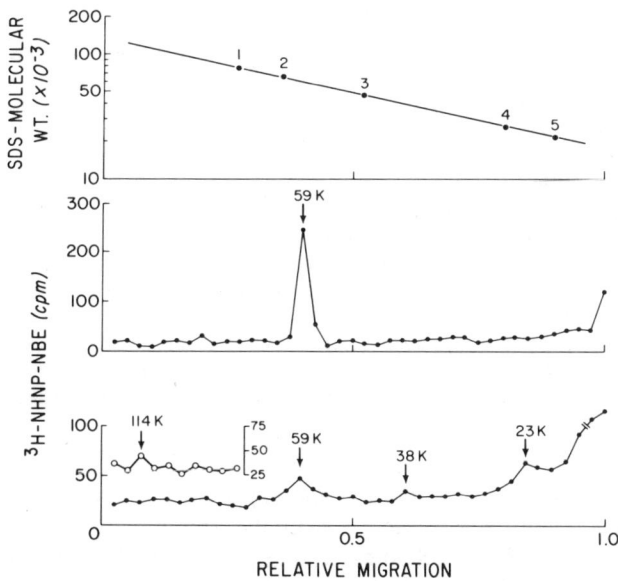

Figure 1 NaDodSO$_4$-polyacrylamide gel electrophoresis of autoantibody-affinity purified calf lung β_2-adrenergic receptors. Triton X-100-solubilized β_2-receptor complexes with the irreversible affinity ligand [^{3}H]NHNP-NBE were purified by gel permeation chromatography and preparative isoelectric focusing and run on NaDodSP$_4$-polyacrylamide gels (lower panel) under reducing (●—●) and nonreducing conditions (○—○). The center panel illustrates the β_2-receptor following an additional purification step consisting of affinity chromatography utilizing autoantibodies to β_2-receptors (23) coupled to Sepharose 4B. The top panel is a linear plot of the molecular-weight standards used for 10% NaDodSO$_4$-polyacrylamide gel electrophoresis; point 1, lactoperoxidase (77,500 MW); 2, bovine serum albumin (66,000 MW); 3, ovalbumin (46,000 MW); 4, chymotrypsinogen A (25,700 MW); and 5, soybean trypsin inhibitor (21,500 MW). These data indicate a subunit molecular weight of 59,000 for the β_2-receptor with a total molecular weight of 114,000 daltons. From Ref. 24.

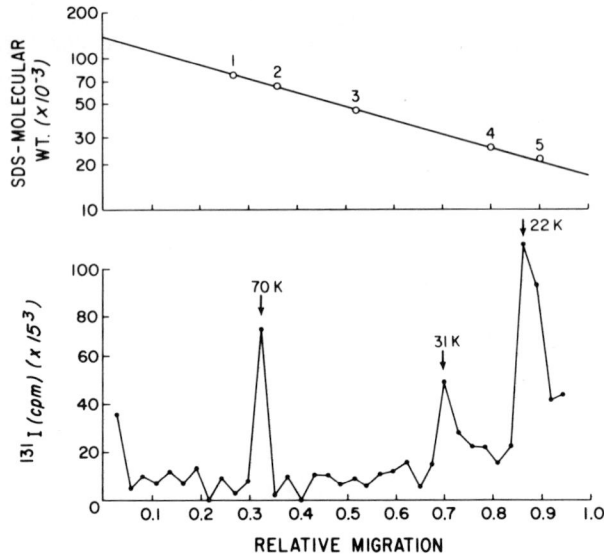

Figure 2 NaDodSO$_4$-polyacrylamide gel analysis of monoclonal antibody-affinity purified turkey erythrocyte β-adrenergic receptors. Turkey erythrocyte ghosts were surface-labeled with [^{131}I]iodine. β-Receptors were solubilized from membranes with 0.5% digitonin and purified by preparative isoelectric focusing to a specific activity of 167 pmol of receptor per milligram of protein. Partially purified receptors were incubated with FV-104 β-receptor monoclonal antibody-Sepharose 4B for 2 hr at 30°C. The Sepharose beads were washed in a column with phosphate buffer, pH 7.4. β-Receptors were eluted from the antibody affinity column with buffer containing 1% NaDodSO$_4$, 2.5% mercaptoethanol, and 5% glycerol and samples were incubated at 100°C for 5 min. Column eluates were analyzed on 10% NaDodSO$_4$-polyacrylamide gels which were sliced and counted for radioactivity. Top panel: molecular-weight calibration curve for 10% NaDodSO$_4$-polyacrylamide gels using point 1, lactoperoxidase (77,500 MW); 2, bovine serum albumin (66,000 MW); 3, ovalbumin (46,000 MW); 4, chymotrypsinogen A (25,700 MW); and 5, soybean trypsin inhibitor (21,500 MW). Bottom panel: [^{131}I]-labeled monoclonal antibody microaffinity column eluates of turkey erythrocyte β-adrenergic receptors. These data illustrate a molecular weight of 70,000 for the turkey β-receptor with a possible subunit of 31,000 daltons. From Ref. 26.

(Figure 2). These data are supportive of the apparent molecular differences from hydrodynamic data, sulfhydryl reagent sensitivity, and immuno-cross reactivity of β$_1$- and β$_2$-receptors (18–23). Studies with monoclonal antibodies (26) and β-receptor reconstitution (27) indicate that although β$_1$- and β$_2$-receptors have major structural differences, a certain degree of molecular homology does exist in these apparently dissimilar molecules (18–27). It is possible that adrenergic receptors evolved from a common molecule, perhaps an α-receptor by gene duplication (26).

2.1 Cellular Localization of Cardiac β-Receptors

The catecholamines (isoproterenol, epinephrine, and norepinephrine) produce characteristic β-adrenergic responses in isolated perfused hearts, cultured heart cells, and papillary muscle preparations when immobilized on glass beads or on soluble amino acid copolymers (28–38). The cardiac effects of the immobilized catechol-amines include positive chronotropic and inotropic responses with a shortening of time to peak tension and elevation of the plateau of the action potential (29,30,38). In contrast to soluble isoproterenol, these effects of immobilized catecholamines are directly and immediately reversible by simply removing the immobilized ligands from the external surface of the cardiac preparation (29,30,33,35,37). The inotropic response decay times following removal of isoproterenol–glass beads from the surface of a cat papillary muscle or the rapid changing of the papillary bath contents to remove isoproterenol covalently attached to a 13,000-MW copolymer are illus-trated in Figure 3. One basic assumption of immobilized drug studies is that the catecholamines by nature of their covalent attachment to a solid support or large macromolecule can only exert their biological action on receptors localized to the external surface of cells.

In 1974, Reuter reported that isoproterenol and norepinephrine produce accel-eration of pacemaker activity and a shift of the plateau level of the action potential

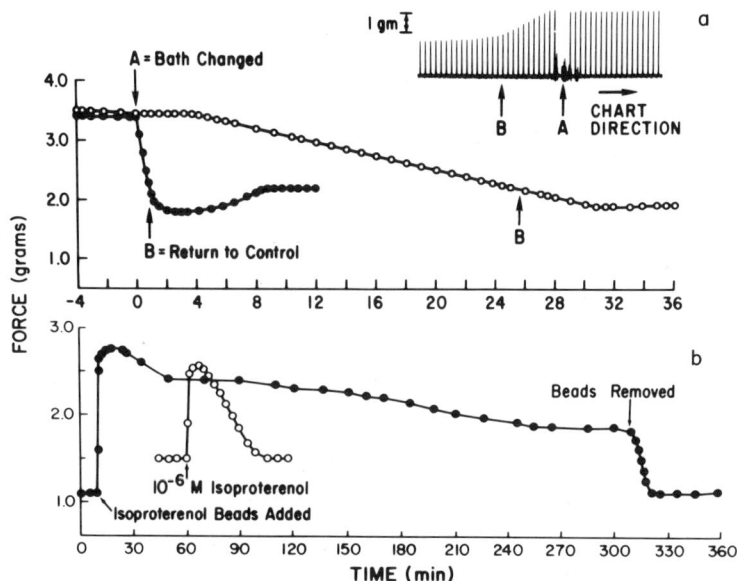

Figure 3 Inotropic response decay times subsequent to isoproterenol and immobilized isoproterenol removal from cat papillary muscles. *(a)* Isolated right ventricular cat papillary muscles were subjected to isoproterenol (O——O) and 13,000-MW polymeric isoproterenol (●——●) for 10 min. The papillary muscle bath was drained (arrow A), refilled with fresh oxygenated Krebs solution, redrained, and again refilled. The force of contraction was monitored until the control inotropic state was obtained (arrow B). The inset is a sample polygraph tracing from a polymeric isoproterenol experiment illustrating the rapid response decay following removal of the immobilized isoproterenol. *(b)* Effect of the addition of glass-bead-immobilized isoproterenol to a cat papillary muscle compared to that of addition of 1 μM isoproterenol to a separate muscle. The rapid decay of the glass bead–isoproterenol inotropic response can be seen from the time the beads were removed from the muscle surface. Redrawn from Ref. 33b with permission of Academic Press, Inc.

when applied iontophoretically to cardiac Purkinje fibers (39). Intracellular injection of the same drugs had no effect (39), further implicating a cell-surface location for cardiac β-receptors. While the effects of glass-bead-immobilized isoproterenol can be potentially explained by a limited release of chemically modified isoproterenol into the microenvironment between the glass bead and muscle surface (33,40), the minute amount of catecholamine released, as well as the properties of drug diffusion, would make the application of catecholamine–glass beads analogous to the ionto-phoretic application of catecholamine (33,40). In contrast, the pharmacological effects of isoproterenol covalently coupled to a high-molecular-weight amino acid polymer have been demonstrated to be due *only* to the covalently coupled species of isoproterenol, which does not penetrate into cells (33–36). The isoproterenol–polymer complex (Copoly-Iso) (Figure 4) can itself produce all the cardiac effects attributable to β-receptor stimulation (33–36)

   These studies taken collectively provide evidence that cardiac β-receptors exist on the cell surface, oriented toward the extracellular space. Similar evidence has recently been provided for the localization of α-adrenergic receptors (41).

3   SITES OF CATECHOLAMINE ACTION IN PRODUCING POSITIVE INOTROPIC RESPONSES IN ISOLATED CARDIAC MUSCLE

The addition of isoproterenol to muscle baths containing isometrically contracting cat papillary muscles produces a positive inotropic response (increased force and increased rate of relaxation) which reaches a maximum or peak effect between 60 and 180 sec subsequent to the catecholamine addition, depending upon the dose

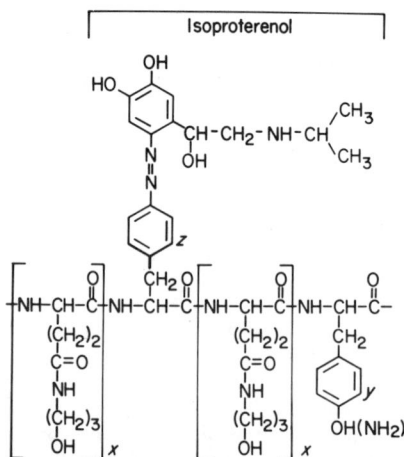

Figure 4   Proposed structure of isoproterenol following diazotization to *p*-amino phenylalanine *(z)* in a random copolypeptide of hydroxypropylglutamine *(x)* with *p*-aminophenylalanine *(y)* (x/y = 4.5:1). The polymer molecular weight with isoproterenol attacted is estimated to be 13,000. From Ref. 34, with permission of Academic Press, Inc.

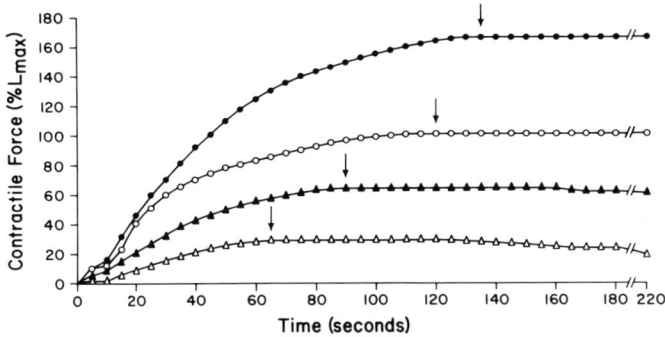

Figure 5   Time course of the positive inotropic response to increasing concentrations of soluble iso-proterenol. The concentrations of isoproterenol were 10 nM (△——△), 20 nM (▲——▲), 50 nM (○——○), and 100 nM (●——●). The arrows indicate points in time when steady-state contraction was achieved with each dose. From Ref. 33, with permission of Academic Press, Inc.

applied (33) (Figure 5). The relationship between isoproterenol concentration and the time course of the inotropic response is not a simple one, and not what one might predict a priori for a diffusion controlled bimolecular reaction. A study of the association rates of [^{125}I]iodohydroxybenzylpindolol ([^{125}I]IHYP), a potent β-receptor antagonist, with cardiac β-receptors (Figure 6) illustrates the expected outcome for such a reaction (i.e., as the drug concentration is increased with the number of β-adrenergic receptors remaining constant, the time to approach equilibrium is shortened). However, in intact cardiac muscle, when the concentration of isoproterenol is increased from 10 to 100 nM, the actual time required to achieve a steady-state contractile response is progressively *increased* from 60 to 130 sec (Figure 5). These data illustrate that cardiac responses cannot be considered to be

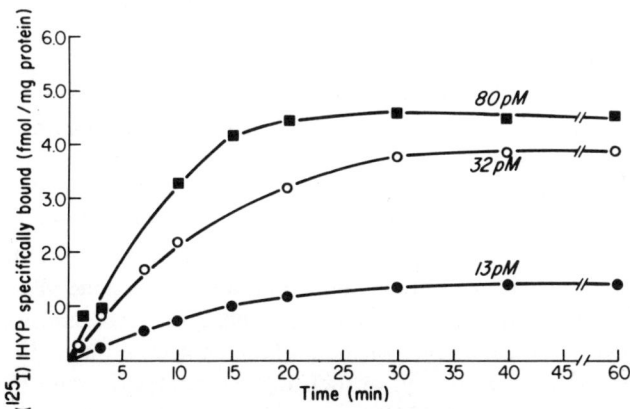

Figure 6   Time course of [^{125}I]IHYP specific binding to dog heart β-adrenergic receptors with increasing concentrations of [^{125}I]IHYP. [^{125}I]IHYP concentrations were 13 pM (●——●), 32 pM (○——○), and 80 pM (■——■). Data from Ref 20.

a direct function of the rate of access of catecholamine to β-receptors throughout the entire muscle (33).

3.1 Time Course of Cyclic AMP Accumulation in Isolated Cardiac Muscle in Response to Catecholamines

The rate of increase of cyclic AMP concentrations in isolated cardiac muscle is extremely rapid. Examples of the times for attainment of maximum cyclic AMP concentrations following the addition of catecholamine are: perfused rat heart, 1–3 sec (42,4) and 10 sec (43,44); rat atrial muscle, 30 sec (45); rat hearts *in situ,* 15 sec (46); rabbit papillary muscles, 30 sec (47); chick embryo hearts, 30 sec (48); guinea pig papillary muscles, 30 sec (49); perfused guinea pig hearts, 15 sec (50); cat papillary muscles, 15 sec (31). Cyclic AMP concentrations, once increased by catecholamines, remain elevated for at least 1.5 min in rat atrium (45), 2 min in rabbit papillary muscles (47), 3 min in cat papillary muscles (31), and 10 min in guinea pig papillary muscles (49); however, in other systems the increased cyclic AMP concentrations are more transient (42,44,4,50). The majority of the transient increases appear to involve isolated perfused hearts and single injections of catecholamine, which would be rapidly washed out of the heart. When isolated rat hearts are perfused continuously with epinephrine, the peak concentration of cyclic AMP is obtained in 60 sec, and the increased cyclic AMP concentration is maintained for at least 4 min (51).

The rapid kinetics of cardiac cyclic AMP formation in response to catecholamines relative to the rate of development of contractile changes provided the first (4) and most consistent "evidence" for a causal relationship between cyclic AMP production and increased cardiac contractility.

Although such a correlation may indeed exist, closer examination of the events actually occurring in cardiac muscle provide us with evidence that prevailing views concerning cardiac inotropic responses to catecholamines are overly simplistic (33).

In studying the rate of catecholamine diffusion into isolated cardiac muscle, I demonstrated that isoproterenol requires 10–15 min to approach an equilibrium (bath) concentration of catecholamine throughout even a small (1-mm-diameter) cat papillary muscle, and that the catecholamine diffusion rates in cardiac muscle are substantially slower than in solution (33). Historically, these findings on their own are not particularly surprising, as in 1928, A. V. Hill described the diffusion of oxygen and lactic acid into and out of muscle tissue and found the diffusion rates of the experimental substances to be substantially slower in muscle than in water (52). The diffusion rate of neurotransmitters in tissue is also generally much slower than in water. Acetylcholine diffusion into rat diaphragm muscle and norepinephrine diffusion through the medial layer of the aorta have been estimated to be approximately 1/7 and 1/10, respectively, of their diffusion rates in water (53,54).

When the kinetics of cyclic AMP formation and cardiac inotropic response development are examined as a function of the distance that a catecholamine such as isoproterenol has penetrated the cardiac muscle, two apparent paradoxes are evident. From Figures 7 and 8, which illustrate the distribution of isoproterenol in a cat papillary muscle based on diffusion theory (Figure 7) and experimental observations

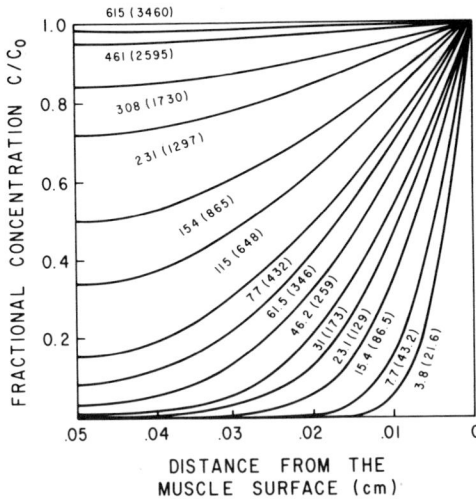

Figure 7 Calculated distribution of isoproterenol and polymeric isoproterenol (13,000 MW) in a 1-mm-diameter isolated cat papillary muscle at 25°C. Concentration distributions at various times with initial surface drug concentration C_O, and C the concentration at different calculated values of r/a in centimeters from the muscle surface. The numbers on the curves indicate the point in time in seconds; numbers without parentheses show the time for soluble isoproterenol diffusion, and those in parentheses show the time for 13,000-MW polymeric isoproterenol diffusion. From Ref 33, with permission of Academic Press, Inc.

(Figure 8), one can readily see that a disparity exists among the time course of the muscle cyclic AMP accumulation, the inotropic response, and the muscle distribution of isoproterenol (33). The maximum cardiac muscle concentration of cyclic AMP is achieved when less than 15% of the equilibrium concentration of isoproterenol is present and the majority of the myocardial muscle cells are exposed to subthreshold levels of drug (Figure 8). The peak inotropic responses are also obtained when the majority of the drug molecules are still at the muscle surface and less than 40% of the equilibrium concentration is present in the tissue (Figures 7 and 8) (33).

3.2 Propagated Inotropic Responses in Heart Muscle

Studies with glass-bead-immobilized catecholamines (31) and with isoproterenol covalently coupled to high-molecular-weight amino acid polymers (34) have demonstrated that the cardiac inotropic responses to these immobilized, and, therefore, diffusion-limited catecholamines are identical in character to the inotropic response obtained with isoproterenol free in solution (31,33). The positive inotropic responses to all three isoproterenol derivatives have the same temporal relationship despite the differences in the amount of tissue exposed to isoproterenol (Figures 8 and 9) (33). When the magnitude of the catecholamine-induced positive inotropic responses (33) are compared directly to an inotropic response produced in the same muscles by paired electrical stimulation (31,34), it becomes clear that even though in some

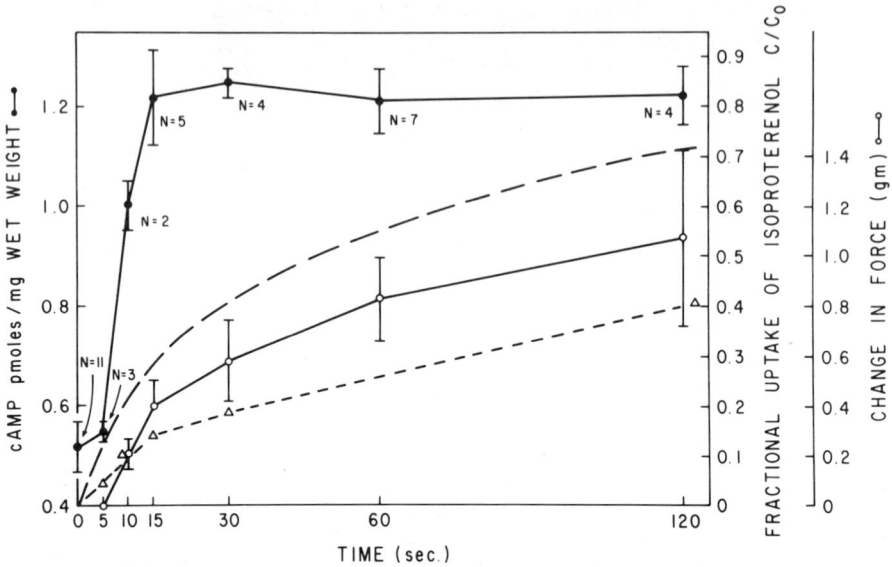

Figure 8 Time course of positive inotropic response and intracellular cyclic AMP response to 1 μM isoproterenol compared with theoretical and experimental fractional uptake of isoproterenol into a cat papillary muscle. The change in isometric force (○) in eight isolated cat papillary muscles in response to 1 μM isoproterenol is compared with the change in intracellular cyclic AMP (cAMP) levels (●). Average contractility for the eight muscles did not increase after 120 sec. Control cyclic AMP levels are shown at zero time. Error bars denote standard errors of the mean. The theoretical fractional uptake of isoproterenol (---) and the experimental fractional uptake (△---△) are presented with respect to time. The data points for fractional uptake of isoproterenol, which represent the average of six muscles, are from Ref. 33 with permission of Academic Press, Inc.

cases as few as 0.01% of the cells in the cardiac muscle are directly exposed to isoproterenol (31), the majority of the cardiac cells in each muscle participate in the increased contractile response (31,33,34). Therefore, these data demonstrate that while the catecholamine stimulation is limited to only relatively few cells, by either diffusion and/or immobilization, all or the clear majority of cardiac muscle cells participate in the inotropic response. These studies provided the first documentation of a basic property of cardiac muscle, the ability to propagate an inotropic response from a site of localized catecholamine stimulation (31). The propagation of inotropic responses is not limited to immobilized isoproterenol, as the diffusion studies indicate that inotropic responses to soluble catecholamines cannot be fully accounted for on the basis of drug diffusion without some myogenic propagation occurring (33).

3.3 Cyclic AMP Responses to Immobilized Isoproterenol

Whereas the contractile responses to soluble isoproterenol and the various immobilized forms of isoproterenol are essentially identical, the cyclic nucleotide responses to these agents appear to differ. The rapid increase in the cardiac muscle cyclic AMP concentration in response to isoproterenol in solution (Figure 8) is not detectable when isoproterenol immobilized on glass beads (31) or soluble polymers (34) is used to stimulate the muscles (Figure 9). The absence of a cyclic AMP increase accompanying the inotropic response to isoproterenol glass beads was confirmed and extended to guinea pig papillary muscles (37). This same laboratory subsequently provided the first electrophysiological evidence for inotropic response propagation, demonstrating action potentials emanating from the site of immobilized catecholamine application to guinea pig papillary muscles (38).

The absence of a detectable cyclic AMP response preceding or during the inotropic response to either polymeric or glass bead-immobilized catecholamines suggests that there is no propagation of a cyclic AMP response in cardiac muscle such as occurs with the contractile response resulting from the stimulation of the superficial muscle cell layers by catecholamines; and that the propagated inotropic responses (involving the majority of cardiac muscle cells) do not appear to depend directly on a concentration change of cyclic AMP in the involved cells (31,33,34).

3.4 Initiation of Cardiac Inotropic Responses in "Initiator Cells": Does Cyclic AMP Play a Role in Response Initiation?

Ventricular cardiac cells can function in dual roles as either "initiator cells" or "propagator cells" based only upon whether or not catecholamine–β-receptor in-

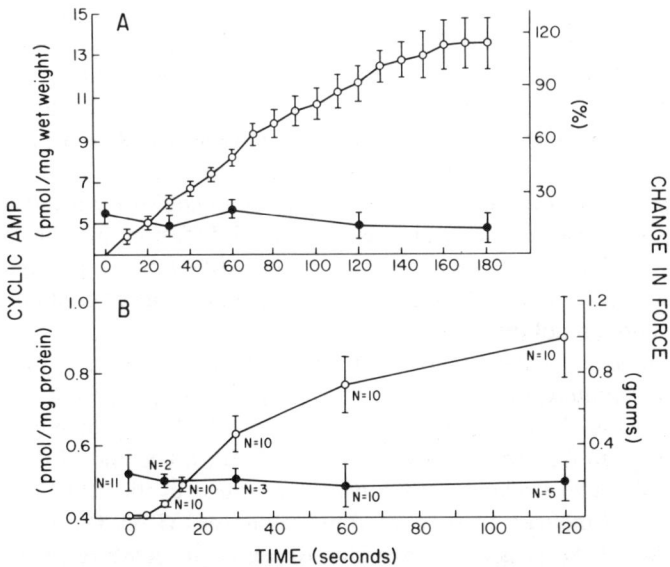

Figure 9 Time course and magnitude of change of isometric force (o———o) and cyclic AMP concentrations (●———●) in response to the 13,000-MW copolyisoproterenol *(A)*; and isoproterenol glass beads *(B)*. Following the addition of immobilized isoproterenol to the muscles baths, the papillary muscles were frozen at the indicated times. Cyclic AMP was determined by radioimmune assay subsequent to muscle homogenization. Control values are indicated at zero time. Vertical bars denote standard errors of the mean. Figure 9*A* was redrawn from Refs. 34 with permission of Academic Press, Inc.

teractions actually occur on the particular cell in question (28). Initiator cells are those cells in actual contact with the catecholamine and which can propagate a stimulus to the propagator cells. Propagator cells, in turn, are those cells that contribute to the increased contractile response without interacting with the catecholamine, by responding to the stimulus received from either initiator cells or from other propagator cells (28). There are no implied anatomical differences between initiator cells and propagator cells, only the presence (initiator cells) and absence (propagator cells) of drug receptor interactions (28). In contrast to intact cardiac muscle, isolated heart cells in tissue culture respond to both soluble and immobilized catecholamines with unequivocal increases in intracellular cyclic AMP concentrations (31). Similarly, thin strips of rat diaphragm muscle show increases in cyclic AMP content in response to both soluble and immobilized isoproterenol (37), demonstrating that when immobilized isoproterenol is exposed to a significant percentage of cells in a preparation, cyclic AMP concentration changes are clearly detectable. These data suggest that cyclic AMP concentrations may increase in the initiator cells in intact cardiac muscle in response to immobilized isoproterenol, but because of the small number of cells involved, the concentration change is undetectable (31,34,37).

The increased fraction of cells that interact with soluble isoproterenol may account for the measurable cyclic AMP concentration changes that are observed in cat papillary muscles (Figure 8). However, it is clear from the isoproterenol diffusion data (Figure 8) (33) that the peak cyclic AMP concentration measured in whole muscle homogenates cannot reflect conditions throughout the muscle. Fifteen seconds subsequent to the addition of isoproterenol to a cat papillary muscle, when muscle cyclic AMP concentrations appear to be maximal (Figure 8), the best distribution of isoproterenol through the papillary muscle would be a 17-fold concentration gradient over 64% of the cells (33). The apparently constant muscle cyclic AMP concentrations after the initial 15 sec (Figure 8) could be a result of continued cyclic AMP synthesis due to further isoproterenol diffusion into the muscle (33). This situation might appear as a wave of cyclic AMP moving through the radius of the muscle with cyclic AMP metabolism equal to the rate of continued synthesis following isoproterenol diffusion (33).

In that there are no apparent quantitative or qualitative differences in the inotropic responses to isoproterenol–glass beads or to polymeric isoproterenol which proceed without detectable changes in cyclic AMP, or to soluble isoproterenol that is preceded by a two- to threefold increase in cyclic AMP, it becomes difficult to assign a definite cause–effect relationship between measured cyclic AMP concentrations and isoproterenol-induced positive inotropic responses (31,33,34). However, given the knowledge of the possible localization of the cyclic AMP response as well as the nature of propagated inotropic response to soluble and immobilized catecholamines, such a concentration dependence would not necessarily be expected under the usual experimental conditions employed in "correlation studies."

What can we conclude, then, about cyclic AMP and cardiac inotropic responses to catecholamines? Tsien (2) proposed that the finding of propagated inotropic responses to catecholamines without apparent cyclic AMP concentration changes should not lead to an abandonment of the cyclic AMP hypothesis but to its modification (2). In modifying the cyclic AMP hypothesis the following should be taken into consideration.

1 Cyclic AMP is not an obligatory mediator of all β-receptor responses in cardiac muscle. This emanates from "dissociation" studies (see, e.g., Refs. 2 and 5) as well as from the finding that cardiac β-receptors can interact with cell membrane components independent of adenylate cyclase (55,56).

2 If cyclic AMP can mediate β-receptor-induced positive inotropic responses, the cyclic AMP effect would be limited to initiator cells (Section 3.3).

3 Cyclic AMP does not appear to mediate the propagation of the inotropic response once initiated (Section 3.3).

4 If cyclic AMP affects the initiation of an inotropic response, it is likely that cyclic AMP acts only as a trigger for contractile events; continued cyclic AMP concentration elevations in a given cell would not be likely to have any additional effects on the inotropic response (Sections 3.1 and 3.4).

Although these might define some of the limits on a potential role for cyclic AMP in cardiac contraction, it is most likely, as proposed by Tsien (2), that the ultimate resolution of the issue of cyclic AMP involvement in cardiac contraction rests with the demonstration of biochemical events, modulated by cyclic AMP upon which contractility changes are dependent.

4 EFFICIENCY OF COUPLING BETWEEN β-RECEPTORS, ADENYLATE CYCLASE, AND CARDIAC CONTRACTILITY: DIRECT EVIDENCE FOR "SPARE" β-RECEPTORS

In 1956, Stephenson introduced the concept of "spare receptors" and drug efficacy to explain differing maximal responses obtained with a series of cholinergic agonists on guinea pig illeum contraction (57). Around the same time, Furchgott (58) and Nickerson (59) using irreversible β-haloalkylamine antagonists, provided evidence for the existence of more α-receptors or histamine receptors than were required for complete tissue activation in certain systems.

The propagated inotropic responses to catecholamines imply the existence of an excess of cardiac β-adrenergic receptors in the tissue in a three-dimensional sense. However, the responses to immobilized catecholamines do not provide information concerning the number or percentage of β-receptors on an individual cell which need to be occupied by catecholamine to achieve the maximum inotropic response (60). The absence of information concerning β-receptor-contractile response stoichiometry has been due in part to the lack of an irreversible β-receptor antagonist. The development by Atlas et al. (61) of the covalent β-blocker N-[2-hydroxy-3-(1-napthoxy)-propyl]-N-bromoacetylethylenediamine (NHNP-NBE) appears to have overcome this deficit. I have recently characterized the nature of the interaction of NHNP-NBE with cardiac β-receptors (60). Formation of the NHNP-NBE–β-receptor complex was found to be irreversible and dependent upon incubation time, temperature, and ligand concentration (60). Occupation of β-receptors by adrenergic ligands prior to NHNP-NBE exposure protected the receptors from inactivation (60). These data are consistent with a covalent modification of β-receptors by NHNP-NBE at a site in or near the catecholamine binding site (60).

4.1 Cardiac β-Receptor Occupation by NHNP-NBE vs. Inotropic Response to Isoproterenol

NHNP-NBE covalently modifies β-receptors in intact cardiac muscle in a dose-related manner (60). Cat cardiac muscle was incubated with various concentrations of NHNP-NBE from 0.1 to 100 μM for 10 min; the heart muscle was then extensively washed, homogenized, and a membrane fraction isolated. The concentration of remaining unoccupied β-receptors was determined by [^{125}I]IHYP binding (60). As summarized in Table 3, NHNP-NBE produced a dose-dependent inactivation of β-receptors in intact heart muscle (60). NHNP-NBE also dramatically affects the concentrations at which isoproterenol produces positive inotropic responses in cat cardiac muscle. As illustrated in Figure 10, increasing the concentration of NHNP-NBE from 0.1 μM to 100 μM produced a progressive rightward shift in the log dose–response curve to isoproterenol. However, there was *no* reduction in the maximum inotropic response to isoproterenol with any of the concentrations of NHNP-NBE tested (60). The percentage of cardiac β-receptors irreversibly inactivated by various concentrations of NHNP-NBE are compared in Table 3 to the shift produced in the ED_{50} for isoproterenol-induced positive inotropic responses.

TABLE 3   Irreversible β-Receptor Occupation by NHNP-NBE vs. Cardiac Inotropic Responses to Isoproterenol[a](60)

NHNP-NBE Concentration (μM)	% Total β-Receptors Occupied	Inotropic Response ED_{50} l-Isoproterenol (nM)	% Control Maximum Response Achieved
Control	0	9.8 ± 2.3 (n = 7)	100
0.1	0	22	100
1.0	43	70	100
10	69	500	100
100	90	5623	100

[a]β-Receptor occupation was determined by [^{125}I]IHYP binding in membranes prepared subsequent to NHNP-NBE treatment (10 min) of intact cat heart muscle (60). The ED_{50}s for isoproterenol-induced inotropic responses were determined from dose–response curves over six orders of magnitude of isoproterenol concentration. Data from Venter (60), with permission of Academic Press, Inc.

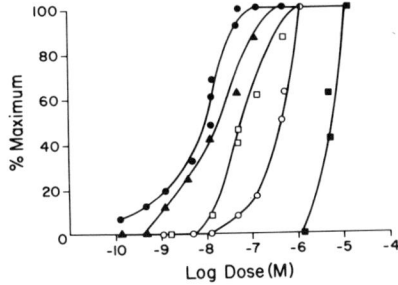

Figure 10 Log dose–response curves for isoproterenol producing positive inotropic responses in cat papillary muscles in the presence of increasing concentration of NHNP-NBE. Cat papillary muscles were treated with 0.1 μM NHNP-NBE (▲———▲); 1.0 μM (□———□); 10 μM (○———○) and 100 μM NHNP-NBE (■———■) for 10 min followed by five bath changes with fresh prewarmed Krebs over a period of 30 min. *l*-Isoproterenol dose–response curves were performed over the ranges indicated in the figure. The control isoproterenol dose–response curve is depicted by (●———●). Each set of experiments was performed sequentially on single papillary muscles. The peak (100%) response was the same for each dose–response curve. Data points represent the mean of at least three experiments. From Ref. 60, with permission of Academic Press, Inc.

The control ED_{50} of 9.8 nM with 100% of the receptors accessible to ligand was shifted to an ED_{50} of 5623 nM if fewer than 10% of the β-receptors were accessible to isoproterenol (60).

These results directly demonstrate the existence of "spare β-receptors" in the heart in relation to cardiac contractility. Maximum inotropic responses to isoproterenol are obtainable even when the β-receptor concentration has been reduced as much as tenfold (60).

4.2 NHNP-NBE Effects on Isoproterenol-Induced Increases in Cardiac Muscle Cyclic AMP Concentrations

With the finding that maximal positive inotropic responses to isoproterenol are elicited with less than 10% of the β-receptors in an accessible form, it was of some interest to assess the cyclic AMP responses under the same conditions (60). Under control conditions isoproterenol increases the cardiac muscle concentration of cyclic AMP in a dose-related manner (Figure 11) (60). The ED_{50} for isoproterenol-induced cyclic AMP changes is 15 nM, a value similar to that of the ED_{50} for inotropic responses (60). NHNP-NBE (100 μM), in addition to effecting β-receptor binding and inotropic responses, also dramatically affects the concentration at which isoproterenol produces cyclic AMP responses (Figure 11) (60). The isoproterenol ED_{50} is shifted to 600 μM as a result of reduced numbers of β-receptors. As with the contractile responses, the same maximum cyclic AMP response is achieved both before and after irreversible β-receptor blockade (60). However, in addition to affecting the dose–response relationship for isoproterenol, NHNP-NBE treatment also affects the time course of the cyclic AMP response. In control muscles, the isoproterenol-induced increase in cyclic AMP is half maximal in 8 sec (a rate of

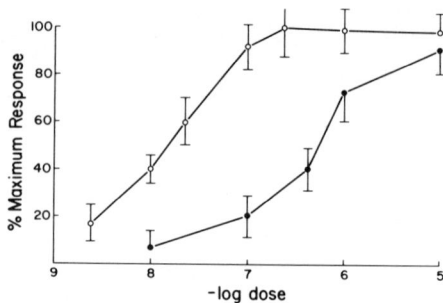

Figure 11 Log dose–response curves for *l*-isoproterenol-induced cyclic AMP concentration increases in the presence and absence of an irreversible β-receptor blocker. Isometrically contracting cat papillary muscles were treated with the indicated concentrations of *l*-isoproterenol for 180 sec prior to freeze clamping. The open circles denote the control dose–response relationship. The closed circles denote the isoproterenol dose–response relationship subsequent to muscle treatment with NHNP-NBE (100 μM) for 10 min, followed by four bath changes with fresh Krebs solution and muscle reequilibration. Frozen papillary muscles were rapidly homogenized in 500 μl of 6% trichloroacetic acid (TCA) at 0°C. The homogenates were centrifuged for 5 min in a microfuge (12,000g). The supernatant was extracted three times with H_2O saturated ether, concentrated to dryness with a stream of nitrogen, redissolved in acetate buffer pH 4.0, and assayed for cyclic AMP using a radioimmunoassay. Each point represents the mean ± SE of at least three determinations. From Ref. 60, with permission of Academic Press, Inc.

36.2 pmol of cyclic AMP per minute per milligram of muscle protein). Following a 10-min treatment of muscles with 100 μM NHNP-NBE, the rate of cyclic AMP formation is reduced almost eightfold, to 4.6 pmol of cyclic AMP per minute per milligram of protein (60). These results indicate that isoproterenol, which can produce full (100%) inotropic responses with less than 10% of the β-receptors, also under these same conditions produces a maximal cyclic AMP response, indicating that cardiac β-receptors couple to adenylate cyclase and produce inotropic responses with equally high efficiency (60).

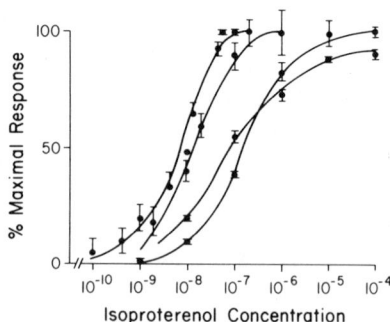

Figure 12 Isoproterenol dose–response curves for inotropic response, cyclic AMP accumulation, adenylate cyclase activation, and β-receptor binding in cat heart. The curves from left to right represent isoproterenol induced positive inotropic responses in cat papillary muscles; cyclic AMP accumulation in cat papillary muscles; adenylate cyclase activation in purified cat heart membranes; and displacement of [^{125}I]IHYP specific binding in purified cat heart membranes. From E.H. Hu and J.C. Venter, unpublished data.

A composite study of the dose effects of isoproterenol on cardiac positive inotropic responses, cyclic AMP accumulation, adenylate cyclase activation, and β-receptor specific binding is illustrated in Figure 12. The isoproterenol-induced contractility and cyclic AMP changes have nearly identical dose–response relationships, again demonstrating the similarity in the coupling of β-receptors to these two possibily related parameters. The dose–response relationship for isoproterenol displacement of [^{125}I]IHYP specific binding is shifted substantially to the right, supporting the notion that only a small percentage of β-receptors need be occupied by isoproterenol to produce maximal muscle responses (60). These data also confirm that the ED_{50} of isoproterenol in producing contractility changes is not the same as the actual affinity of isoproterenol for the β-receptor.

4.3 Irreversible β-Receptor Occupation by NHNP-NBE and Cyclic AMP Responses to Isoproterenol in Cultured Cells

VA$_2$ cells, an SV40 transformed clone of human lung WI-38 cells, and C6 rat glioma cells provide well-characterized systems with regard to [^{125}I]IHYP binding and cyclic AMP production mediated by β-receptors (62). As in cardiac muscle and heart membranes (60), the covalent β-receptor antagonist, NHNP-NBE, induces a dose-dependent loss of [^{125}I]IHYP specific binding in these cells (60 and unpublished observations). However, unlike the heart, as β-receptors are irreversibly inhibited in VA$_2$ and C6 cells by NHNP-NBE, there is a concomitant, stoichiometric loss of isoproterenol-induced cyclic AMP production (60 and unpublished observations). The data from VA$_2$ cells (Figure 13) illustrates the parallel loss of β-receptors and isoproterenol-induced cyclic AMP production (60). These data, together with the heart data, indicate that there are at least two types of coupling between β-receptors and adenylate cyclase: the high-efficiency coupling found in cardiac muscle and the low-efficiency or stoichiometric coupling in some cultured cell systems (60,63).

4.4 Coupling Mechanisms Between Cardiac β-Receptors and Adenylate Cyclase That Allow for the Expression of "Spare Receptors"

Cardiac cells possess β-adrenergic receptors that differ in apparent structure and molecular weight from β-receptor molecules found in most other cell types (18,23). In addition, cardiac β-receptors *unlike* human lung (VA$_2$) cell (60), rat glioma cell (C6) (unpublished observations), and rat lymphoma (S49) cell (63) β-receptors, can couple to adenylate cyclase in a nonstoichiometric high-efficiency manner (60). There are at least three known molecular components involved in the coupling of β-receptors to adenylate cyclase: the molecular entity referred to as the β-adrenergic receptor itself, which is estimated in the heart to have a molecular weight of 65,000 (Table 1); a membrane-associated guanosine triphosphatase (GTPase) with a reported molecular weight of 42,000 (64,65); and the catalytic unit of adenylate cyclase, with an estimated molecular weight of 220,000 (66). A guanine nucleotide-binding component exists in cell membranes which can regulate the affinity of the

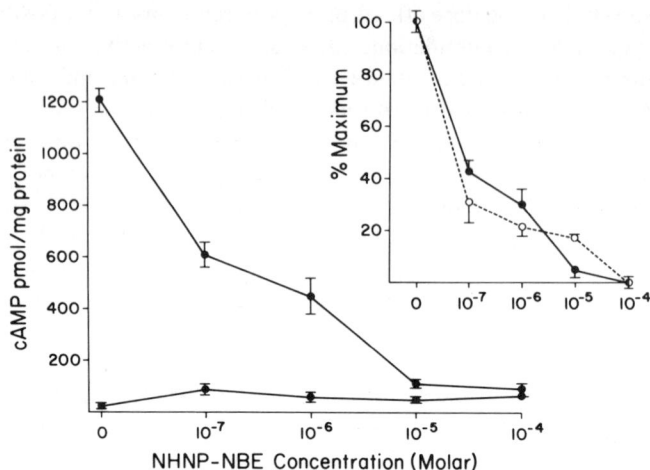

Figure 13 Inhibition of isoproterenol-induced cyclic AMP formation in cultured cells by increasing concentrations of NHNP-NBE. VA_2 cells grown to confluency on 60-mm culture dishes were assayed for cyclic AMP in trichloroacetic acid (TCA) extracts, subsequent to treatment with 10 μM l-isoproterenol for 10 min (upper curve). Cells were preincubated for 10 min with the indicated concentrations of NHNP-NBE followed by three washes with PBS. The cells were then subjected to isoproterenol for 10 min or no treatment, followed by TCA extraction. Cyclic AMP was determined by radioimmunoassay. Error bars represent standard errors from two experiments performed in quadruplicate. The lower curve represents "basal" levels of cyclic AMP (0 concentration) and cyclic AMP produced in response to NHNP-NBE. The inset illustrates the percent inhibition of the isoproterenol-induced cyclic AMP production by NHNP-NBE (●——●) and of IHYP specific binging to VA_2 membranes (○——○). From Ref. 60, with permission of Academic Press, Inc.

receptor for agonists such as isoproterenol as well as regulate the receptor coupling to adenylate cyclase (67). It is generally assumed that the guanine nucleotide regulatory site and the GTPase reside in the same macromolecule. Studies by Cassel and Selinger indicate that the hydrolysis of GTP may be responsible for the inactivation of hormone-stimulated adenylate cyclase (68).

Tolkovsky and Levitski (69) have shown that irreversible β-receptor blockade in turkey erthrocytes by NHNP-NBE affected the time course of adenylate cyclase activation by β-receptor stimulation but that the same maximal enzyme activity was eventually achieved when guanylylimidodiphosphate was included in the reaction mixture to irreversibly activate adenylate cyclase once stimulated by the β-receptor. The guanylylimidodiphosphate is thought to activate adenylate cyclase irreversibly by inhibition of the membrane GTPase (68). In these studies with guanylylimidodiphosphate, the NHNP-NBE-induced reduction in β-receptor binding produced a proportional decrease in the rate of enzyme activation by isoproterenol (69). Tolkovsky and Levitski concluded from these and other data that collisions between β-receptors and adenylate cyclase resulted in enzyme activation (69). In the absence of guanylylimidodiphosphate and with a reduced concentration of β-receptors obtained with NHNP-NBE treatment, the turkey erythrocyte appears to change to a stoichiometrically coupled system with a reduced maximum attainable adenylate cyclase activity. In a single system we can, therefore, shift from a state of stoichiometric coupling between β-receptors and adenylate cyclase (apparently the "normal" state in these cells), to conditions where there is an apparent high-efficiency coupling with limited numbers of β-receptors providing maximal enzyme stimulation, although at a reduced rate. The only apparent change from the stoi-chiometric state to the high-efficiency state is with the presence or absence of guanylylimidodiphosphate, which presumably alters GTPase activity.

Cholera toxin, which inhibits the membrane GTPase in a dose-dependent manner by adenosine diphosphate (ADP) ribosylation of the GTPase (64), prolongs the active state of hormone-stimulated adenylate cyclase (68). In turkey erythrocytes the addition of an excess of propranolol to membranes pretreated with isoproterenol, cholera toxin, and GTP produces evidence for prolonged active state of adenylate cyclase upon isoproterenol β-receptor dissociation (69). Cholera toxin increases the enzyme active state from a half-life of 2–5 sec (69) to 2–3 min (68). The prolongation of the active state of adenylate cyclase produced by cholera toxin is apparently due to the partial inhibition of the GTPase by cholera toxin (68), leading to a reduced rate of GTP hydrolysis and a resultant reduced rate of adenylate cyclase inactivation (68).

In attempting to explain the basis of the spare β-receptor phenomenon in the heart, the duration of the active state of cardiac adenylate cyclase was investigated. The half-life of isoproterenol activated adenylate cyclase upon excess propranolol addition was found to be on the order of 120 sec (Figure 14), similar to that found in turkey erythrocytes in the presence of cholera toxin (68). These data indicate that the cardiac adenylate cyclase in its "normal" physiological state behaves like turkey erythrocytes that have been treated with cholera toxin.

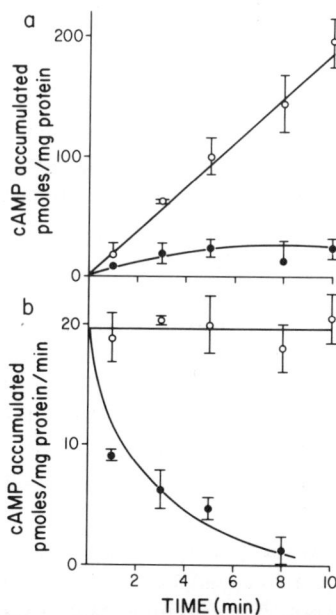

Figure 14 Cardiac adenylate cyclase active state decay as a consequence of isoproterenol-β-receptor dissociation induced by propranolol. Cat heart membranes were incubated for 5 min in the presence of l-isoproterenol (1 μM), GTP (50 μM), and an ATP regenerating system. At the end of the 5 min incubation (time 0 on the figure) a mixture of l-propranolol (10 μM) and [α-^{32}P]ATP was added to one set (open circles). Aliquots were withdrawn at the time indicated in the figure and assayed for cyclic AMP. As parts a and b illustrate, cyclic AMP production proceeded at a constant velocity in the absence of propranolol. In the presence of propranolol the cardiac adenylate cyclase activity decayed with a half life of approximately 120 sec. From E.H. Hu and J.C. Venter, unpublished data.

The VA_2 cell experiments (60, Figure 13) are consistent with a collision coupling mechanism for adenylate cyclase activation; however, the apparent stoichiometry between β-receptor number and cyclic AMP production argues that the half-life of the activated adenylate cyclase is no greater than the half-life of the agonist–receptor–GTPase–catalytic unit interaction (60). Therefore, a single β-receptor could not in this system activate multiple catalytic units and have them remain active. If a β-receptor dissociated from one catalytic unit and collided with a second, the first enzyme would deactivate while the second enzyme unit was in the process of being "found" and activated. The net effect would be essentially the same if one receptor remained coupled to any one enzyme unit. The cardiac β-receptor data argue, in agreement with Tolkovsky and Levitski (69), that single β-receptors cannot *simultaneously* activate multiple units of adenylate cyclase, but that with time, single β-receptors can activate multiple units of enzyme, owing to the existence of a mechanism for prolonging the active enzyme state. A reduced or altered GTPase activity presents a mechanism for obtaining a prolonged active enzyme state, consistent with our present knowledge (68). However, it is also possible that the cardiac β-receptor membrane GTPase interaction may differ from the β-receptor–GTPase coupling in other tissues. Increased molecular resolution of the components involved is clearly warranted.

5 RELATIONSHIP AMONG SPARE β-RECEPTORS, β-RECEPTOR DENSITY, AND PHYSIOLOGICAL REGULATION OF THE HEART

The high-efficiency coupling between cardiac β-receptors and cardiac contractility which allows for the expression of spare β-receptors also provides the heart with a response regulatory mechanism with a considerable safety margin. The data obtained with the irreversible β-receptor antagonist NHNP-NBE have demonstrated that even with a substantial reduction in β-receptor density, the maximal obtainable β-receptor-mediated responses are not diminished, provided that sufficient catecholamine is available to the receptors (60). However with a reduced β-receptor density, increasing concentrations of catecholamines are needed to achieve the same level of contractile responsiveness (60). These data indicate that the density of cardiac β-adenergic receptors will determine the sensitivity of a given portion of cardiac muscle to catecholamines. The data in Figure 10 best illustrate this point. As β-receptor density was effectively reduced from 100% to approximately 10% of the normal receptor complement, the concentration of isoproterenol required to produce a response 50% of the maximum (ED_{50}), shifted from 10 nM to over 5μM (60, Section 4).

Although the concept of β-receptor concentrations controlling tissue sensitivity to catecholamines in the heart may appear obvious, all the physiological ramifications of this concept may not. It is also important to keep in mind how the cardiac β-receptor system differs from, for example, the VA_2 cell system (60). In a stoichiometrically coupled system such as in VA_2 cells, a reduction in the number of β-receptors reduces the maximal obtainable response (Figure 13) as well as the sensitivity to catecholamines, by the law of mass action (60). The heart's ultimate or maximal responsiveness is not altered by changes in β-receptor density; only the sensitivity to catecholamine varies.

5.1   Cardiac Cellular Differentiation

The mammalian heart is a complex integrated organ composed of a variety of cell types, most of which have highly specialized functions. For example, the sinoatrial node, which functions as the cardiac pacemaker, contains two principal cell types: small round P cells, which have few myofibrils and are probably the pacemaker cells, and elongated cells called transitional cells, which are intermediate in appearance between P cells and myocardial cells and probably conduct the impluse within the node and to the nodal margins (70). The cardiac action potential originating in the sinoatrial node arrives via internodal pathways at the atrioventricular (AV) node. The AV node contains both P cells and transitional cells, but the P cells are more sparse (70). As one progresses down the conducting system to the bundle of His and on to the Purkinje system, three or more different cell types have been described (71). Automaticity declines as one moves down the conducting system, and disappears at the level of ventricular myocardial cells. The specialized cells of the conducting system possess little mechanical activity, and there are major differences in the force loads upon the various myocardial cells of the atrium and the right and left ventricles (69,70,73). Distinctive differences have also been detailed by McNutt and Fawcett (72,73) between atrial and ventricular myocardial cells. For example, cat atrial cells are smaller in diameter (5–6 μm) than ventricular cells (10–12 μm) and have substantially fewer T tubules (72,73).

5.2   Sympathetic Innervation of the Heart

Sympathetic fibers contribute to the beat-to-beat regulation of the heart rate and contractile force (74). Therefore, one means of controlling the adrenergic responsiveness of the heart rests in the structure and function of the sympathetic innervation of the myocardium. The course of sympathetic nerves to the heart has been described in detail (75,76), and it has been documented that substantial variation exists in the source and density of innervation throughout the heart (74–76). This variation in innervation is of functional significance. Independent stimulation of individual cardiac nerves can produce discrete changes in heart rate and/or the contractile force of specific cardiac areas (see Ref. 74 for a review).

The extent of sympathetic innervation at the microscopic level also varies dramatically throughout the heart, from at least one adrenergic varicosity per cell in the sinus node to a single nerve terminal per many cells in the ventricles (74–76). However, unlike sympathetic innervation and neuroeffector junctions, which have been characterized in detail in smooth muscle (see, e.g., Refs. 77 and 78), not as much detailed information is available concerning the heart. Most descriptions of sympathetic neuroeffector junctions of ventricular cells report wide gaps of 600–3000 Å, with the widest junctions in the majority (7). Reviewing this situation,

Rolett (7) concluded that "in view of this lack of postjunctional structure special-
ization and the width of the junctional space it is possible to speculate that the
sympathetic neurotransmitter diffuses widely after release to interact with receptors
distributed over the surface of neighboring cardiac muscle fibers."

Although the neuroeffector synaptic cleft in the sinus node and atrial tissue may
be more narrow than in the ventricles, the number or ratio of varicosities per
myocardial cell in both the atria and ventricles suggest that adrenergic nerve activation of the myocardium may depend upon similar inotropic response propagation
mechanisms, as has been described in isolated cardiac muscle (31,33,34; see Section
3.2). Support for the concept of catecholamine-induced propagated inotropic responses occurring in the intact heart comes from a recent study by Matsuda et al.
(79), demonstrating that neuronal uptake is the principal mechanism of terminating
the action of neurally released norepinephrine in the intact canine myocardium.
Extraneuronal uptake participates in the termination of norepinephrine action but
to a lesser extent (79). These findings suggest that norepinephrine released upon
stimulation of cardiac sympathetic nerves has its primary action in the *vicinity* of
the varicosities from which it was released. If ventricular responses to released
transmitters were a direct result of diffusion of norepinephrine to noninnervated
cells (7) rather than to a propagated response, extraneuronal uptake and general
diffusion processes would be expected to play a much greater role in the termination
of norepinephrine action.

5.3   Myocardial Distribution of β-Adrenergic Receptors and Variations in Adrenergic Responsiveness

Given the diversity of cell types, cellular function, and adrenergic responsiveness
throughout the heart, a study was undertaken to assess the density of β-receptors
in principal areas of the canine myocardium (80,81). We found that β-receptor
density varies substantially from one section of the myocardium to another (Table
4). Of the cardiac sections investigated, the sinoatrial node and left ventricles contain
the highest receptor density, 75 fmol/mg and 66 fmol/mg of membrane protein,
respectively. The Purkinje fibers (46 fmol/mg) also contain a high receptor density.
The right ventricle, septum, atrium, and the AV node ranged from 9 to 19 fmol/
mg. While the β-receptor density varies as much as eightfold from area to area,
adenylate cyclase activity is more uniform throughout the heart, varying only two-
to threefold. The ratio of β-receptors to adenylate cyclase activity ranges from
3 : 1 in the septum and atrium to 60 : 1 in the sinus node (Table 4).

The nonstoichiometry of β-receptor number with adenylate cyclase activity is not
surprising in view of the demonstrated existence of spare β-receptors (60) and may
be related to the sensitivity of that portion of the heart to catecholamines (80). As
one might predict from the shifts in the catecholamine sensitivity of isolated cardiac
muscle brought about by altering the number of β-receptors with NHNP-NBE (60),
it appears that the naturally occurring variations in β-receptor density (Table 4) may
affect adrenergic responsiveness. The log dose–response curves for isoproterenol
activation of adenylate cyclase from canine atria (receptor/adenylate cyclase ratio

TABLE 4   Adenylate Cyclase Activity and [^{125}I]IHYP Binding to Cardiac Membranes from 12 Principal Regions of the Canine Myocardium[a]

Cardiac Section	Adenylate Cyclase Activity pmol/min/mg protein (total activity, pmol/min)	[^{125}I]IHYP Specific Binding, fmol/mg protein (total fmol)	Ratio Number of β-Receptors to Isoproterenol-Stimulated Adenylate Cyclase Activity
1. Left ventricle	1.5 ± 0.1 (405)	66 ± 7 (17,820)	44
2. Right ventricle	1.5 ± 0.2 (190.5)	19 ± 5 (2445)	13
3. Septum	2.7 ± 0.15 (172.8)	11 ± 2 (540)	3
4. Left atrium	1.3 ± 0.3 (28.6)	14 ± 2 (308)	11
5. Right atrium	1.3 ± 0.1 (61)	10 ± 3 (475)	8
6. Left atrial appendage	1.5 ± 0.2 (37.5)	19 ± 3 (469)	12
7. Right atrial appendage	0.5 ± 0.4 (5)	15 ± 1 (157)	31
8. Left papillary muscles	1.2 ± 0.3 (12)	11 ± 1 (126)	11
9. Right papillary muscles	1.7 ± 0.1 (15.3)	10 ± 1 (89)	6
10. Sinus node	1.4 ± 0.2 (7)	75 ± 15 (426)	60
11. Atrioventricular node	1.7 ± 0.1 (11.9)	9 ± 1 (64)	5
12. Purkinje fibers	1.9 ± 0.3 (22.8)	46 ± 10 (548)	24

[a]Canine hearts were rapidly dissected into anatomically distinct sections, the tissue homogenized, and sarcolemma-enriched fractions isolated. Adenylate cyclase activity was determined in the presence of 10 μM isoproterenol and GTP. [^{125}I]IHYP specific binding was determined from saturation isotherms in the presence and absence of 10 μM ($-$)ℓ-propranolol (80,81).

of 8 : 1) and canine left ventricle (ratio 44 : 1) show that left ventricular adenylate cyclase is approximately 12-fold more sensitive to isoproterenol ($ED_{50} = 0.2$ μM) than atrial adenylate cyclase ($ED_{50} = 3$ μM) (80). These data clearly support the concept that cardiac β-receptor density has control over the sensitivity of the heart to catecholamines. The variations in β-receptor density can be of functional significance; for example, the sinoatrial node, which is contained in or is surrounded by atrial myocardial cells, has eight times the density of β-receptors as the surrounding atrial muscle cells. Thus the sinoatrial node will be able to respond to relatively low catecholamine concentrations. In fact, the dose–response data (Section 4.1) indicate that based upon β-receptor density alone, the sinoatrial node could possibly elicit a maximal response to catecholamines before the threshold for the atria was exceeded.

Mechanistically, these regional sensitivity differences could play an important role in maintaining the heart in a normally functioning condition. Randall discusses the effect of adrenergic responses, including maintenance of the synchrony of contraction of the heart, the pacemaker location, and cardiac rhythm (74). It is not difficult to surmise a role for altered β-receptor concentrations in the development of cardiac arrythmias. It is also possible that decreased β-receptor concentrations in the ventricles may contribute to heart failure. In contrast, the increased contractile responsiveness of the myocardium and the tachycardia occurring with hyperthyroidism could be readily explained on the basis of increased numbers of cardiac β-receptors as a function of thyroid hormone exposure (see Section 6.2).

6   FACTORS INFLUENCING THE DENSITY OF β-ADRENERGIC RECEPTORS

The concentration of β-receptors on the surface of cells both in terms of the number of actual receptor molecules and the number of functional receptor molecules is determined by a number of contributing factors. β-Receptor density has been reported to be affected by the cell cycle (82), the rate of receptor synthesis and turnover (83,84), cell density and cell-to-cell contact (83–85), by self-regulation in terms of desensitization phenomena (85–87), hormones such as thyroid hormone and glucocorticoids (84,88,89), and possibly by circulating autoantibodies to β-receptors (23).

6.1   Regulation of β-Receptor Synthesis and Turnover Rates

Fraser and Venter (83,84) have recently measured the rate of β-receptor synthesis and incorporation into plasma membranes of rat glioma cells (C6 cells), and human embryonic lung cells (VA$_2$, VA$_4$, and WI-38 cells), following the irreversible blockade of preexisting β-receptors. As Figure 15 illustrates, new β-receptors are incorporated at a relatively constant rate for approximately 20 hr, after which a plateau is achieved, possibly due to the turnover or degradation of some of the newly synthesized receptors (83,84). The rate of appearance of new receptors is on the order of 2% of the initial density per hour. Assuming that each receptor in the membrane has an equal probability of being removed from the membrane, the half-life of the β-receptors in this system can be estimated to be on the order of 20–30 hr (83,84). A similar half-life was found for C6 cell β-receptors. The effects of protein synthesis inhibitors puromycin and cycloheximide demonstrate that new β-receptor incorporation into cell membranes is directly dependent upon protein synthesis (83,84).

These β-receptor synthesis and turnover rates indicate that the half-life of β-receptors in cell membranes is relatively short, requiring continual β-receptor synthesis to maintain a constant receptor concentration in the membrane. This active process provides the prerequisite conditions for a dynamic regulation of β-receptor density.

Kebabian and co-workers in 1975 first suggested that β-receptors were macromolecules that may undergo a rapid turnover, by demonstrating that the density of β-receptors in rat pineal glands could change twofold within a 12- to 24-hr period, depending on the exposure of the rat to light or darkness (90).

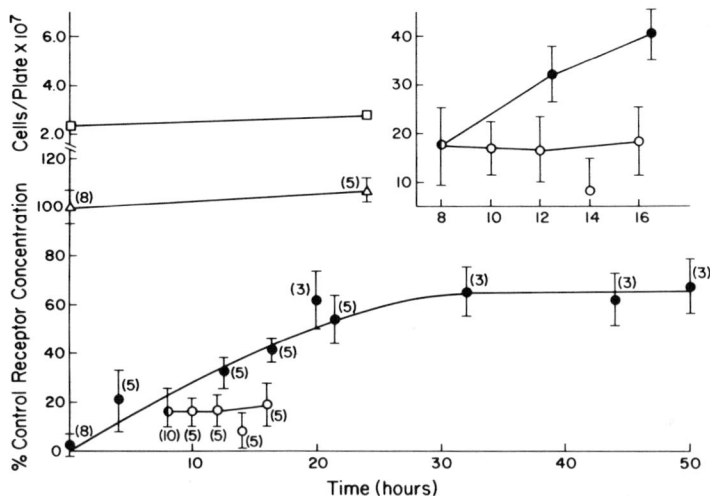

Figure 15 β-Adrenergic receptor incorporation rates into VA$_2$ cell membranes. Cells were incubated twice with NHNP-NBE (100 μM) to inactivate irreversibly existing β-receptors at time 0. Cells were incubated at 37°C in minimal essential media with 10% calf serum for the times indicated. Cells were harvested and membranes isolated. β-Receptor concentrations were determined by assessing [^{125}I]IHYP (87 pM) specific binding. At 87 pM [^{125}I]IHYP occupied 71% of the VA$_2$ β-receptors. Assays were performed in quintuplet using 70 μg of protein per assay. The control synthesis rates are indicated by the open triangles. The closed circles represent β-receptor incorporation rates subsequent to NHNP-NBE treatment. The inset shows the β-receptor incorporation in cells treated with NHNP-NBE (closed circles) or NHNP-NBE plus puromycin (0.1 mg/ml) 8 hours subsequent to NHNP-NBE treatment (open circles). From ref. 84, with permission of Academic Press, Inc.

The physiological state of a tissue has been shown to influence the rate at which specific membrane proteins turn over. For example, the turnover rate of the nicotinic acetylcholine receptor is increased by muscle denervation (91,92), and this effect can be inhibited by direct electrical stimulation of the muscle (91,93,94). The presence of antibodies against the acetylcholine receptor has also been shown to enhance degradation of this receptor (95,96).

In addition, cell density and cell-to-cell contact in tissue culture has been shown to effect β-receptor density in human astrocytoma cells (85) and C6 cells (83,84) but not in VA$_2$ cells. In the astrocytoma/glial cells, β-receptor density was highest in low-density cells which lacked cell-to-cell contacts (84,85). As cell density increased, the number of β-receptors per cell decreased, reaching a plateau at confluency (84). These and other studies led Harden et al. (85) to conclude that "responsiveness to catecholamines increases or decreases during culture as a consequence of an increase or decrease in the number of β-adrenergic receptors per cell."

Like most cellular proteins, the β-receptor also appears to be synthesized at a specific time during the cell cycle (82). Charlton and Venter have shown that the concentration of C6 cell β-receptors is lowest during mitosis and the G$_1$ phase of the cell cycle (82). Receptor synthesis occurred during the S phase, followed by a decrease during G$_2$ to the levels found at mitosis (82). In nonsynchronized C6 cells, mitotic cells were found to have one-half of the β-receptor concentration of

6.2 Hormonal Influence over β-Receptor Concentrations

Hormones are known to regulate the concentration of several membrane receptors. Low levels of insulin downregulate the concentration of insulin receptors in lymphocytes (97). This phenomenon of insulin regulation of insulin receptor concentrations has since been observed in several species and is believed to involve an energy-dependent increase in insulin receptor degradation (98). Both thyrotropin-releasing hormone (TRH) (99) and thyroid hormone (100) at low concentrations reduce the number of TRH receptors in cultures of rat pituitary cells. In this same line of cells, hydrocortisone causes an increase in TRH receptor density (101). Lefkowitz and co-workers have demonstrated that the administration of thyroid hormone to rats results in an increased number of cardiac β-adrenergic receptors (88). This result has also been obtained by others (89). This cardiac affect of thyroid hormone on increasing the concentration of myocardial β-receptors has particular significance to the increased sensitivity of the heart to catecholamines in hyperthyroid states. Decreased catecholamine responsiveness of the heart has been noted with hypothyroidism (102) and supports the hypothesis that β-receptor density is a controlling influence on cardiac function.

Hydrocortisone and other glucocortoids produce a 100% increase in the β-receptor concentration of human lung VA_2 cells in culture within a 24-hr period (83,84). In contrast, hydrocortisone produces no changes in C6 cell β-receptor density (Fraser and Venter, unpublished observations), although responsiveness is increased, owing apparently to increased amounts of adenylate cyclase (103).

6.3 Other Potential Regulators of β-Receptor Density

While desensitization of β-receptors by the presence of agonists has been demonstrated to play a regulatory role over adrenergic responsiveness of a number of isolated cell systems (86,87) by an apparent decrease in the concentration of β-receptors, there is little evidence for a major role of the receptor desensitization phenomenon in the regulation of cardiac adrenergic responses. Isolated cardiac muscle such as cat papillary muscles display little, if any, tachyphylaxis (decreased responsiveness to catecholamines) when exposed repeatedly to isoproterenol during dose–response studies (Venter, unpublished observations). This phenomenon has been observed by others (G. A. Robison, personal communication). The β-receptor excess or "spareness" (60) in the myocardium, however, would supply a buffering capacity to the effects of desensitization and therefore offers additional safety to the heart, protecting from a loss of adrenergic responsiveness from prolonged activation of the sympathetic system. Preliminary experiments suggest that the myocardial β-receptors may be more susceptible to desensitization under conditions of excess receptor depletion due to NHNP-NBE treatment (Venter, unpublished observations).

There is increasing evidence that autoantibodies to membrane receptors can have substantial influence over receptor function. Autoantibodies to cell membrane receptors have recently been documented in a number of disease states in humans.

For example, autoantibodies to the nicotinic acetylcholine receptor are associated with myasthenia gravis (104–107); to the thyrotropin receptor with Graves' disease (108,109); and to the insulin receptor with certain types of insulin-resistant diabetes (110–112). Antibodies to certain receptors can increase the degradation of the receptors, thereby reducing the membrane receptor density (95,96). Venter, Fraser and Harrison recently discovered the existence of autoantibodies to β_2-adrenergic receptors in a series of asthmatic and allergic rhinitis patients (23). It is conceivable that β-receptor antibodies may have a profound influence on β-receptor concentrations and function (23).

Temperature affects adrenergic responsiveness of various heart preparations by altering the relative contribution of α- and β-adrenergic receptors to force and rate responses (113–115).

A final type of regulation of cardiac β-receptors is that proposed by Watanabe et al. (116) to explain the attenuation of β-receptor responses by muscarinic–cholinergic receptor agonists. These workers presented results which indicate that the affinity of catecholamines for cardiac β-receptors may be modulated by the β-receptor–adenylate cyclase-associated GTPase regulatory protein, as a consequence of acetylcholine–muscarinic receptor interactions. This observation, which clearly deserves further investigation, suggests that complex membrane protein interactions may be involved in autonomic regulatory phenomenon in the heart.

7 SUMMARY AND CONCLUSION

New insight into the relationship among β-adrenergic receptors, adenylate cyclase, and cardiac contractility has resulted from recent advances in our knowledge concerning β-receptors both at the molecular and tissue levels.

It is proposed here that the catecholamine activation of the heart mediated by β-adrenergic receptors occurs in part by response propagation mechanisms from sites of localized catecholamine stimulation. This appears to be true whether the catecholamine is of exogenous or adrenal origin or is present due to the stimulation of sympathetic nerves. Although it seems likely that propagated inotropic responses are not directly mediated by cyclic AMP, the initiation of the contractile responses in "initiator cells" by β-receptor stimulation is, for the most part, consistent with a role for cyclic AMP in response initiation, although it is by no means necessarily an obligatory one. Adrenergic control over cardiac contractility is regulated by the density of β-adrenergic receptors in the myocardium and by the efficiency of coupling of β-receptors to adenylate cyclase. β-Adrenergic receptor concentrations in the heart appear to be under dynamic regulation. Thus any alterations in β-receptor regulatory factors could lead to alterations in cardiac β-receptor density and therefore the adrenergic responsiveness of the heart. Variations in cardiac β-receptor concentrations and the resultant hyper- or hyporesponsiveness of the heart to catecholamines may be a major contributing factor to the etiology of clinical manifestations associated with cardiovascular disease.

ACKNOWLEDGMENTS

The unpublished data presented in this chapter were supported by the National Institutes of Health Grant HL-21329, and by Grants 77-693 and 79-688 from the American Heart Association, with funds contributed in part by the Heart Association of Western New York.

REFERENCES

1   S. E. Mayer, *Fed. Proc.*, **29**, 1367 (1972).
2   R. W. Tsien, *Ad. Cyclic Nucleotide Res.*, **8**, 363 (1977).
3   F. Murad, Y.-M. Chiu, T. W. Rall, and E. W. Sutherland, *J. Biol. Chem.*, **237**, 1233 (1962).
4   G. A. Robison, R. W. Butcher, I. Øye, H. E. Morgan, and E. W. Sutherland, *Mol. Pharmacol.*, **1**, 168 (1965).
5   B. E. Sobel and S. E. Mayer, *Circ. Res.*, **32**, 407 (1973).
6   M. L. Entman, *Adv. Cyclic Nucleotide Res.*, **4**, 163 (1974).
7   E. L. Rolett, in G. A. Langes, and A. J. Brady, Eds., *The Mammalian Myocardium*, Wiley, New York, p. 219.
8   G. I. Drummond, and D. L. Severson, *Circ. Res.*, **44**, 145 (1979).
9   A. Wollenberger, in W. G. Nayler, Ed., *Contraction and Relaxation in the Myocardium*, Academic Press, New York, 1975, p. 113.
10 J. R. Williamson, in *Handbook of Physiology*, Vol. 6, *Endocrinology*, American Physiological Society, Washington, D.C., 1976, p. 605.
11 G. A. Robison, R. W. Butcher, and E. W. Sutherland, *Cyclic AMP*, Academic Press, New York, 1971.
12 E. G. Krause and A. Wollenberger, in H. Cramer and J. Schultz, Eds., *Cyclic Nucleotides: Mechanisms of Action*, Wiley, New York, 1976.
13 J.-B. Osnes, Cyclic AMP Dependent and Independent Effects of Adrenergic Amines, Doctoral dissertation, University of Oslo, 1978.
14 R. P. Ahlquist, *Am. J. Physiol.*, **153**, 586 (1948).
15 A. M. Lands, A. Arnold, J. P. McAuliff, F. P. Ludvena, and R. G. Brown, Jr., *Nature (Lond.)*, **214**, 597 (1967).
16 D. H. Jenkinson, *Br. Med. Bull.*, **29**, 142 (1973).
17 R. F. Furchgott, in H. Blaschko and E. Muschell, *Catecholamines*, Handbook of Experimental Pharmacology, Vol 33, Springer-Verlag, Berlin, 1972, p. 283.
18 C. M. Fraser, G. Ghai, A. L. Cave, and J. C. Venter, *Fed. Proc.*, **37**, 684 (1978).
19 W. L. Strauss, C. M. Fraser, G. Ghai, and J. C. Venter, *Fed. Proc.*, **38**, 843 (1979).
20 W. L. Strauss, G. Ghai, C. M. Fraser, and J. C. Venter, *Arch. Biochem. Biophys.*, **196**, 566 (1979).
21 W. L. Strauss, G. Ghai, C. M. Fraser, and J. C. Venter, submitted for publication.
22 S. Wrenn and E. Haber, *J. Biol. Chem.*, **254**, 6577 (1979).
23 J. C. Venter, C. M. Fraser, and L. Harrison, *Science*, **207**, 1361 (1980).
24 A. I. Soiefer, R. Greguski, D. Triggle, C. M. Fraser, and J. C. Venter, submitted for publication.
25 A. I. Soiefer and J. C. Venter, *Fed. Proc.* **39**, 243 (1980).
26 C. M. Fraser and J. C. Venter, *Proc. Natl. Acad. Sci. U.S.A.*, **77**, 7034 (1980).
27 D. R. Jeffery, R. R. Charlton, and J. C. Venter, *J. Biol. Chem.*, **255**, 5015 (1980).

28 J. C. Venter, Immobilized and Insolubilized Drugs, Hormones and Enzymes: Characterizations and Applications to Physiology and Medicine, Doctoral dissertation, University of California, San Diego (Xerox Univ. Microfilm 76-10, p. 131), 1975.

29 J. C. Venter, J. E. Dixon, P. R. Maroko, and N. O. Kaplan, *Proc. Natl. Acad. Sci. U.S.A.*, **69**, 1141 (1972).

30 J. C. Venter, J. Ross, J. E. Dixon, S. E. Mayer, and N. O. Kaplan, *Proc. Natl. Acad. Sci. U.S.A.*, **70**, 1214 (1973).

31 J. C. Venter, J. Ross Jr., and N. O. Kaplan, *Proc. Natl. Acad. Sci. U.S.A.* **72**, 824 (1975).

32 J. C. Venter, L. J. Arnold, Jr., and N. O. Kaplan, *Mol. Pharmacol.*, **11**, 1 (1975).

33 J. C. Venter, *Mol. Pharmacol.*, **14**, 562 (1978).

34 E. H. Hu and J. C. Venter, *Mol. Pharmacol.*, **14**, 237 (1978).

35 M. S. Verlander, J. C. Venter, M. Goodman, N. O. Kaplan, and B. Saks, *Proc. Natl. Acad. Sci. U.S.A.*, **73**, 1009 (1976).

36 J. C. Venter, M. S. Verlander, N. O. Kaplan, M. Goodman, J. Ross, and S. Sesayama, in R. S. Kostelnik, Ed., *Polymeric Delivery Systems: Midland Macromolecular Monographs*, Vol. 5, Gordon and Breach, London, 1978, p. 237.

37 W. R. Ingebretsen Jr., E. Becker, W. F. Friedman, and S. E. Mayer, *Circ. Res.*, **40**, 474 (1977).

38 E. Becker, W. R. Ingebretsen, Jr., and S. E. Mayer, *Cir. Res.*, **41**, 653 (1977).

39 H. Reuter, *J. Physiol. (Lond.)*, **242**, 429 (1974).

40 J. C. Venter, N. O. Kaplan, M. S. Yong, and J. B. Richardson, *Science*, **185**, 459 (1974).

41 J-P. Dehayne, P. F. Blackmore, J. C. Venter, and J. H. Exton, *J. Biol. Chem.*, **255**, 3905 (1980).

42 G. Drummond, L. Duncan, and E. Hertzman, *J. Biol. Chem.*, **241**, 5899 (1966).

43 J. G. Dobson, Jr., *Am. J. Physiol.*, **234**, H638 (1978).

44 W. Y. Cheung and J. R. Williamson, *Nature (Lond.)*, **207**, 979 (1965).

45 T. T. Martinez and J. H. McNeill, *J. Pharmacol. Exp. Ther.*, **203**,457 (1977).

46 D. H. Namm and S. E. Mayer, *Mol. Pharmacol.*, **4**, 61 (1968).

47 H. J. Schüman, M. Endoh, and O. E. Brodde, *Pharmacology*, **289**, 291 (1975).

48 J. B. Polson, N. D. Goldberg, and F. E. Shideman, *J. Pharmacol. Exp. Ther.*, **200**, 630 (1977).

49 J. G. Dobson, J. Ross, Jr., and S. E. Mayer, *Circ. Res.*, **39**, 388 (1976).

50 W. R. Kukovetz, G. Pöch, and A. Wurm, *Naunyn-Schmiedebergs Arch. Pharmacol.* **278**, 403 (1973).

51 S. L. Keely and J. D. Corbin, *Am. J. Physiol.*, **233**, 269 (1977).

52 A. V. Hill, *Proc. R. Soc. Lond. Ser. B.*, **104**, 39 (1928).

53 K. Krnjevic and J. F. Mitchell, *J. Physiol. (Lond.)*, **153**, 562 (1960).

54 J. A. Bevan and J. Török, *Circ. Res.*, **27**, 325 (1970).

55 S. Wrenn, C. Homey, and E. Haber, *J. Biol. Chem.*, **254**, 5708 (1979).

56 F. Horata, W. J. Strottmatter, and J. Axelrod, *Proc. Natl., Acad. Sci. U.S.A.*, **76**, 368 (1979).

57 R. P. Stephenson, *Br. J. Pharmacol.*, **11**, 379 (1956).

58 R. F. Furchgott, *Pharmacol. Rev.*, **7**, 183 (1955).

59 M. Nickerson, *Nature (Lond.)*, **178**, 697 (1956).

60 J. C. Venter, *Mol. Pharmacol.*, **16**, 429 (1979).

61 D. Atlas, M. L. Steer, and A. Levitzki, *Proc. Natl. Acad. Sci. U.S.A.*, **73**, 1921 (1976).

62 M. E. Maguire, R. A. Wilkund, H. J. Anderson, and A. G. Gilman, *J. Biol. Chem.*, **251**, 1221 (1976).

63 G. L. Johnson, H. R. Bourne, M. K. Gleason, P. Coffino, P. A. Insel, and K. L. Melmon. *Mol. Pharmacol.*, **15**, 16 (1979).

64 D. M. Gill, and R. Meren, *Proc. Natl. Acad. Sci. U.S.A.*, **75**, 3050 (1978).

65   T. Pfeuffer, *J. Biol. Chem.*, **252**, 7224 (1977).

66   T. Haga, K. Haga, and A. G. Gilman, *J. Biol. Chem.*, **252**, 5776 (1977).

67   M. E. Maguire, E. Ross, and A. G. Gilman, *Adv. Cyclic Nucleotide Res.* **8**, 1 (1977).

68   D. Cassel and Z. Selinger, *Proc. Natl. Acad. Sci. U.S.A.*, **74**, 3307 (1977).

69   A. M. Tolkovsky and A. Levitski, *Biochemistry*, **17**, 3795 (1978).

70   R. M. Berne and M. N. Levy, *Cardiovascular Physiology*, C. V. Mosby, St. Louis, Mo., 1972.

71   S. Virágh and C. E. Challice, in C. E. Challice and S. Virágh, Eds., *Ultrastructure of the Mammalian Heart*, Academic Press, New York, 1973, p. 43.

72   D. W. Fawcett and N. S. McNutt, *J. Cell Biol.*, **42**, 1 (1969).

73   N. S. McNutt and D. W. Fawcett, *J. Cell Biol.*, **42**, 46 (1969).

74   W. C. Randall, in W. C. Randall, Ed., *Neural Regulation of the Heart*, Oxford University Press, New York, 1977, p. 42.

75   A. Yamauchi, in C. E. Challice and S. Virágh, Eds., *Ultrastructure of the Mammalian Heart*, Academic Press, New York, 1973, p. 127.

76   W. C. Randall and J. A. Armour, in W. C. Randall, Ed., *Neural Regulation of the Heart*, Oxford University Press, New York, 1977, pp. 13.

77   J. A. Bevan, *Circ. Res.*, **45**, 161 (1979).

78   G. Burnstock, in E. Buelging, A. F. Brading, A. W. Jones, and T. Tomita, Eds., *Smooth Muscle*, Williams & Wilkins, Baltimore, Md., 1970, p. 1.

79   Y. Matsuda, Y. Matsuda, and M. Levy, *Circ. Res.*, **45**, 180 (1979).

80   G. Ghai, E. H. Hu, and J. C. Venter, submitted for publication.

81   G. Ghai and J. C. Venter, *Fed. Proc.*, **37**, 685 (1978).

82   R. R. Charlton and J. C. Venter, *Biochem. Biophys. Res. Commun.*, **94**, 1221 (1980).

83   C. M. Fraser and J. C. Venter, *Fed. Proc.*, **38**, 362 (1979).

84   C. M. Fraser and J. C. Venter, *Biochem. Biophys. Res. Commun.*, **94**, 390 (1980).

85   T. K. Harden, S. J. Foster, and J. P. Perkins, *J. Biol. Chem.*, **254**, 4416 (1979).

86   R. J. Lefkowitz and L. T. Williams, *Adv. Cyclic Nucleotide Res.*, **9**, 1 (1978).

87   T. K. Harden, Y-F. Su, and J. P. Perkins, *J. Cyclic Nucleotide Res.* **5**, 99 (1979).

88   L. T. Williams. R. J. Lefkowitz, A. M. Watanabe, D. R. Hathaway, and H. R. Besch Jr., *J. Biol. Chem.*, **252**, 2787 (1977).

89   T. Ciaraldi and G. V. Marinetti, *Biochem., Biophys. Res. Commun.*, **74**, 984 (1977).

90   J. W. Kebabian, M. Zatz, G. A. Romero, and J. Axelrod, *Proc. Natl. Acad. Sci. U.S.A.*, **72**, 3735 (1975).

91   T. Lømo and J. Rosenthal, *J. Physiol.*, **221**, 493 (1972).

92   D. M. Fambrough, *Science*, **168**, 372 (1970).

93   R. Jones and G. Vrbova, *J. Physiol.*, **236**, 517 (1974).

94   R. Gruener, N. Baumbach, and D. Coffee, *Nature (Lond.)*, **248**, 68 (1974).

95   I. Kao and D. B. Brachman, *Science*, **196**, 527 (1977).

96   S. H. Appel, R. Anwyl, M. W. McAdams, and S. Elias, *Proc. Natl. Acad. Sci. U.S.A.*, **74**, 2130 (1977).

97   J. T. Gavin III, J. Roth, D. M. Neville, P. MeMeytes, and D. M. Buell, *Proc. Natl. Acad. Sci. U.S.A.*, **71**, 84 (1974).

98   C. R. Kahn, *J. Cell. Biol.*, **70**, 261 (1976).

99   P. M. Hinkle and A. H. Tashjian, Jr., *Biochemistry*, **14**, 3845 (1975).

100   M. H. Perrone and P. M. Hinkle, *Prog. 59th Meet. Endocr. Soc., 1977*, p. 193 (Abstr. 274).

101 A. H. Tashjian, Jr., R. Osborne, D. Mainn, and A. Knaian, *Biochem. Biophys. Res. Commun.*, **79**, 333 (1977).

102 G.Kunos, I. Vermes-Kunos, and M. Nickerson, *Nature (Lond.)*, **250**, 779 (1974).

103 M. A. Brostrom, C. Kon, D. R. Olson, and B. Breckenridge, *Mol. Pharmacol.*, **10**, 711–720 (1974).

104 J. Patrick, J. Lindstrom, B. Culf, and J. McMillan, *Proc. Natl. Acad. Sci. U.S.A.*, **70**, 334 (1973).

105 R. R. Almon. C. G. Andrew, and S. H. Appel, *Science*, **186**, 55 (1974).

106 O. Abramsky, A. Aharonov, C. Webb, and S. Fuchs, *Clin. Exp. Immunol.*, **19**, 11 (1975).

107 J. A. Lindstrom, M. E. Seybold, V. A. Lennon, S. Whittingham, and D. D. Duane, *Neurology*, **26**, 1054 (1976).

108 B. R. Smith and R. Hall, *Lancet*, **2**, 427 (1974).

109 S. W. Manley, J. R. Bourke, and R. W. Hawker, *J. Endocrinol.*, **61**, 437 (1974).

110 J. S. Flier, C. R. Kahn, J. Roth, and R. S. Bar, *Science*, **190**, 63 (1975).

111 C. R. Kahn, J. S. Flier, R. S. Bar, J. A. Archer, P. Gordon, M. M. Martin, and J. Roth, *N. Engl. J. Med.* **294**, 739 (1976).

112 L. C. Harrison, J. S. Flier, C. R. Kahn, D. B. Jarrett, M. Muggeo, and J. Roth in N. R. Rose, P. E. Bigazzi, N. L. Warner, Eds., *Genetic Control of Autoimmune Disease*, Elsevier/North-Holland, New York, 1978, p. 61.

113 G. Kunos, M. S. Yong, and M. Nickerson, *Nature New Biol. (Lond.)*, **241**, 119–120 (1973).

114 B. G. Benfey, G. Kunos, and M. Nickerson, *Br. J. Pharmacol.*, **51**, 253–257 (1974).

115 G. Kunos, and M. Nickerson, *J. Physiol. (London)* **256**, 23 (1976).

116 A. M. Watanabe, M. M. McConnaughey, R. A. Strawbridge, J. W. Fleming, L. R. Jones, and H. R. Besch, Jr., *J. Biol. Chem.*, **253**, 4833 (1978).

117 R. J. Lefkowitz, *Biochem. Pharmacol.*, **24**, 583 (1975).

118 K. P. Minneman, L. R. Hegstrand, and P. B. Molinoff, *Mol. Pharmacol.*, **16**, 21 (1979).

119 C. Mukerjee, M. G. Caron, D. Mullikin, and R. J. Lefkowitz, *Mol. Pharmacol.*, **12**, 16 (1976).

120 E. L. Rugg, D. B. Barnett, and S. R. Nahorski, *Mol. Pharmacol.*, **14**, 996 (1978).

121 G. Vauquelin, S. Bottari, L. Kanarek, and A. D. Strosberg, *J. Biol. Chem.*, **254**, 4462 (1979).

CHAPTER EIGHT

DIRECT BINDING STUDIES OF THE β-ADRENOCEPTOR

Charles J. Homcy and Edgar Haber

Cardiac Unit, Department of Medicine, Massachusetts General Hospital, Boston, Massachusetts

1 *IN VITRO* IDENTIFICATION OF THE β-RECEPTOR

No other pharmacological system has engendered more interest at all levels of investigation than has the β-adrenergic receptor. Historically, the sympathetic nervous system has provided a major research focus for physiologists and pharmacologists since the classic experiments of Cannon and Uridil (1) established that a chemical mediator is liberated at neuroeffector junctions. In 1948, Ahlquist (2) provided the conceptual basis for defining the actions of catecholamines with the introduction of a classification system that broadly divided their effects into α, or excitatory, and β, or inhibitory. It followed that a potency series for various agonists was developed, with isoproterenol being greater than epinephrine, which was more potent than norepinephrine as regards β effects, in contrast with the α system, where epinephrine and norepinephrine were equipotent and isoproterenol was essentially without effect. In 1958, Moran and Perkins (3) validated the system for a larger series of agonists, partial agonists, and antagonists. More recently, a subdivision of the beta system into type 1 and type 2 effects has occurred following the initial observations of Lands et al. (4). For β_1 receptors, epinephrine is equipotent with norepinephrine. For the β_2 receptor, epinephrine is far more potent than norepinephrine. Functionally, the heart is most characteristic of the β_1 receptor, whereas the liver and arterial and bronchial smooth muscles can be considered examples of β_2 receptors.

The *in vitro* identification of a binding site has depended upon and been validated by these physiological and pharmacological characterizations, which have keenly defined the functions of the β-adrenergic receptor. One could thus predict its specificity, ligand-binding properties, and even its binding constant or affinity based on these known parameters. It simply remained for acceptable radiolabeled ligands to be synthesized and for investigators to develop the necessary technology before direct binding studies could be performed successfully.

2 REQUIREMENTS FOR A BINDING SITE TO BE A RECEPTOR

One can simply list the requisites that a ligand binding site must meet for it to be accepted as the physicochemical equivalent of the functional β-adrenergic receptor. Cuatrecasas (5) and Clark (6) have previously emphasized these. A first requirement is that the properties of ligand binding, such as reversibility and affinity, correlate not only with well-defined receptor-mediated physiologic effects, but also with the activation of adenylate cyclase in the target organ. This effector enzyme has provided a powerful tool for assessing the validity of the pharmacologically defined binding site. Thus any particular agonist or antagonist must bind with an affinity to the receptor which correlates with its ability to stimulate or inhibit adenylate cyclase. The relative binding characteristics of a series of analogs, including agonists, partial agonists, and antagonists, should parallel their biologic potency as well as their capacity to stimulate this enzyme. Stereospecificity would be predicted in that the levo-isomers of β-receptor agonists and antagonists are 50- to 100-fold more potent than the dextrorotatory forms. These qualitative considerations can be more rigorously tested by applying the same quantitative approaches that have been classically employed for describing enzyme–substrate interactions, either in terms of equilibrium interactions or based on the kinetics of ligand binding. It is the aim of such an approach to demonstrate that the affinity for a receptor–ligand interaction as determined by binding experiments parallels the value determined by measuring functional parameters. Clearly, such an analysis requires that these binding sites are saturable and demonstrate appropriate dissociation and association rates. A brief consideration of this will serve as an introduction and as a means of criticizing the design of certain experimental approaches.

3 QUANTITATIVE APPROACHES TO RECEPTOR IDENTIFICATION

Each of the following derivations has been well worked out for enzyme–substrate interactions. Applications of the law of mass action to the dose–response relationship upon which the following are based can be attributed to Clark (7). More recent reviews by Cuatrecasas and Hollenberg (8) and Williams and Lefkowitz (9) have more formally applied these considerations to measurements of receptor–ligand interaction.

3.1 Occupancy Theory

It will be assumed that a single class of equivalent independent binding sites exists. Another assumption will be that the biologic effect directly correlates with the fractional occupancy of available binding sites. In a later section, it will become evident that the second assumption may be incorrect in view of the spare receptor phenomenon. This general approach, however, will still be valid.

If a ligand L combines with a receptor site R to yield a complex RL, then the biologic effect is a function of the concentration of RL. This can be written

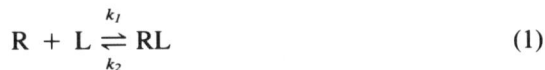

$$R + L \underset{k_2}{\overset{k_1}{\rightleftharpoons}} RL \tag{1}$$

and

$$K_D = \frac{k_2}{k_1} = \frac{[R] [L]}{[RL]} \quad \text{at equilibrium} \tag{2}$$

where K_D is the dissociation constant. Let $[R_t]$ be the total receptor concentration and $[R_t] = [R] + [RL]$. Substituting for $[R] = [R_t] - [RL]$, we obtain

$$K_D = \frac{[R_t - RL] [L]}{[RL]}$$

$$= \frac{[L]R_t}{[RL]} - [L]$$

$$K_D + [L] = \frac{[R_t] [L]}{[RL]}$$

$$\frac{[RL]}{[R_t]} = \frac{[L]}{K_D + [L]} \tag{3}$$

As stated above, any graded response Δ will be proportional to the fractional receptor occupancy. Therefore, $\Delta = \alpha[RL]$, where R represents ligand-occupied receptors and α is a proportionality constant. Therefore, any fractional response at ligand concentration [L] is

$$\frac{\Delta}{\Delta_{max}} = \frac{\alpha[RL]}{\alpha[R_t]} = \frac{[RL]}{[R_t]}$$

and substituting directly into Equation 3 yields

$$\Delta = \frac{\Delta_{max}[L]}{K_D + [L]} \tag{4}$$

This is identical to the classical Michaelis–Menten equation, which relates the velocity V to the substrate concentration [S], the K_m (substrate concentration producing one-half maximal activity) and the maximal velocity V_{max}:

$$v = \frac{V_{max}\,[S]}{K_m + [S]}$$

Both equations result in hyperbolic functions where Δ or $V = 0$, when [L] or [S] $= 0$ and Δ or V approximate Δ_{max} or V_{max} when [L] or [S] become large in relation to the K_D or K_m. At $\Delta/\Delta_{max} = \frac{1}{2}$ (the half-maximal response),

$$\frac{[L]}{K_D + [L]} = \frac{1}{2}$$

and

$$[L] = K_D \qquad \text{(as does } K_m = [S] \text{ for } V = \frac{V_{max}}{2}\text{)}$$

A direct derivation from Equation 3 includes the double-reciprocal plot of Lineweaver–Burke applied to a receptor–ligand interaction producing a biologic effect in a manner analogous to enzyme–substrate reactions. Furthermore, assessment of competitive and noncompetitive antagonists can be performed in this manner and calculation of their dissociation constants can be measured by well-defined methods (8,9). It should be obvious that the foregoing treatment can be applied in quantitatively assessing β-receptor–ligand interactions by measuring the adenylate cyclase response. However, under certain conditions, as pointed out by Goldstein et al. (10) and reemphasized by Cuatrecasas and Hollenberg (8), the derivation of the K_D will be incorrect. For example, let the fractional response

$$f = \frac{\Delta}{\Delta_{max}} = \frac{[RL]}{[R_t]} = \frac{[L]}{K_D + [L]}$$

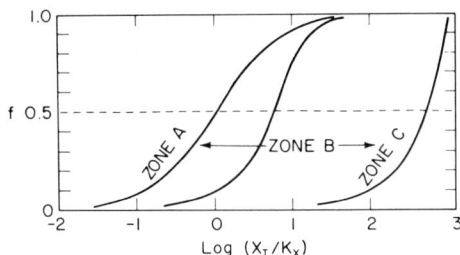

Figure 1 Theoretical log dose–response curves for the three zones of behavior. Ordinal values f are fractions of maximal response or fractional occupancy of receptors. Abscissal values are logarithms of drug concentration (dose) normalized by expressing in units of the drug–receptor dissociation constant. In zone A practically all the drug is free. In zone C practically all the drug is combined with receptor sites. In zone B neither free nor combined drug can be neglected; only a single representative curve is shown. Adapted from Strauss and Goldstein. Reprinted by permission from Ref 10.

and by rearranging, we obtain

$$f = \frac{[L]}{K_D + [L]}$$

$$[L] = K_D \left(\frac{f}{1 - f} \right)$$

and by substituting $[L_t] - [RL]$ for $[L]$ and $f[R_t]$ for $[RL]$, we obtain

$$[L_t] = K_D \left(\frac{f}{1 - f} \right) + f[R_t] \tag{5}$$

This equation has been widely applied to all varieties of ligand–receptor interactions in which the fractional response is assumed to be proportional to receptor occupancy. If $f[R_t]$ is quite small (typically if $[R_t]$ is small compared with the K_D), then even at receptor saturation, essentially all ligand molecules are free and

$$[L_t] = [L] = K_D \left(\frac{f}{1 - f} \right)$$

When these conditions are met, this has been referred to as the zone A approximation for a drug–receptor interaction based on the occupancy assumption. The log dose–response curve for zone A is given by the familiar symmetrical sigmoid, which inflects at $f = 0.5$ (with $K_D = [L_t]$). It becomes apparent, then, at quite high R_t concentrations, where essentially no drug molecules are free, that $[L_t] = f[R_t]$. However, it is the intermediate zone that can present practical problems in attempting to calculate the K_D of an agonist for the β-receptor by measuring the adenylate cyclase response, for example. Figure 1 depicts the log dose–response for each of the foregoing zones. One can see for a representative zone B, where $[R_t]/K_D$ is not $< 1/10$, that at $f = 0.5$ the log $[L_t]/K_D$ can be significantly greater than zero, and therefore the $[L_t]$ at $f = 0.5$ would be a significant overestimation of the K_D. In fact, in most situations one is working with receptor concentrations significantly less than the K_D although this problem has been encountered in measuring the true K_D for certain receptor systems (11).

It will be noted later that in the competitive radioligand displacement assay, the hormone concentration at which 50% displacement of the radioligand occurs is a valid measure of the K_D only when this condition is met. Furthermore, since Scatchard analysis of binding data is based on the same relationships, it will be valid only for zone A approximations. Figure 2 demonstrates this effect for the

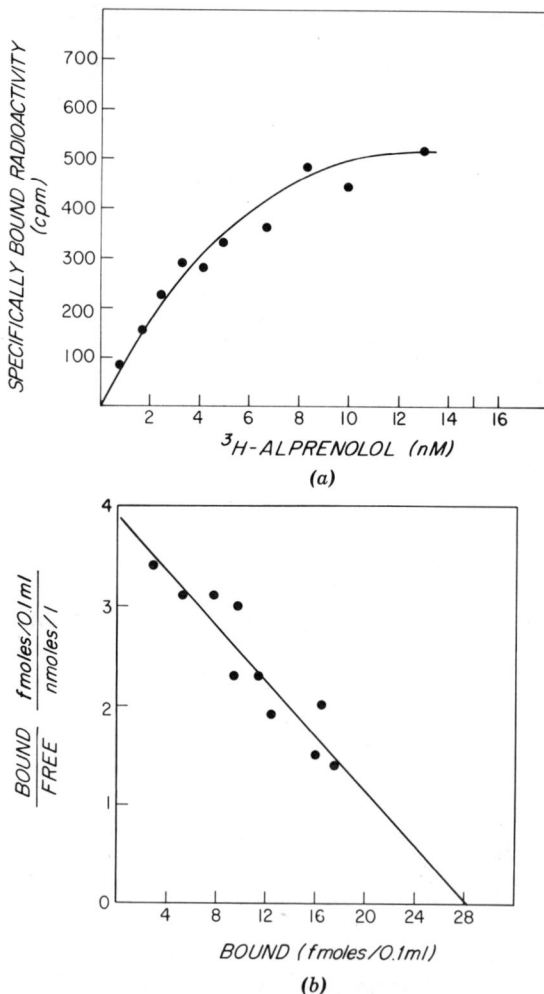

Figure 2 (*a*) Saturation assay of β-adrenergic binding sites in a particulate membrane preparation from canine myocardium. Total binding of the radioligand at each concentration of [³H]alprenolol is corrected for the nonspecific binding observed in the presence of 1 μM *l*-propranolol. (*b*) Scatchard analysis of the saturation binding data from Figure 2A. The linear transformation of the data indicates a single class of binding sites, with a K_D of 7.3 nM and a density of receptors in the preparation of 28.4 fmol/0.1 ml (the membrane protein concentration was 0.37 mg/ml).

cardiac muscarinic cholinergic receptor, which has an extremely low K_D, in the range 20–30 pM. It can be seen that the apparent K_D as determined by saturation isotherms progressively decreases as the receptor concentration is lowered.

3.2 Scatchard Analysis

In a manner analogous to the preceding example, where the K_D for a ligand was derived from the simple consideration of mass action and the occupancy principle, direct ligand binding studies can be evaluated. The similarity of the equations is obvious. One simply determines the amount of bound ligand as the hormone concentration is increased until saturation of available receptors is approached. Both the receptor concentration and K_D can then be derived.

Referring back to Equation 2, at any ligand concentration [L], a fraction [RL] of the total receptor concentration [R$_t$] will be occupied and by analogy to Equation 3,

$$\frac{[RL]}{[R_t]} = \frac{[L]}{K_D + [L]} \quad \text{and} \quad [RL] = \frac{[R_t]\,[L]}{K_D + [L]}$$

$$\frac{[RL]}{[L]} = R_t \left(\frac{1}{K_D + [L]} \right)$$

$$K_D \frac{[RL]}{[L]} + [RL] = [R_t]$$

$$\frac{]RL]}{[L]} = [R_t - RL]\,K_D^{-1}$$

$$\frac{[RL]}{[L]} = \frac{\text{bound hormone}}{\text{free hormone}} = [R_t]\,K_D^{-1}(\text{a constant}) - [RL]\,K_D^{-1} \tag{6}$$

If the bound/free (B/F) hormone ratio is plotted vs. bound hormone, then for B/F = 0, $[R_t]K_D^{-1} = [RL]K_D^{-1}$, or the X intercept equals the receptor concentration. The slope is equal to $-K_D^{-1}$. A saturation isotherm for the cardiac β-receptor utilizing [³H]alprenolol and the derived Scatchard plot is shown in Figure 2.

3.3 Kinetic Analysis of Binding Data

Finally, one may approach the determination of receptor affinity by utilizing a kinetic approach, that is, by determining the on and off rates for hormone binding. The K_D is therefore simply the ratio k_2/k_1, where k_1 is the association rate constant and k_2 the dissociation rate constant, and should agree with the K_D as determined by equilibrium methods.

It is obvious that for the interaction $R + L \rightleftharpoons RL$, the change in the concentration of RL versus time will be defined by

$$\frac{d[RL]}{dt} = k_1\,[R]\,[L] - k_2[RL] \tag{7}$$

The approach to solving this second-order differential equation can be simplified by realizing that [L] is typically $\gg$ than R_t and thus remains essentially constant, allowing us to call $k_1[L]$ a new constant k'_1 and then $d[RL]/dt = k'_1[R] - k_2[RL]$. At equilibrium, let us term ligand-occupied R as RL_{eq}. Since $[RL_{eq}]$ can be determined, let us solve for $[R]$ in the following equation:

$$[R] = [R_t] - [RL_{eq}] = [RL_{eq}] + [R_{eq}] - [RL]$$

or since the ratio $[R_{eq}][L]/[RL_{eq}] = k_2/k_1$ (when $d[RL]/dt$ is 0) or $[R_{eq}] = k_2[RL_{eq}]/k_1[L] = k_2[RL_{eq}]/k'_1$, then $[R] = [RL_{eq}] + [RL_{eq}](k_2/k_1') - [RL]$, and substituting, then $d[RL]/dt = (k'_1 + k_2)([RL_{eq}] - [RL])$ and $[RL] = [RL_{eq}](1 - e^{-(k'_1 + k_2)t})$ by simple integration. The natural logarithm of this is

$$\ln\{[RL_{eq}]/([RL_{eq}] - [RL])\} = (k'_1 + k_2)t \qquad (8)$$

This then provides an obvious method for solving for $k'_1 + k_2$. After first determining $[RL]$ at various times for a given $[L]$ until equilibrium is reached, one can plot $[RL_{eq}]/([RL_{eq}] - [RL])$ versus time. The slope of this line will be equal to $k'_1 + k_2$. k_2 can be readily determined as indicated below. Since k_2 is known, k_1 can be calculated since $k_1 = k'_1/L$.

Another approach (12) is to consider the time $(t_{1/2})$ at which $[RL]$ approaches one-half of its value at equilibrium, and by substituting into Equation 8, we have

$$\ln 2 = (k_1[L] + k_2)t_{1/2} \qquad (9)$$

and after solving $0.63/t_{1/2} = k_1[L] + k_2$. This is a simple linear equation of the form $y = ax + b$, allowing determination of $k_1 + k_2$ when appropriately plotted at various concentrations of L. An application of this derivation is given in Figure 3 for the turkey erythrocyte β-receptor.

k_2, the rate constant for dissociation of the complex RL, is readily derived if conditions are set that prevent significant reassociation of R and L. Thus $d[RL]/dt = -k_2[RL]$ and with simple integration $[RL]/[RL]_0 = e^{-k_2t}$, or $\ln([RL]/[RL]_0) = -k_2t$, where $[RL]_0$ is the concentration of bound radioligand at $t = 0$. Thus a simple plot of $\ln([RL]/[RL_0])$ vs. t would provide a slope equal to $-k_2$. This kind of experiment is typically designed so that upon initiation the complex $[RL]_0$ is infinitely diluted or a large excess of unlabeled competing ligand is added to prevent reassociation with the membrane-bound receptors; however, it is difficult to ensure that effective dilution at the membrane–solute interface, particularly with lipid-soluble ligands, has been achieved. Furthermore, addition of a large excess of ligand can give spurious results if ligand–ligand interaction is promoted (8) (as in the case of certain polypeptide hormones, for example) or if cooperativity among receptor sites exists (13). Cooperativity can be demonstrated by calculating the off rate for a labeled ligand in the presence and absence of varying concentrations of added unlabeled ligand. When the off rate changes in parallel with the fractional receptor occupancy, a cooperative interaction is suggested.

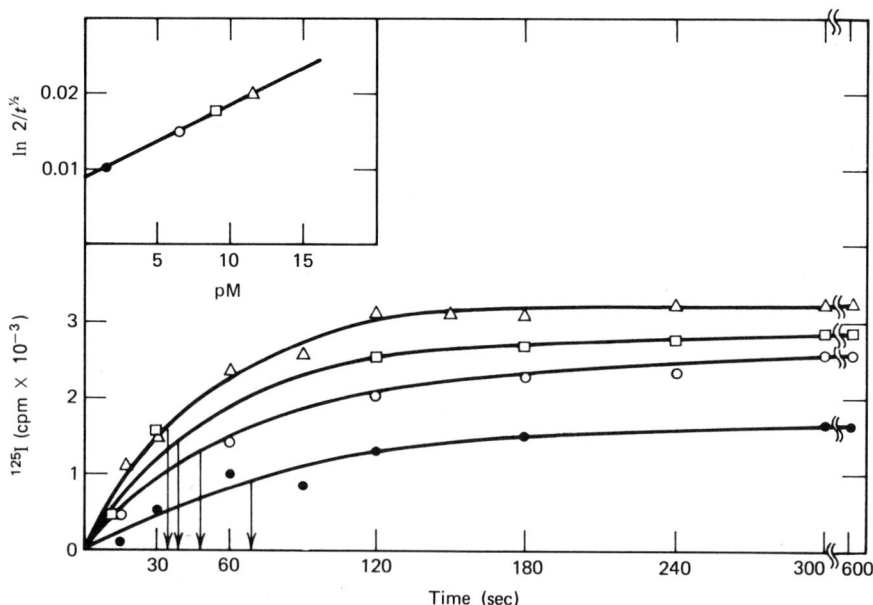

Figure 3 Binding reaction for [^{125}I]hydroxybenzylpindolol ([^{125}I]HYP) at 37°C for varying concentrations of [^{125}I]HYP. Varying concentrations of [^{125}I]HYP were added to 0.02 mg/ml of membrane protein in 0.05 M hepes, pH 7.5, and 0.001 M phenol at 37°C at the times indicated. Aliquots were rapidly passed over glass fiber filters, and ^{125}I was determined on triplicate samples as described in the text. Equilibrium was reached within 10 min in each experiment. [^{125}I]HYP bound in the presence of 10^{-7} M HYP was considered nonspecifically bound and did not change with time. Inset, plot of ln 2/ $t_{1/2}$ versus [L] for each of the curves shown below (see the text). $t_{1/2}$, time at which [RL] reaches one-half of equilibrium value. Reprinted by permission of the *Journal of Biological Chemistry* from Ref. 12, p. 1236.

3.4 Determination of Affinity Constants for Competing Ligands

From a consideration of the following conditions, two important relationships can be defined that allow one to determine the K_d for a competing ligand.

1 When an unlabeled ligand A is allowed to compete with the radiolabeled ligand L for the same receptor R and one is assumed to competitively inhibit the binding of the other, then

$$R + I \rightleftharpoons RI \quad \text{and} \quad R + L \rightleftharpoons RL$$

and the mass-action equation defining these simultaneous equilibria is

$$RL = \frac{[L][R_t]}{K_L(1 + K_A) + L} \quad \text{analogous to Equation 3} \quad (10)$$

Let us examine this relationship when the competing ligand is added at a concentration (I_{50}) which reduces the binding of the radiolabeled ligand L by R to 50% of its initial value; then

$$\frac{2[L]\ [R_t]}{K_L(1\ +\ I_{50}/K)\ +\ [L]} = \frac{[L]\ [R_t]}{(K_L)\ +\ [L]} \qquad \text{(in the absence of A)}$$

and

$$K_A\ =\ \frac{I_{50}}{1\ +\ [L]/K_D} \qquad\qquad (11)$$

This then allows one to calculate the equilibrium dissociation constant for any nonradiolabeled ligand by simply determining its I_{50}. As was shown to be the case when the fractional response was related to receptor occupancy based on equilibrium considerations, this relationship is valid only when R_t is significantly less than the K_D (i.e., $R_t < 0.1\ K_D$).

2   Equation (10) can also be employed to describe the relationship between an agonist and antagonist, as first derived by Schild (14). Let us assume now that A is a competitive antagonist. Then if for any given L,

$$\frac{\Delta}{\Delta_{\max}} = \frac{L}{K_L\ +\ L}$$

When a certain concentration of the antagonist A is added, the agonist concentration must be increased to a new value L′ to produce the same response, $\Delta/\Delta_{\max}$; then

$$\frac{[L]}{K\ +\ [L]} = \frac{\Delta}{\Delta_{\max}} = \frac{[L']}{K_L(1\ +\ [A]/K_A)\ +\ [L']}$$

and rearranging,

$$\frac{L'}{L\ -\ 1} = \frac{A}{K_A} \qquad\qquad (12)$$

Therefore, log $(L'/L\ -\ 1)\ =\ -$log $K_A\ +\ A$. Thus by simply determining the concentration ratio L′/L for several different concentrations of A, we can obtain the value K_A by simply plotting log $(L'/L\ -\ 1)$ vs. $-$log A. With such an analysis, there is no constraint on the receptor concentration. The X intercept is then $-$log K_A.

4   ASSAY TECHNIQUES

The method for determination of bound counts can influence the sensitivity and specificity of any assay system and was initially a problem in the identification of β-adrenergic receptor binding. Since labeled β-receptor antagonists are hydrophobic in nature, a fraction of their binding is simply due to dissolution within the lipid milieu of the membrane. It has been shown, for example, using nuclear magnetic resonance techniques, that naphthalene, which constitutes the ring structure of propranolol, inserts into detergent micelles quite effectively (15). It should be clear from analysis of binding isotherms that approximately 99% of the binding of any labeled ligand to its receptor when present in trace quantities ($\leq 0.1K_D$) will be inhibited when unlabeled ligand is added at a concentration at least two log units greater than the K_D. Specific binding to the β-receptor then is simply the difference between the amount of labeled ligand bound in the presence and absence of 10^{-6} M l-propranolol. Low-affinity sites for certain labeled ligands, such as hydroxy-benzylpindolol, have been detected in certain membrane preparations. These can usually be detected by careful attention to the shape of binding isotherms and the use of Scatchard analysis to ensure the unique affinity of the defined binding site. More significant difficulties have appeared when [^{3}H]agonists have been used for direct binding studies. First, nonspecific binding of such agents to membrane preparations can occur via the catechol ring, either through biological interactions such as metabolism of the label by the membrane-bound form of the enzyme catechol-O-methyltransferase (16) or by chemical oxidation of the catechol ring to the quinone and the subsequent formation of covalent linkages through Schiff base formation with available membrane proteins containing lysine residues (16,17). Second, the binding affinity of agonists is generally 10- to 100-fold less than that of antagonists (12,18,19). Thus binding of these compounds to low-affinity nonspecific sites may be more difficult to satisfactorily discriminate from actual receptor binding. Recently, several groups have appeared to surmount these problems by employing a large excess of catechol to flood nonspecific binding sites as well as ascorbic acid to protect the label from chemical oxidation. Thus Lefkowitz and Williams (20) have successfully used [^{3}H]hydroxybenzylisoproterenol to quantitate receptor binding and affinity directly. U'Pritchard et al. (21) have done the same with [^{3}H]epinephrine. As will be pointed out, the use of an agonist in direct binding studies allows one to compare binding affinity and cyclase responsiveness directly.

With regard to the actual technique for separation of bound and free ligand, certain points deserve emphasis. In the ideal situation, the method should preserve equilibrium conditions. This then immediately adds the constraints that the method be rapid and gentle. Since the β-receptor typically demonstrates a low K_D for labeled ligands, the calculated off rate for bound ligand is quite slow. Therefore, several techniques are applicable and afford rapid separation of particulate-bound from free ligand. The most extensively employed procedure involves rapid filtration of the particular preparation on any of a variety of synthetic membranes, with glass fiber filters typically providing the lowest background. Centrifugation provides an alternative rapid means of separation, although removing unbound labeled ligand involves more delay. Equilibrium dialysis allows the assay to be performed entirely under equilibrium conditions and is particularly useful for assaying relatively low-affinity binding sites, where the dissociation rate of the ligand–receptor complex may be so rapid as to preclude either filtration or centrifugation. The technique, however, requires that a relatively high concentration of receptor be present in order to detect the difference between the compartment containing bound ligand and that containing free ligand. Furthermore, it is difficult to apply this technique to large numbers of samples.

More recently, Caron and Lefkowitz (22) and Vauquelin et al. (23) have solubilized the β-receptor from frog and turkey erythrocytes, respectively. Both groups were able to successfully demonstrate receptor binding in digitonin-dispersed preparations. Other detergents either resulted in such high degrees of nonspecific binding of the radiolabel, or physically disrupted the binding interaction, that specifically bound ligand could not be detected. Caron and Lefkowitz (22) utilized small gel filtration columns to separate bound from free ligand. Although the authors reported that the results obtained by this technique were similar to those obtained by equilibrium dialysis, Levitzky (24) has pointed out that at the extremely low receptor concentration present during these experiments, less than a 10% difference could be expected to occur at saturating ligand concentrations. This small difference would be further obscured by the pool of nonspecific binding created by the distribution of the tritiated ligand in detergent micelles. Vauquelin et al. (23) precipitated the receptor–ligand complex with polyethylene glycol, as has been done with other proteins, including antibodies and the insulin receptor. This technique also provided results comparable with those obtained with gel filtration. An additional method for measuring receptor–ligand interactions in solution depends on the use of a two-phase system in which the ligand distributes equally in both but the receptor is excluded from one (24). Typically, ligand, receptor protein, and a porous bead such as Sephadex G-50, which will exclude receptor but include ligand, are incubated until equilibrium is reached. An aliquot of the supernatant not containing gel is then counted and compared with a control containing only ligand without receptor protein. The increase in counts over control is then a measure of protein-bound ligand. The difficulty with this technique is the same as that associated with equilibrium dialysis, in that a relatively high receptor concentration must be present in order to detect a difference between the two compartments. However, this technique has been applied successfully in the characterization of a partially purified and thereby concentrated preparation of canine cardiac β-receptor (25).

5   APPLICATION OF RADIOLIGAND BINDING METHODS TO RECEPTOR IDENTIFICATION: VALIDATION AND QUANTITATION

Several groups have demonstrated a close correlation between β-receptor-mediated physiologic effects and the interaction of a variety of β-active drugs with the enzyme adenylate cyclase in membrane particulate preparations. Receptor affinity as determined by Schild analysis has shown a close agreement with the biologic potency for a large series of antagonists and partial agonists in several different tissue preparations. However, this agreement has generally not been found for full agonists. Kaumann and Birnbaumer (26) reported that catecholamines were several orders of magnitude less potent in increasing adenylate cyclase activity in membrane preparations from kitten heart than in producing a given inotropic or chronotropic response. The same disparity was not observed in the case of partial agonists in that their K_D's in membrane preparations were determined on the basis of their relative antagonist properties. The apparent higher affinity elicited for the physiologic response of full agonists has been attributed to a spare-receptor phenomenon (26–28). In some systems, only a fraction of the receptor sites may need to be occupied for maximal response to be produced. Although occupancy relationships described earlier still hold, Δ_{max} is now achieved when RL < R_t. A direct consequence is that the EC_{50} or ligand concentration necessary to produce a half-maximal response for an agonist will be less than the K_D as determined by radioligand-binding techniques. In particular, in these experiments, described by Kaumann and Birnbaumer (26), receptor sites may be spare as related to the maximal biological effect, but not in relationship to adenylate cyclase stimulation, suggesting that the former may be achieved before adenosine $3',5'$-cyclic monophosphate (cyclic AMP) production is maximal. A similar disparity between adenylate cyclase stimulation in broken-cell preparations and biologic effect in the intact organ has been observed for other systems (29,30). An alternative hypothesis proposed that the receptor acts via mechanisms unrelated to cyclase activation. Recently a direct interaction between membrane-bound catechol-O-methyltransferase and the β-receptor of cardiac microsomal preparations has been demonstrated. The stimulatory effect of β-agonists on this enzyme could not be mimicked by the addition of exogenous cyclic AMP (31).

Having shown a generally close correlation between biological potency and adenylate cyclase effect for a variety of β-active drugs, the same approaches have been employed to validate direct receptor-binding studies. Levitzki et al. (32) were the first group to successfully demonstrate this relationship in the turkey erythrocyte utilizing both [^{3}H]isoproterenol and [^{3}H]propranolol. In that specific binding represented only a small fraction of total binding, it was remarkable that a meaningful apparent K_D could be generated for both compounds ($\sim 1.5 \times 10^{-7}$ M for isoproterenol and approximately 2.5×10^{-9} M for propranolol) which correlated with values generated from cyclase activation data. Saturation binding experiments performed with both ligands placed the number of receptors per cell at approximately 1000. Over the past 5 years, analogous experiments have been carried out in a variety of tissues, for which both equilibrium and kinetic data utilizing a variety of radiolabeled ligands have been accumulated. There has been excellent agreement in general, and several specific examples will be cited which elucidate these points.

Brown et al. (12,33) have characterized in great detail the receptor–cyclase system of the turkey erythrocyte employing the ligand hydroxybenzylpindolol, which can be iodinated to the extremely high specific gravity of 2000 Ci/mmol. With this ligand and by employing rapid filtration for separation of bound from free ligand, approximately 80–90% of total binding was identified as receptor-specific. They utilized saturation experiments in which increasing concentrations of hydroxybenzylpindolol were incubated under equilibrium conditions and the amount of bound ligand determined. The K_D was then calculated utilizing Scatchard analysis. They then also determined the same value based on the kinetics of receptor binding at several different ligand concentrations, as illustrated earlier (Equation 8; Figure 3). There was relatively good agreement between the two methods with a K_D in the range 10^{-10}–10^{-11} M. Furthermore, the K_D determined by competitive binding curves (Equation 10) showed close agreement with those calculated by cyclase inhibition experiments for a variety of antagonists. However, in the absence of added guanylyl nucleotides, a consistent disparity in predicted receptor affinity was found for agonists when cyclase activation data were compared with binding-inhibition experiments. From binding data, a K_D of 10^{-7} was calculated for iso-proterenol, whereas the K_{act} for cyclase was approximately ten-fold greater. Similar results were observed by Wrenn and Haber (unpublished observations) utilizing a canine cardiac microsomal preparation wherein the EC_{50} for adenylate cyclase activation of the membrane preparation by agonists was approximately 100-fold greater than the apparent K_D as determined by [^{3}H]propranolol binding. In the turkey erythrocyte, the presence of guanylyl nucleotides appears to shift the K_m for catecholamine activation of adenylate cyclase to a lower value and abolish this discrepancy (33). In the frog erythrocyte (34), the ability of either an agonist or antagonist to compete with a labeled high-affinity ligand, [^{3}H]alprenolol, in direct receptor-binding studies showed close agreement with its ability to stimulate or inhibit adenylate cyclase in the absence of added guanylyl nucleotides. However, in both the frog erythrocyte (35) and in the cultured mammalian lymphoma line S49 (36), guanylyl nucleotides can be shown to shift the agonist dose–response curve for cyclase activation to the left with an apparent decrease in K_{act}, an effect similar to that observed in the turkey erythrocyte. Paradoxically, a simultaneous decrease in receptor affinity for agonists can be demonstrated by direct binding studies in the presence of guanylyl nucleotides in both the frog erythrocyte and in the lymphoma cell line (Figure 4). In both cases, the ability of a particular nucleotide [Gpp(NH)p > Gpp(CH)p > GTP] to produce an increase in the K_D for agonist binding was paralleled by its capacity to effect a decrease in the K_{act} for cyclase. In contrast, neither Brown et al. (33) nor Tolkovsky and Levitzki (37) could detect a guanylyl nucleotide-induced alteration in receptor affinity when determined by direct binding studies in the turkey erythrocyte. Antagonist binding has been shown by all groups to be unaltered in the presence of guanylyl nucleotides. The reason for the discrepancies among these different systems is unclear. They may result from the unique characteristics of the guanoside triphosphate (GTP)-regulatory subunit in a particular cell type. Clarification of these findings should be forthcoming with a better understanding of the relationship among receptor, catalytic unit, and the guanylyl nucleotide-binding subunit. It appears however, that guanylyl nucleo-tides not only mediate receptor–cyclase coupling but directly alter receptor–agonist interaction. These findings suggest that the GTP–regulatory subunit influences the isomerization of the agonist-occupied receptor.

More recently, direct radiolabeled agonist binding studies have been carried out utilizing [^{3}H]epinephrine (20) and a new ligand, [^{3}H]hydroxybenzylisoproterenol (21), which has an approximately tenfold higher affinity for frog erythrocyte β-receptor than does the parent compound, isoproterenol. As noted earlier, the use of excess catechol to block nonreceptor sites with an affinity for the ring structure of the agonist molecule and, just as important, the development of a greater so-phistication with binding studies of this kind have allowed investigators to succeed

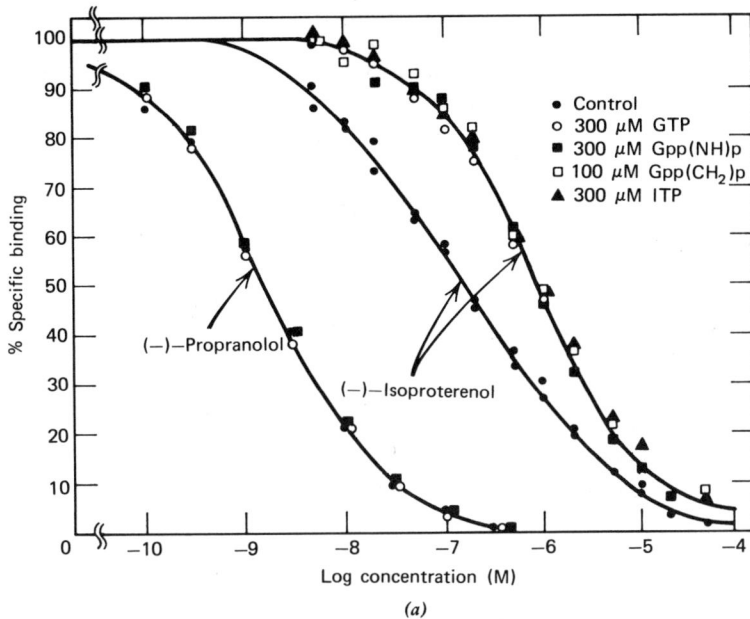

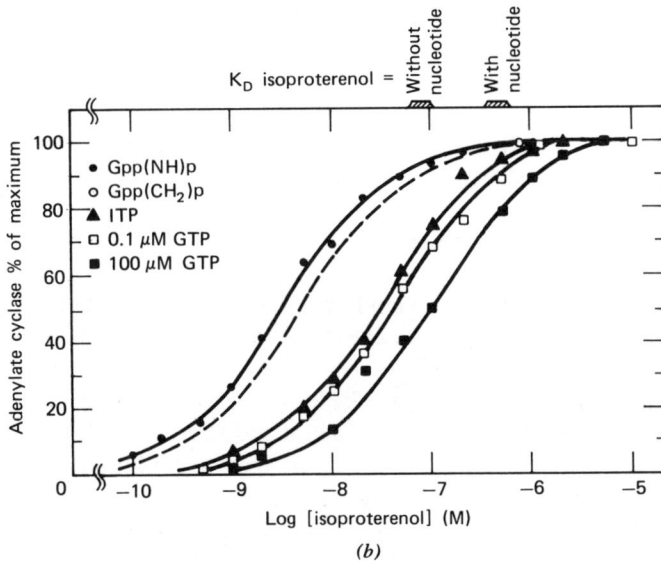

Figure 4 (a) Effects of purine nucleotides on (-)-propranolol and (-)-isoproterenol binding to plasma membrane receptors. Specific binding of each ligand was assayed by competition for [^{125}I]iodohydroxybenzylpindolol ([^{125}I]IHYP) binding sites as described in the text. Data are shown from several experiments in which protein concentrations were approximately 0.2 mg/ml; [^{125}I]IHYP concentrations were 130 pM for experiments with propranolol or 60–75 pM for experiments with isoproterenol. The incubation time was 30 min. Reprinted by permission of the *Journal of Biological Chemistry* from Ref. 36, p. 5766. (b) Effect of purine nucleotides on the activation of adenylate cyclase by (-)-isoproterenol. Enzyme activity was assayed over the range of isoproterenol concentrations shown in the presence of 50 μM Gpp(NH)p (●), 100 μM Gpp(CH$_2$)p (○), 100 μM ITP (▲), 0.1 μM GTP (□), or 100 μM GTP (■). Data are expressed as a percent of maximal stimulation by the agonist during a 20-min assay. The range of calculated values of the K_D for isoproterenol in the presence of nucleotides (Table 2) is shown for comparison. Reprinted by permission of the *Journal of Biological Chemistry* from Ref. 36, p. 5769.

where earlier attempts had failed. The availability of these ligands has allowed direct comparison of agonist receptor binding and adenylate cyclase stimulation. The kinetics of agonist binding were extensively investigated by Williams and Lefkowitz (38). In a manner analogous to that shown for glucagon binding to liver membranes by Lin et al. (39), these investigators have noted that there is an almost irreversible (extremely slow off rate) binding of agonist to β-receptors when guanylyl nucleotides are excluded from the incubation. Their addition promoted rapid dissociation. As had been observed in competitive inhibition studies (35), the addition of guanylyl nucleotides directly and specifically reduced the affinity of the receptor for agonists.

A more quantitative approach for characterizing the receptor-GTP subunit interaction has recently been introduced by Stadel, DeLean and Lefkowitz (39a). This technique utilizes a computer program to model the competitive binding curves for agonist and antagonist ligands. Agonist binding curves, in the absence of guanylyl nucleotides, are best explained by two binding states of different affinities in contrast to antagonist curves which are well described by a single state of receptor binding. However, the presence of guanylyl nucleotides uniquely alters the competitive binding curves for agonist ligands so that they can be modeled by a single low affinity class of receptors suggesting that the nucleotide mediates a transition between low and high affinity receptor states.

6 MOLECULAR CHARACTERIZATION OF THE β-RECEPTOR-ADENYLATE CYCLASE SYSTEM

6.1 Solubilization of the Receptor

The recent ability to detect receptor binding in detergent preparations has initiated attempts to probe the structure and character of the receptor molecule.

Caron and Lefkowitz (22) initially described a solubilization procedure employing digitonin, a sterol nonionic detergent, for turkey erythrocyte membranes which preserved receptor activity. Thereafter, Limbird and Lefkowitz (40) reported physical separation of the receptor and cyclase on an agarose gel filtration column using direct binding techniques to identify the elution profile of the receptor and thus demonstrated that the cyclase and receptor were unique molecules. More recently, the availability of direct binding techniques has enabled biophysical studies of the molecule to proceed. First, these groups have reported that the soluble form of the receptor has binding properties almost identical to that for the particulate form in terms of ligand specificity with an affinity in the nanomolar range. Preliminary gel filtration and equilibrium sedimentation experiments using radioligand binding to identify receptor elution patterns have estimated the molecular weight of the receptor to be in the range of 130,000 (41). More recently, Limbird and Lefkowitz (42) detected a shift in the apparent Stokes radius of the receptor to a larger form if agonist occupancy is accomplished prior to solubilization. The implication of these observations is unclear, although it has been postulated that this may be the physical parallel of receptor–GTP regulatory subunit coupling in the membrane during agonist occupancy of the receptor.

6.2   Use of Irreversible Inhibitors—Affinity and Photoaffinity Labeling

Affinity labeling techniques for the characterization and isolation of cellular receptors have met with varied success. This probably relates to the large extent to which other membrane proteins are labeled nonspecifically and to the extremely low concentration at which receptors are typically found. Thus it is essential that the chemical modification of a ligand which renders it capable of covalent attachment not significantly alter its affinity or specificity. Atlas et al. (43) have introduced a bromoacetyl derivative of propranolol which reacts with any available nucleophile to form a covalent linkage. Exposure of membrane preparations to this compound resulted in a decrease in the total receptor number as calculated from subsequent [I^{125}]hydroxybenzylpindolol binding. They were then able to show that the initial velocity of adenylate cyclase activation upon incubation with any given concentration of agonist was reduced but that peak activity was eventually reached (37). As Nickerson (44) had shown many years previously for the histamine receptor in guinea pig ileal smooth muscle, the dose–response curve for an agonist is shifted to the right, but the maximal response is still attainable following irreversible blockade of a large fraction of available receptors. It was calculated from dose–response curves that occupancy of 1% of available receptors was adequate to effect a maximal response. Thus there appear to be spare receptors in the sense that with time all available effector units can be recruited despite depletion of the total receptor number. Tolkovsky and Levitzki (37) suggested certain models for β-receptor–cyclase interactions based upon these observations—in particular, the collision-coupling model—as best explaining the experimental results. This states that the receptor and the cyclase molecule are functionally uncoupled in the membrane but form a short-lived complex during agonist occupancy leading to an activated state of the enzyme.

Finally, by introducing tritium into their affinity label, it was possible to track labeled receptors following solubilization and SDS gel electrophoresis through the use of autoradiography. Binding principally to a 40,000-molecular weight peptide was identified which could be prevented if the initial incubation was carried out in the presence of saturating concentrations of hydroxybenzylpindolol (45).

More recently, Daifler and Marinetti (46) have introduced a photoaffinity label that was shown to have a moderately high affinity for the receptor. It is hypothesized that this label would provide more specificity since removal of unbound material could be accomplished prior to covalent attachment during photolysis.

Finally, Wrenn and Homcy (46a) have recently introduced a new photoaffinity label for the beta adrenergic receptor wherein the chemically active group is not linked to the ethanolamine side chain, but to the primary ring structure of the molecule. Prior to photolysis, this derivative of acebutolol behaved as a competitive antagonist of isoproterenol-stimulated adenylate cyclase in the rat reticulocyte. Following light activation, however, labeling of the beta adrenergic receptor was shown to be irreversible. This labeling could be prevented stereospecifically with l-propranolol but not d-propranolol. There was no inhibition either of fluoride- or guanylyl nucleotide-mediated cyclase stimulation. Furthermore, no effect was observed on the glucagon-mediated stimulation of adenylate cyclase. Further experiments utilizing this ligand have demonstrated in the rat reticulocyte, at least, that spare receptors do not exist (46b). With increasing inactivation of the receptor pool, there is progressive loss in maximal cyclase stimulation, but no alteration in the apparent K_D for this effect. This observation is consistent with a model in which there is a close coupling of receptor and catalytic subunits. In contast, Venter has shown that there is evidence for spare receptors in myocardial cells. Utilizing the affinity label first introduced by Atlas and Levitski, he was able to demonstrate that maximal cyclase stimulation could still be achieved following inactivation of a large percentage of the receptor pool (46c). In conclusion, these data suggest that there may be significant differences in the way that a tissue responds to catecholamines based both on the mechanism of receptor-cyclase coupling as well as the stoichiometry of the interaction.

6.3   Cell Culture Systems, Fusion Experiments, and Genetic Reconstitution

The recent explosion in cell biology has provided unique opportunities to study the relationship of the individual components of the receptor–cyclase system. The use of cultured cell lines has shown that independent mutations can alter receptor binding without affecting maximal cyclase activity. Furthermore, genetic reconstitution experiments have demonstrated that such defects are complementary, indicating that the receptor and cyclase are distinct gene products. Several groups (47,48) have shown by direct binding studies that mutant cell lines can lose receptor sites but maintain maximal fluoride or GppNHp-stimulated cyclase activity, or, conversely, receptors can be identified in clones without appreciable cyclase activity. Furthermore, these investigators have solubilized components from mouse L cells which possess cyclase but lack receptor and have reconstituted a mutant S49 lymphoma line which possessed receptor but lacked cyclase (49). A completely active receptor–cyclase system was recovered. Similar studies have demonstrated the unique molecular identity of the guanylyl nucleotide-binding subunit as distinct from either the receptor or the catalytic unit of the cyclase (50,51).

In a parallel approach, Orly et al. (52,53) successfully fused turkey erythrocytes (whose cyclase activity had previously been destroyed by sulfhydryl modification with N-ethylmaleimide) with erythroleukemia cells which possessed cyclase activity but lacked receptor sites, as defined by [I^{125}]hydroxybenzylpindolol-binding studies. The resultant hybrids were replete with a catecholamine-sensitive adenylate cyclase, suggesting that the two proteins were distinct entities. Furthermore, as has been alluded to previously, it suggested that the receptor and cyclase molecules can interact freely within the lipid milieu of the cell membrane in a manner analogous to the floating receptor model proposed by Cuatrecasas (5).

6.4 Chemical Modification of the Receptor Molecule

The protein character of the receptor is indicated by its sensitivity to proteolytic enzymes as well as to agents that alkylate specific amino acid side chains (54). Definition of key chemical structures within the active site of the receptor molecule has been carried out utilizing a variety of reagents that are relatively substrate specific. Limbird and Lefkowitz (54) have shown that alkylation of free sulfhydryl groups has no effect on alprenolol binding. Vauquelin et al. (55) have shown recently that reduction of a presumed disulfide linkage within the active site of the receptor with dithiothreitol significantly reduces alprenolol binding in the turkey erythrocyte. This occurred both in the particulate and solubilized preparation and equally affected agonist and antagonist binding. No relationship to guanylyl nucleotide binding was observed. Alterations in the lipid milieu by treatment with phospholipases (54) have been shown to reduce the number of available receptor sites and to produce a parallel decrease in maximal catecholamine-stimulated adenylate cyclase activity.

Finally, the role of divalent cations on ligand binding has also been approached using direct binding experiments. It had previously been shown that Ca^{2+} will allosterically inhibit catecholamine stimulation of adenylate cyclase either by specific interaction with Ca^{2+} binding sites (56) or by competition for specific Mg^{2+} binding sites (57). The locus through which these cations exert their effect has been thought to be primarily on the cyclase molecule. More recently, Williams et al. (58) have shown that Mg^{2+} specifically increases the affinity of the frog erythrocyte β-receptor for agonists but not antagonists. This effect was not observed in solubilized preparations, leading the authors to suggest that the increase in affinity requires functional coupling of the receptors to the cyclase.

6.5 Immunologic Probes of Adrenergic Receptor Binding

The development of specific antibodies to hormone receptors has provided a great deal of information regarding the kinetics of ligand binding, as well as about those steps distal to binding that are critical in the activation process. Thus studies utilizing antibodies to the insulin receptor have suggested that cross-linking of receptors may be important for activation by an agonist. This was based on the observation that intact antibody molecules could mimic hormone effects, such as stimulating glucose uptake, whereas Fab fragments, still capable of binding to the receptor, were without biological effect (59).

Only one group has thus far successfully raised anti-β-receptor antibodies. Wrenn and Haber (25) have reported the harvesting of an antiserum that inhibited stereospecific propranolol binding following immunization of rabbits with an affinity-purified receptor preparation from canine cardiac microsomes (Figure 5). This component that was present in the IgG fraction also inhibited isoproterenol-mediated adenylate cyclase stimulation. Significant antireceptor activity was observed at a dilution of antiserum of approximately 1 : 10,000. Furthermore, there appeared to

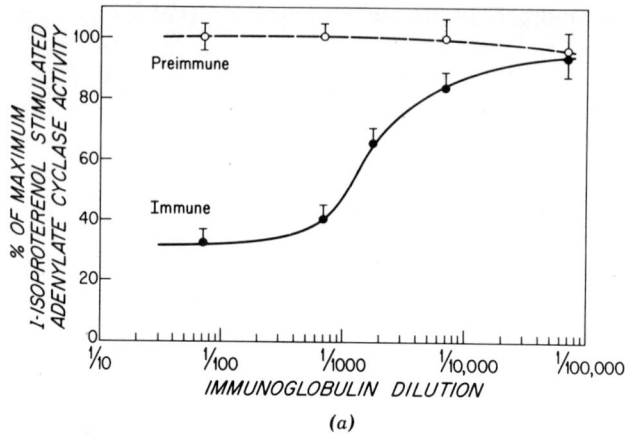

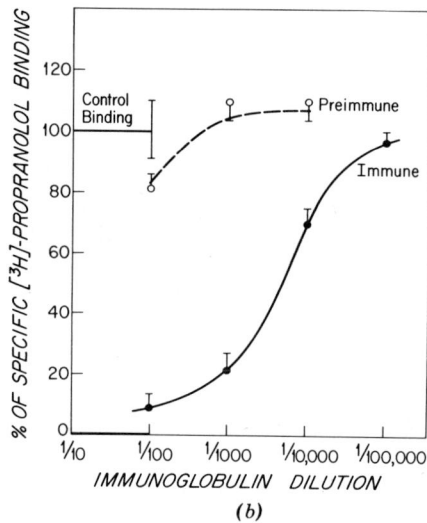

Figure 5 *(a)* Effect of globulin fractions on isoproterenol-stimulated adenylate cyclase activity. Membranes were added to the adenylate cyclase assay mixture containing 1 μM (-)-isoproterenol and globulin fractions from either immune or preimmune sera and incubated for 10 min at 37°C. Each point is the mean of triplicate determinations; bars indicate standard deviations. *(b)* Effect of immune and preimmune globulin fraction I on [³H]propranolol binding to cardiac membranes. Specific binding is defined as the fraction of total binding displaceable by 10 μM (-)-isoproterenol. Immunoglobulin dilution refers to its concentration in serum (320 μg/ml). Specific binding as defined in legend of Figure 1. Reprinted by permission of the *Journal of Biological Chemistry* from Ref. 25, p. 6579.

be a high degree of specificity, in that the antiserum did not inhibit guanylyl nucleotide-mediated adenylate cyclase stimulation or basal enzymatic activity. Furthermore, ouabain binding and membrane-bound catechol-*O*-methyltransferase activity were not affected. The antireceptor antibody did not block glucagon-mediated cyclase stimulation in the same cardiac microsomal preparation. Finally, structural differences between β_1- and β_2-receptors were apparently recognized by the immune sera in that β_2-receptor-mediated adenylate cyclase activity in liver membranes was not altered to the same extent as was β_1-linked activity in the heart. Thus this approach was able to define the unique molecular identity of the receptor as separate from the cyclase as well as to suggest that there is an actual structural heterogeneity among receptor subtypes.

6.6   Direct Binding Studies in Understanding Receptor Regulation

Homologous hormone regulation appears, in part, to be controlled by β-receptor density. In most cells, the effects of catecholamines are transient, with a rapid decline in intracellular cyclic AMP despite continued receptor occupancy by an agonist. Several investigators (59–61), either following *in vivo* catecholamine exposure or by employing intact cell systems *in vitro,* have shown that this decrease in adenylate cyclase responsiveness is accompanied by a loss of β-adrenergic receptors as assessed by labeled antagonist binding. Wessels et al. (62) have shown that in the intact frog erythrocyte the rate of loss of binding sites for [^{3}H]agonist binding correlated well with the rate of loss of adenylate cyclase responsiveness, but neither correlated well with the loss of [^{3}H]antagonist binding, suggesting that the desensitization process possibly involves a conformational change in the receptor, altering its binding specificity. More recently, it has been shown that in the intact astrocytoma cell, the initial rapid loss of cyclase responsiveness occurs prior to significant depletion of receptor sites (63,64). These investigators concluded that the desensitization process involves an initial uncoupling of receptor-linked cyclase, leading eventually to the loss of receptor number. Recent studies using variant cultured cell lines that possess receptor binding sites and cyclase catalytic activity but are functionally uncoupled, in that catecholamines do not stimulate the enzyme, have provided further insight into the problem. Shear et al. (65) have shown that exposure to agonists in this type of cell does not result in decreased receptor number, indicating that functional coupling is required for homologous hormonal desensitization to occur.

6.7   Other Mechanisms Regulating Receptor Activity

Numerous mechanisms regulating β-receptor number have now been cataloged.

Heterologous hormonal control has been best characterized for the hormone thyroxine. (This topic is discussed in more detail in Chapter 10.) The hearts of hyperthyroid rats (66,67) have a twofold greater concentration of β-receptors than do their euthyroid counterparts without a change in receptor affinity. It has been difficult, however, to demonstrate that isoproterenol-stimulated adenylate cyclase activity is altered in tissue from hyperthyroid animals. This, together with the confusing body of literature concerning the *in vivo* effects of thyroid hormone on heart action, has made the significance of increased receptor number less clear. Furthermore, thyroxine can be shown to directly affect other membrane proteins; in particular, an increase in renal cortical adenosine triphosphatase (ATPase) occurs following thyroxine administration (68).

In contrast to these results, adrenalectomy has been shown to lead to an increase in β-receptor number as well as in adenylate cyclase responsiveness to isoproterenol in rat liver (69). These effects can be reversed by the *in vivo* administration of cortisone. This regulatory mechanism may play an important physiologic role in mediating the effects of cortisol on intermediary metabolism in the liver.

More recently, Watanabe et al. (70) have suggested that a direct interaction between muscarinic and adrenergic innervation may occur in the heart. They have shown that cholinergic activation may directly alter β-receptor affinity in cardiac tissue without affecting receptor number, through the modulation of the effects of guanine nucleotides.

The β-receptor–adenylate cyclase system of the rat pineal gland (71,72) has also been well characterized. In this system, a diurnal variation coincident with exposure to light occurs. Thus at 6 p.m., when afferent sympathetic activity is minimal, receptor number is greatest, and adenylate cyclase responsiveness is maximal; the converse is true at 6 A.M. This appears to be analogous to the phenomenon of downregulation, wherein receptor–cyclase activity varies inversely with the ambient level of sympathetic activity.

Finally, the phenomenon of chemical denervation and resulting supersensitivity can be shown to have a biochemical counterpart. Thus, after treatment with either 6-hydroxydopamine or guanethidine, an increase in receptor number in brain (73) and heart (74) could be demonstrated by direct *in vitro* binding studies. This correlates with an increase in cyclic AMP generation upon exposure to catecholamines.

6.8   New Approaches for Studying Receptor Activity

A major goal for this area in the future will be the purification and isolation of the individual components of the receptor–cyclase system. High degrees of purification using classical affinity methods have already been achieved (23,25,75,76). This area deserves major research emphasis, as eventual reconstitution experiments employing each of the purified components of the system should provide information about transmembrane hormonal signaling not available with our present technology. Lipid requirements and the role of lipid–protein interactions in these events can be defined only through the use of artificial membrane systems.

In addition to understanding the mechanism through which a receptor activates its effector arm, defining why a particular ligand is recognized as agonist, partial agonist, or antagonist remains a key question. Clearly, a purified receptor–cyclase system would provide the ideal system for investigating the mode of agonist-induced receptor isomerization and the molecular mechanism underlying structure–function relationships. Since the receptor protein is present in such minute quantity and thus does not lend itself to such studies, investigators have begun to employ model systems. Rockson et al. (77) have shown that a subset of antibodies raised to various β-ligands have a specificity and affinity for both agonists and antagonists which mirror receptor binding. Furthermore, an affinity fractionation technique employing *l*-propranolol elution resulted in the separation of an antibody class which recognized the levoisomer of propranolol with a 200-fold higher affinity than did the dextroform. The availability of such binding sites in relatively large quantity will allow a variety of physicochemical approaches to be used to define the determinants of binding specificity. The development of cell fusion technology should allow monoclonal antibodies to be raised with any of a desired spectra of ligand affinities. The potential for understanding binding-site specificity through the use of primary structure determination and x-ray crystallographic techniques becomes even greater.

REFERENCES

1 W. B. Cannon and J. E. Uridil, *Am. J. Physiol.*, **58**, 353 (1921).

2 R. P. Ahlquist, *Am. J. Physiol.*, **153**, 586 (1948).

3 N. C. Moran and M. E. Perkins, *J. Pharmacol. Exp. Therap.*, **124**, 223 (1958).

4 A. M. Lands, A. Arnold, J. P. McAuliff, F. P. Luduena, and T. G. Brown, Jr., *Nature (Lond.)*, **214**, 597 (1967).

5 P. Cuatrecasas, *Biochem. Pharmacol.*, **23**, 2353 (1974).

6 E. Haber and S. Wrenn, *Physiol. Rev.*, **56**, 317 (1976).

7 A. J. Clark, General Pharmacology, in *Handbuch der Experimentellen Pharmakologie*, Vol. 4, Springer-Verlag, Berlin, 1937.

8 P. Cuatrecasas and M. D. Hollenberg, *Adv. Protein Chem.*, **30**, 252 (1976).

9 L. T. Williams and R. J. Lefkowitz, in *Receptor Binding Studies in Adrenergic Pharmacology*, Raven Press, New York, 1978, Chap. 4.

10 A. Goldstein, L. Aronow, and S. M. Kalman, *Principles of Drug Action: The Basis of Pharmacology*, Harper & Row, New York, 1968, Chap. 1.

11 J. Z. Fields, W. R. Roeske, E. Morkin, and H. I. Yamamura, *J. Biol. Chem.*, **253**, 3251 (1978).

12 E. M. Brown, D. Hauser, F. Troxler, and G. D. Aurbach, *J. Biol. Chem.*, **251**, 1232 (1976).

13 P. DeMeyts, A. R. Bramo, and J. Roth, *J. Biol. Chem.*, **251**, 1877 (1976).

14 H. O. Schild, *Br. J. Pharmacol.*, **4**, 277 (1949).

15 F. M. Menger, R. U. Rhee, and L. Mandell, *J. Chem. Soc. Chem. Commun.*, **23**, 918 (1973).

16 P. Cuatrecasas, G. P. E. Tell, V. Sica, I. Parikh, and K. J. Chang, *Nature (Lond.)*, **247**, 92 (1974).

17 B. B. Wolfe, J. A. Zirrolli, and P. B. Molinoff, *Mol. Pharmacol.*, **10**, 582 (1974).

18 D. Atlas, M. L. Steer, and A. Levitzki, *Proc. Natl. Acad. Sci. U.S.A.*, **71**, 4246 (1974).

19 C. Mukherjee, M. G. Caron, M. Coverstone, and R. J. Lefkowitz, *J. Biol. Chem.*, **250**, 4869 (1974).

20 R. J. Lefkowitz and L. T. Williams, *Proc. Natl. Acad. Sci. U.S.A.*, **74**, 515 (1977).

21 D. C. U'Pritchard, D. B. Bylund, and S. H. Snyder, *J. Biol. Chem.*, **253**, 5090 (1978).

22 M. G. Caron and R. J. Lefkowitz, *J. Biol. Chem.*, **251**, 2374 (1976).

23 G. Vauquelin, P. Geynet, J. Hanoune, and A. D. Strosberg, *Proc. Natl. Acad. Sci. U.S.A.*, **74**, 3710 (1977).

24 A. Levitzki, in P. Cuatrecasas and M. F. Greaves, Eds., *Receptors and Recognition*, Ser. A, Vol. 2. Chapman & Hall, London, 1976.

25 S. Wrenn and E. Haber, *J. Biol. Chem.*, **254**, 6577 (1979).

26 A. J. Kaumann and L. Birnbaumer, *J. Biol. Chem.*, **249**, 7874 (1974).

27 R. P. Stephenson, *Br. J. Pharmacol.*, **11**, 379 (1956).

28 M. Nickerson and N. K. Hollenberg, in W. S. Root and E. G. Hofman, Eds., *Physiological Pharmacology*, Vol. 4, Academic Press, New York, 1967, pp. 129–178.

29 F. Murad, Y. M. Chi, T. W. Rall, and E. W. Sutherland, *J. Biol. Chem.*, **237**, 1233 (1962).

30 J. E. Birnbaum, P. W. Abel, G. I. Amidon, and C. K. Buckner, *J. Pharmacol. Exp. Ther.*, **194**, 396 (1975).

31 S. Wrenn, C. Homcy, and E. Haber, *J. Biol. Chem.*, **254**, 5708 (1979).

32 A. N. Levitzki, N. Selville, D. Atlas, and M. L. Stern, *Proc. Natl. Acad. Sci. U.S.A.*, **71**, 2773 (1974).

33 E. M. Brown, S. A. Fedak, C. J. Woodard, G. D. Aurbach, and D. Rodbard, *J. Biol. Chem.*, **251**, 1239 (1976).

34 C. Mukherjee, M. G. Caron, D. Mullikin, and R. J. Lefkowitz, *Mol. Pharmacol.*, **12**, 16 (1976).

35 R. J. Lefkowitz, D. Mullikin, and M. G. Caron, *J. Biol. Chem.*, **251**, 4686 (1976).

36 E. M. Ross, M. E. Maguire, T. W. Sturgill, R. L. Biltoven, and A. G. Gilman, *J. Biol. Chem.*, **252**, 5761 (1977).

37 A. M. Tolkovsky and A. Levitzki, *Biochemistry*, **17**, 3795 (1978).

38 L. T. Williams and R. V. Lefkowitz, *J. Biol. Chem.*, **252**, 7207 (1977).

39 M. C. Lin, S. Nicosie, P. M. Lad, and M. Rodbell, *J. Biol. Chem.*, **252**, 2790 (1977).

39a J. M. Stadel, A. DeLean, and R. J. Lefkowitz, *J. Biol. Chem.*, **255**, 1436 (1980).

40 L. E. Limbird and R. J. Lefkowitz, *J. Biol. Chem.*, **252**, 799 (1977).

41 T. Haga, K. Haka, and A. G. Gilman, *J. Biol. Chem.*, **252**, 5776 (1977).

42 L. E. Limbird and R. J. Lefkowitz, *Proc. Natl. Acad. Sci. U.S.A.*, **75**, 228 (1978).

43 D. Atlas, M. L. Steer, and A. Levitzki, *Proc. Natl. Acad. Sci. U.S.A.*, **73**, 1921 (1976).

44 M. Nickerson, *Nature (Lond.)*, **178**, 697 (1956).

45 D. Atlas and A. Levitzki, *Nature (Lond.)*, **272**, 370 (1978).

46 F. J. Daifler and G. V. Marinetti, *Biochem. Biophys. Res. Commun.*, **79**, 1 (1977).

46a S. M. Wrenn, Jr., and C. J. Homcy, *Proc. Natl. Acad. Sci.*, **77**, 4449 (1980).

46b S. M. Wrenn, Jr., and C. J. Homcy, *Trans. Assoc. Am. Phys.*, (in press).

46c J. C. Venter, *Mol. Pharmacol.*, **16**, 429 (1979).

47 P. A. Insel, M. E. Maguire, A. G. Gilman, P. Coffins, H. Bourne, and K. Melman, *Mol. Pharmacol.*, **12**, 1062 (1976).

48 L. L. Brunton, M. E. Maguire, H. J. Anderson, and A. G. Gilman, *J. Biol. Chem.*, **252**, 1293 (1977).

49 E. M. Ross and A. G. Gilman, *Proc. Natl. Acad. Sci. U.S.A.*, **74**, 3715 (1977).

50 E. M. Ross and A. G. Gilman, *J. Biol. Chem.*, **252**, 6966 (1977).

51 E. M. Ross. A. C. Howlett, K. M. Ferguson, and A. G. Gilman, *J. Biol. Chem.*, **253**, 6401 (1978).

52 J. Orly and M. Schramm, *Proc. Natl. Acad. Sci. U.S.A.*, **73**, 4410 (1976).

53 M. Schramm, J. Orly, S. Eimerl, and M. Korner, *Nature (Lond.)*, **268**, 310 (1977).

54 L. E. Limbird and R. J. Lefkowitz, *Mol. Pharmacol.*, **12**, 559 (1976).

55 G. Vauquelin, S. Bottan, L. Kanarek, and A. D. Strosberg, *J. Biol. Chem.*, **254**, 4462 (1979).

56 M. L. Steer and A. Levitzki, *J. Biol. Chem.*, **250**, 2080 (1975).

57 G. I. Drummond and L. Duncan, *J. Biol. Chem.*, **245**, 976 (1970).

58 L. T. Williams, D. Mullikin, and R. J. Lefkowitz, *J. Biol. Chem.*, **253**, 2984 (1978).

59 E. Remold-O'Donnell, *J. Biol. Chem.*, **249**, 3615 (1974).

60 C. Mukherjee, M. G. Caron, and R. J. Lefkowitz, *Proc. Natl. Acad. Sci. U.S.A.*, **72**, 1945 (1975).

61 J. V. Mickey, R. Tate, and R. J. Lefkowitz, *J. Biol. Chem.*, **250**, 5727 (1976).

62 M. R. Wessels, D. Mullikin, and R. J. Lefkowitz, *J. Biol. Chem.*, **253**, 3371 (1978).

63 G. L. Johnson, B. B. Wolfe, T. K. Hardin, P. B. Molinoff, and J. P. Perkins, *J. Biol. Chem.*, **253**, 1472 (1978).

64 Y-F. See, T. K. Hardin, and J. P. Perkins, *J. Biol. Chem.*, **254**, 38 (1979).

65 M. Shear, P. A. Insel, K. L. Melman, and P. Coffins, *J. Biol. Chem.*, **251**, 7572 (1976).

66 T. Ciaraldi and G. V. Marinetti, *Biochem. Biophys. Res. Commun.*, **74**, 984 (1977).

67 L. T. Williams, R. J. Lefkowitz, D. R. Hathaway, A. M. Watanabe, and H. R. Besch, *J. Biol. Chem.*, **252**, 2787 (1977).

68 C. S. Lo, T. R. August, U. A. Liberman, and I. S. Edelman, *J. Biol. Chem.*, **251**, 7826 (1977).

69 B. B. Wolfe, T. K. Harden, and P. B. Molinoff, *Proc. Natl. Acad. Sci. U.S.A.*, **73**, 1343 (1976).

70 A. W. Watanabe, M. M. McConnaghey, R. A. Strawbridge, J. W. Fleming, L. R. Jones, and H. R. Besch, Jr., *J. Biol. Chem.*, **253**, 4833 (1978).

71 J. A. Romero, M. Zatz, J. W. Kelabran, and J. Axelrod, *Nature (Lond.)*, **258**, 435 (1975).

72 J. W. Kalabran, M. Zatz, J. A. Romero, and J. Axelrod, *Proc. Natl. Acad. Sci. U.S.A.*, **72**, 3735 (1975).

73 J. R. Sporn, T. K. Harden, B. B. Wolfe, and P. B. Molinoff, *Science*, **194**, 624 (1976).

74 G. Glaubriger, B. S. Tsai, R. J. Lefkowitz, E. M. Johnson, and B. Weiss, *Nature (Lond.)*, **273**, 240 (1978).

75 G. Vauquelin, P. Geyneyt, J. Hanoune, and A. D. Strosberg, *Eur. J. Biochem.*, **98**, 543 (1979).

76 M. G. Caron, Y. Srinivasan, J. Pitha, K. Kociolok, and R. J. Lefkowitz, *J. Biol. Chem.*, **254**, 2923 (1979).

77 S. Rockson, C. Homcy, and E. Haber, *Clin. Res.*, **27**, 441A (1979).

CHAPTER NINE

(Na$^+$,K$^+$)ATPase AND ADRENOCEPTORS

John W. Phillis and Peter H. Wu

Department of Physiology, College of Medicine, University of Saskatchewan, Saskatoon, Canada

1   INTRODUCTION

From an evolutionary point of view, one of the principal functions of cell membranes was probably the maintenance of appropriate sodium and potassium concentrations within the cell body. Among the first enzymatic reactions to be associated with the cell membrane must have been the various membrane pumps which transport sodium, potassium, and calcium ions across the membrane against their electrochemical gradients. The concept of sodium pumping as a means of removing sodium from the cell against its concentration gradient is a relatively recent development. A sodium pump in muscle membrane was hypothesized by Dean (1) when he realized that if skeletal muscle membrane was permeable to sodium, there must be an active sodium-excreting mechanism to control intracellular sodium levels. Later studies by Schatzmann (2) clearly showed that the sodium and potassium fluxes across red cell membranes were adenosine triphosphate (ATP)-dependent and that the "pump" mechanism could be inhibited by cardiac glycosides. Skou (3) extended this finding by showing that in crab nerve, (Na$^+$,K$^+$)ATPase activity was closely associated with sodium and potassium ion fluxes and that the activity of the enzyme was inhibited by cardiac glycosides. It is now generally accepted that the distribution of sodium and potassium ions across nerve and muscle cell membrane is regulated by (Na$^+$,K$^+$)ATPase. Active calcium transport across red cell membranes was originally described by Schatzmann (4) and there has been increasing evidence for such transport in a variety of excitable cell membranes (5). Membrane-bound enzyme Ca^{2+}-activated Mg^{2+}-dependent ATPase (Ca^{2+}-ATPase), the biochemical avatar of the calcium pump, is not inhibited by the cardiac glycosides.

Because diffusion potentials generated across membranes that were considerably more permeable to K$^+$ than to Na$^+$ could adequately account for the resting potential of excitable cells, the sodium pump was not generally considered to affect the membrane potential directly (6). It was later discovered that in some situations the Na$^+$ pump could be made electrogenic. The first method discovered was to increase the internal sodium concentration and then allow it to recover. During excretion of the extra sodium, the membrane potential exceeded that predicted by diffusion potential theory (7). This technique has since been used frequently to demonstrate electrogenic pumping in many tissues (8). Until recently, sodium loading was the only known means of inducing a sodium pump to become electrogenic. This fact is probably responsible for the commonly held belief that the sodium pump does not contribute to the membrane potential except under exotic nonphysiological conditions. Recent evidence has given rise to some questioning of this point of view.

It is now apparent that β-adrenergic agonists can cause the sodium pump to contribute significantly to the membrane potential. In frog sartorius muscle (9), noradrenaline increases the rate of ouabain-sensitive sodium efflux. In addition, it causes a hyperpolarization that can be blocked by ouabain (10). The hyperpolarization of frog skeletal muscle fibers seen during the excretion of extra sodium after sodium loading is also enhanced by adrenaline (11). The soleus muscles of the guinea pig (12) and rat (13) hyperpolarize in the presence of β-adrenergic agonists, and this hyperpolarization is blocked by ouabain. Catecholamines accelerate ouabain-sensitive sodium efflux in soleus muscle as in the frog sartorius muscle (13).

These data suggest that catecholamines are able to activate an electrogenic pump in skeletal muscle membrane or to induce a nonelectrogenic pump to become electrogenic. Similar suggestions have been made regarding the action of monoamines on central (14) and peripheral (15) neurons, and the ensuing hyperpolarization is thought to account for the inhibitory actions of the monoamines on such cells.

(Na$^+$,K$^+$)ATPase may also play a physiological role in the release mechanism of transmitters from nerve endings (16). According to this concept, the influx of Ca^{2+} ions which accompanies the nerve impulse inhibits membrane (Na$^+$,K$^+$) ATPase and thus triggers the release of neurotransmitter. Substances that inhibit enzyme activity would have an action comparable to that of Ca^{2+} and elicit release. Agents such as the catecholamines, which are known to be able to inhibit the release of transmitter from nerve terminals, may achieve this by stimulating sodium pumping.

Although the electrophysiological studies are consistent with the concept that catecholamines are able to activate an electrogenic pump and that such an action might be the underlying mechanism for catecholamine action in the excitable membrane, the findings of biochemical studies on catecholamine stimulation of the sodium-pump (Na$^+$,K$^+$)ATPase, have not been conclusive. Enzyme activation can be achieved consistently in various preparations, but explanations for this activation have varied. Some investigators postulate a non-receptor-mediated activation of the enzyme; others are in favor of a receptor-mediated activation of (Na$^+$,K$^+$)ATPase. Subsequent sections of this chapter will describe the results of attempts to characterize the monoamine-elicited stimulation of brain (Na$^+$,K$^+$)ATPase and will attempt to relate these findings to the observed actions of monoamines on excitable cells.

2   ACTION OF NORADRENALINE ON EXCITABLE MEMBRANES

Noradrenaline can hyperpolarize a variety of excitable tissues, including neurons in various regions of central nervous system, smooth (17), skeletal (10) muscles, and sympathetic ganglion cells (18). Although the effects of noradrenaline on mammalian cerebral neurons have been extensively documented (19–21), there is a limited understanding as to the mechanisms by which the amines exert their effects on neuronal excitability. There are two observed effects of noradrenaline on central neurons: depression of cell firing and excitatory actions. The depressant effect of noradrenaline is commonly observed in most regions of the central nervous system. It is characterized by a rapid and reversible depression of cell firing. Excitatory actions of the noradrenaline have been observed, in some instances on cells that were also depressed by the noradrenaline, raising the possibility that both excitatory and inhibitory receptors may be present on the same neuron. Although some of the excitatory actions of noradrenaline have been found to be the results of hydrogen ions released during the experiment from the iontophoretic microelectrode (22), the excitatory responses of neurons in the medulla, pons, and hypothalamus suggest the existence of specific excitatory receptors for noradrenaline (19,23–26). Adrenergic agonists and antagonists have been used to identify the type of adrenergic receptors. Attempts to categorize the noradrenaline response by application of either α- or β-blocking agents or α- and β-agonists have not yielded unequivocal evidence for the existence in the brain of pharmacologically distinct adrenergic receptors. Phillis et al. (27) reported that noradrenaline hyperpolarized spinal motoneurons, with an associated reduction in membrane excitability. Engberg et al. (28) have extended these findings to include dopamine and isoprenaline, and have shown that the hyperpolarizing action of the amines is associated with an increase in membrane resistance, is enhanced by a conditioning hyperpolarization, and reverses to a depolarization at a membrane potential of approximately -20 mV. This evidence was taken to suggest that the increase in membrane resistance reflects a decrease in permeability of the membrane to one or more ions, especially sodium ions. Studies with intracellular recording electrodes have shown that noradrenaline hyperpolarizes neurons in other regions of the central nervous system, including cerebral cortical neurons (29), cerebellar Purkinje cells (30), and hippocampal pyramidal cells (31). These hyperpolarizations are generated in the absence of any decrease in membrane resistance (Figure 1), and in some cases there may actually be a small increase in

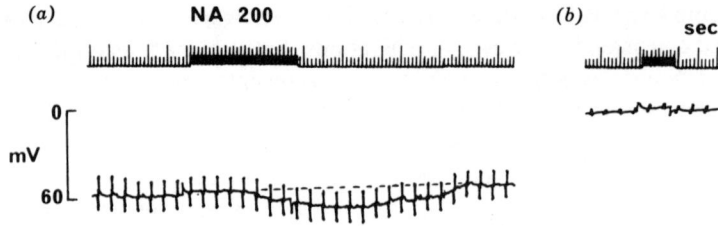

Figure 1 *(A)* Intracellular potential of a cerebral cortical neuron recorded on a chart recorder. Norepinephrine (NA, 200 nA) hyperpolarized the neuron. 0.5-nA pulses of inward and outward current lasting 250 msec were used to test the neuron's input impedance. This did not change during the norepinephrine-induced hyperpolarization. *(B)* Recording made after the microelectrode had been withdrawn from the neuron, showing the almost negligible effects of electrode polarization and bridge imbalance. Time calibrations: 1 and 5 sec. From Ref. 29.

membrane resistance. Phillis et al. (27) have suggested that postsynaptic hyperpolarization accounts for a major portion of the depressant action of this amine since it is an effective depressant against the glutamate-evoked and spontaneous firing of cerebral cortical neurons. However, presynaptic actions of noradrenaline should not be excluded, since it is known that noradrenaline reduces transmitter release from nerve terminals in the rabbit superior cervical ganglion (18). Vizi (32) has demonstrated that noradrenaline can reduce acetylcholine output from cerebral cortical slices, raising the possibility of an inhibitory presynaptic action of noradrenaline in the central nervous system.

The isolated amphibian spinal cord is a useful preparation with which to further explore the presynaptic actions of noradrenaline. When synaptic transmission in this preparation is blocked by exposure to a high Mg^{2+}/low Ca^{2+} containing perfusion solution or to tetrodotoxin, substances can be tested for their direct effects on motoneurons and dorsal root terminals. In such preparations, noradrenaline and other catecholamines have a depolarizing action on dorsal root terminals (Figure 2) which can be blocked by the α-adrenergic antagonist phentolamine. Depolarization of the terminals of primary afferent fibers would be expected to reduce transmitter release by reducing the amplitude of action potentials and could thus account for the presynaptic inhibitory action of noradrenaline (33).

Experiments on the guinea pig large intestine had shown that noradrenaline is an inhibitory transmitter released by sympathetic nerve fibers onto smooth muscle cells of the taenia coli (34). The inhibitory action of noradrenaline (and other catecholamines) on the smooth muscle is, in part, brought about by hyperpolarization due to increased potassium conductance of the membrane (α-receptor activation) and in part by the suppression of the pacemaker potential (β-receptor activation) (34,35). These studies also revealed that the inhibitory actions of noradrenaline (and other catecholamines) are calcium-dependent, which simply means that noradrenaline (and other catecholamines) are ineffective in a calcium-free solution, whereas the inhibitory action of adrenaline is potentiated by an excess of calcium (36,37). Calcium antagonists such as manganese and ruthenium red abolish the action of noradrenaline on smooth muscle cells of the taenia coli, even in the

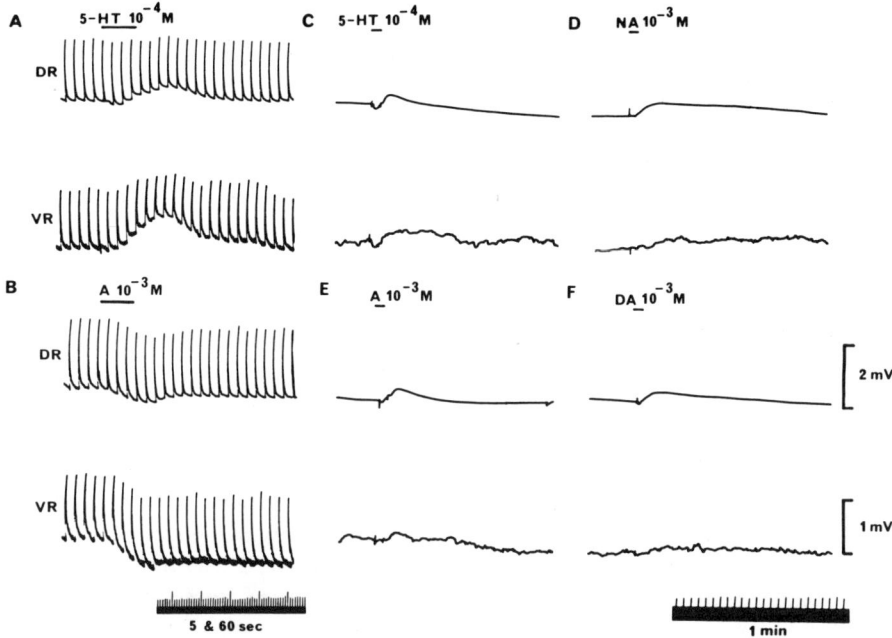

Figure 2 Direct-coupled recordings from the dorsal (DR) and ventral (VR) roots of an isolated hemisected toad spinal cord. In curves A and B the adjacent dorsal root was stimulated once every 20 sec. The responses in C–F were recorded during perfusion of the cord with a solution containing 0 mM Ca^{2+} and 10 mM Mg^{2+}. 5-Hydroxytryptamine (5-HT; 10^{-4} M) depolarized both roots in the normal preparation (A) and adrenaline (A; 10^{-3} M) hyperpolarized both roots, the ventral root hyperpolarization having a prolonged duration. Both substances, together with noradrenaline (NA; 10^{-3} M) and dopamine (DA; 10^{-3} M) depolarized the dorsal roots of the magnesium-treated preparations, and 5-HT evoked a small depolarization in the ventral root. From Ref. 33.

presence of calcium (34,38). Calcium itself produced an effect similar to that of adrenaline, causing a hyperpolarization and stabilization of the membrane with an increase in membrane resistance (34). Thus it was proposed that adrenaline (and noradrenaline) act by increasing calcium binding in the membrane, which would enhance membrane potassium fluxes (α-action) and by immobilizing membrane-bound calcium, prevent depolarization and the generation of pacemaker potentials (β-action). Noradrenaline (10^{-5}~10^{-2}M) hyperpolarizes frog sartorius and rat soleus muscle fibers, and this effect is abolished by ouabain (10,39). Replacement of sodium in the perfusion fluid with lithium also abolished the hyperpolarizations. In both types of muscle, as the K^+ concentration was raised, the hyperpolarization disappeared. For the soleus muscle this occurred with 15 mM K^+. Membrane potential recording from frog sartorius muscle fibers showed that with external K^+ concentrations in excess of 7 mM, the noradrenaline effect reversed into a ouabain-sensitive depolarization. Studies on rat soleus muscle have indicated that the hyperpolarization of the membrane by noradrenaline is sensitive to the β-antagonist propranolol, but not to the α-antagonist phentolamine (39). These results are consistent with the findings obtained from central neurons, namely that noradrenaline-induced hyperpolarization can be blocked by adrenergic antagonists. Questions then arose regarding the relationship of noradrenaline action and the receptor-mediated activation of sodium pumping by the cell membrane, which is intimately associated with the transmembrane potentials of excitable cells.

3   NORADRENALINE AND A MEMBRANE-ION PUMP

It has been proposed that the slow inhibitory postsynaptic potential (IPSP) generated by noradrenaline in sympathetic ganglion cells is produced by activation of an electrogenic sodium pump (40) rather than by inactivation of the sodium conductance of the postsynaptic membrane, as suggested by Weight and Padjen (41). Supportive evidence was obtained from experiments in which perfusion of bullfrog sympathetic ganglion with sodium-free lithium or hydrazinium solutions, which might be expected to reduce the intracellular sodium concentration and thus block the activity of the electrogenic sodium pump, abolished both the slow IPSP and the hyperpolarization evoked by catecholamines (42). Ouabain, sodium cyanide, 2,4-dinitrophenol, and low-temperature selectively block the slow IPSP and the hyperpolarization induced by adrenaline, suggesting that some of these metabolic or $(Na^+,K^+)ATPase$ inhibitors interfere with the functioning of the electrogenic cation pump (43,44). The $(Na^+,K^+)ATPase$ inhibitors ouabain and azide antagonize the depressant actions of noradrenaline on cerebral cortical neurons (45), indicating that the depressant action of noradrenaline here may also result in part from stimulation of an electrogenic membrane ion pump. Other agents, such as barbiturates and ethanol, which have been shown to inhibit the activity of $(Na^+,K^+)ATPase$, also antagonize the depressant action of noradrenaline on central neurons (46–48). Fluoride ions reduce the availability of ATP and thus inhibit the pump. Although fluoride had a weak excitation action of its own when applied onto central neurons, it reduced or abolished the depressant action of noradrenaline on 85% of the neurons tested (45). Lithium, which displaces sodium from access to the pump, can antagonize noradrenaline-induced depression of cerebral cortical neurons (49).

It has long been apparent that the activity of cerebral cortical neurons driven by L-glutamate is more readily suppressed by noradrenaline than is spontaneous firing (48). Glutamate is known to increase the uptake of sodium by brain cells (50), and it is possible that the increase in intracellular sodium levels by glutamate and enhanced electrogenic activity of the sodium pump might increase the sensitivity of these cells to noradrenaline-induced depression. The greater sensitivity of glutamate-evoked firing to noradrenaline depression thus provides further evidence to support the contention that noradrenaline depresses central neurons by activating an electrogenic sodium pump. Other supporting evidence has been forthcoming from experiments which showed that catecholamines have a hyperpolarizing action on skeletal muscle (10,13) and that this hyperpolarization is a result of stimulation of $(Na^+,K^+)ATPase$, thus greatly increasing electrogenic sodium efflux. Experiments on membrane potential changes recorded before and during the exposure of frog sartorius muscle fibers to noradrenaline in the presence or absence of ouabain demonstrate that noradrenaline hyperpolarizes frog muscle fibers by a ouabain-sensitive process in low external potassium solution (10). This hyperpolarization can be reversed to a ouabain-sensitive depolarization when the external potassium concentration is in excess of 7 mM. The most parsimonious explanation for this decrease or reversal of potential change evoked in skeletal muscle fibers by noradrenaline is that the amine stimulates a membrane pump whose electrogenicity is dependent on the concentration of the ions on either side of the membrane. A comparable finding has been made with the smooth muscle fibers of guinea pig taenia coli (36). Exposure of the latter to a 24 mM potassium solution abolished the adrenaline-induced hyperpolarization. This was not a voltage-dependent event, for when the muscle was depolarized to the same extent by acetylcholine, the hyperpolarization induced by adrenaline was greatly increased. If the pump in central neurons is similarly ion concentration-dependent, it would offer an explanation for the observed reversal potential of noradrenaline-evoked responses at -20 mV (28). A neuron depolarized to this extent would have a high membrane potassium permeability but a low sodium permeability, resulting from inactivation of sodium channels. Hence extracellular potassium levels would rise in conjunction with a diminishing of the electrochemical gradient to drive sodium into the cell. Therefore, the ratio of Na$^+_{(in)}$/K$^+_{(out)}$ across the membrane might change significantly in favor of potassium, and with this alteration in the ion ratios, the electrogenicity of the pump could be reduced or reversed, as apparently happens in skeletal muscle.

The various physiological studies mentioned above give a good indication that the action of noradrenaline on excitable membranes involves a receptor-mediated sodium pumping mechanism. The receptor associated with sodium pumping by skeletal muscle is clearly β-adrenergic in type. There has not been, however, any clear evidence regarding the existence of a pharmacologically distinct receptor for catecholamines associated with the sodium pump in central neurons. Since it is accepted that (Na$^+$,K$^+$)ATPase is closely associated with the cell membrane sodium pump, regulation of (Na$^+$,Ks$^+$)ATPase activity in brain tissue by catecholamines has been studied in some detail.

4 BIOCHEMICAL BASIS OF THE SODIUM PUMP—(Na$^+$,K$^+$)ATPase

The cell membrane contains many functional and structural entities which regulate immunological, hormonal, and pharmacological receptor-mediated responses. Among these, one of the most important functions of the cell membrane is the maintenance of appropriate intracellular concentrations of sodium and potassium. Most mammalian cells maintain an intracellular environment containing low sodium and high potassium, even though they are surrounded by an extracellular space that contains high sodium and low potassium. Therefore, the cell membrane is characterized by a low permeability to sodium, a high permeability to potassium, and a negatively charged inner transmembrane potential. A sodium-pumping mechanism is involved in the maintenance of these transmembrane cation and potential gradients. Skou (51) demonstrated a fraction of crab nerve ATPase activity in microsomes which was stimulated by the addition of sodium plus potassium in the presence of magnesium. The mechanism was proposed as the basis for sodium pumping, and the enzyme operating the pump was identified as (Na$^+$,K$^+$)-activated magnesium-dependent ATPase (E.C. 3.6.1.3). The essential function of the "pump" is to remove sodium ions from the cell against an electrochemical gradient and to transport potassium ions into the cell down an electrical but against a chemical gradient, with energy derived from the hydrolysis of ATP molecules (52). The requirement for sodium on the Na$^+$-activated site is absolute (3); however, the potassium ions can be replaced by Rb$^+$, NH$_4$$^+$, Cs$^+$, and Li$^+$. The ability of these ions to replace potassium is in the order in which they are presented. The stoichiometry of sodium pumping is believed to be three sodium ions pumped out for every two potassium ions pumped in, per hydrolysis of each ATP molecule (53). This suggests that there is a net outflow of sodium that is not chemically coupled to an inward transport of potassium; thus the pump may be electrogenic. When the effects of varying conditions on the fluxes of sodium and potassium by intact transporting systems are compared to the results obtained with (Na$^+$,K$^+$)ATPase, the data strongly suggest that (Na$^+$,K$^+$)ATPase activity is the catalytic expression of the enzyme-transporting complex (3,54). (Na$^+$,K$^+$)ATPase is embedded in the cell membrane, and attempts to purify the enzyme have met with many difficulties. However, it is generally accepted from studies on purified preparations that the enzyme consists of large polypeptides (molecular weight approximately 100,000) and has 3–4 nmol of ouabain binding sites per milligram of protein. There are two polypeptides in purified kidney (Na$^+$,K$^+$)ATPase preparations, one with a molecular weight of about 84,000 and the other with a molecular weight of 57,000 (55). The smaller molecule appears to be a sialoglycoprotein and the large molecule is possibly a "catalytic" subunit that can be phosphorylated by ATP in the native state (56).

Cross-linking experiments have yielded results suggesting that two polypeptides are present in purified (Na$^+$,K$^+$)ATPase preparations in a mass ratio for the large to small chains of 1.7 : 1, which probably corresponds to a molar ratio of one large chain to two small chains. [For a more detailed discussion on the purification and characterization of (Na$^+$,K$^+$)ATPase, see the review article (52).] More recent findings on the molecular structure of (Na$^+$,K$^+$)ATPase made by using different high-affinity ligands suggest that native (Na$^+$,K$^+$)ATPase, as found in pig kidney microsomal preparations from the outer medulla, seems to be a molecule containing eight large catalytic peptides with four ouabain-binding sites and four nucleotide-binding sites. Among the four nucleotide-binding sites, only two are available for binding substrate (ATP) at any one time. Thus there is total negative cooperativity between the two pairs of such sites. The nucleotide-binding sites and ouabain-binding sites would be most reasonably distributed on four identical subunits, containing two large peptides each (57).

Since the (Na$^+$,K$^+$)ATPase is associated with the active transport of Na$^+$ and K$^+$, and since these cations activate the enzyme, it is possible that the cation activator sites are also the carrier sites for the transport process (58). Therefore, the sequence of enzyme reactions with the cations may be the interaction of Na$^+$

and K$^+$ with one class of monovalent cation site on the (Na$^+$,K$^+$)ATPase, which alternately prefers Na$^+$ and then K$^+$ for enzyme activation (transport), or there may be distinct coexisting classes of sites for each ion. Several studies (59,60) have led to the suggestion the Na$^+$ and K$^+$ compete for a common site and that the binding of each ion at its respective site influences binding of the other ion at its site.

In order to explain the phenomena of sodium and potassium movement, pump models were proposed to reconcile the reaction scheme of (Na$^+$,K$^+$)ATPase to the vectorial movement of Na$^+$ and K$^+$ across the membrane. There are many types of proposed models, of which a pump driven by undulating peptide chains undergoing conformational change (61) is particularly interesting. Other models include one which proposes that the pump resembles a hole surrounded by charged walls of component enzymes catalyzing the ATPase reaction (62), or that the pump contains a membrane with sequential oxidation–reduction reactions of sulfhydryl groups induced by ATP (63).

In general, it is agreed upon that the sequence of pumping action may be divided into two main steps, similar to the sequence of ATPase chemical reaction: namely, phosphorylated intermediate(s) is (are) formed and Na$^+$ ions move outward; then inorganic phosphate is released inside the cell and K$^+$ ions move inward (64).

5   AMINE STIMULATION OF (Na$^+$,K$^+$)ATPase

Herd et al. (65) showed that (Na$^+$,K$^+$)ATPase from brown adipose tissue was maximally activated with 6 mM noradrenaline and that this effect could be blocked by 4 mM of propranolol. Subsequent to this finding, there were many other reports concerning biogenic amine stimulation of (Na$^+$,K$^+$)ATPase from various tissue preparations (Table 1). Schaefer et al. (66) showed that a soluble factor in brain homogenates could promote catecholamine stimulation of particle-bound (Na$^+$, K$^+$)ATPase from rat brain synaptosomal membranes. Catecholamines that activate the enzyme include noradrenaline, dopamine, adrenaline, and isoprenaline at 2×10^{-5}M concentrations. The phenolic amines tyramine, α-methyltyramine, metaraminol, amphetamine, and phenylephrine do not stimulate (Na$^+$,K$^+$)ATPase. The involvement of a tissue-soluble factor in catecholamine stimulation of (Na$^+$,K$^+$)ATPase was also reported by Rodriguez de Lores Arnaiz and Mistrorigo de Pacheco (67), who showed that noradrenaline stimulated rat cerebral cortical membrane (Na$^+$,K$^+$)ATPase in the presence of a soluble brain fraction, but that in its absence noradrenaline inhibited membrane-bound (Na$^+$,K$^+$)ATPase. They concluded that noradrenaline has a direct inhibitory effect on the enzyme which is not mediated by adrenergic receptors. Coffey et al. (68) reported that membrane ATPase activity in human lymphocyte and neutrophil membranes was stimulated by noradrenaline. Yoshimura (69) confirmed that activation of (Na$^+$,K$^+$)ATPase by catecholamines occurred in brain tissue with effective catecholamine concentrations of 5×10^{-5} M. Catecholamines and other biogenic amines have been reported to stimulate (Na$^+$,K$^+$)ATPase in cat, rabbit, and rat brain homogenates and synaptosomal preparations (70–75). Sulakhe et al. (76) showed that catechol-

TABLE 1   Adrenergic Agonists Inducing Stimulation of (Na$^+$,K$^+$) ATPase

Tissue	Adrenergic Agonist	Reference
Brown adipose	Noradrenaline	65
Rat brain	Noradrenaline, dopamine, adrenaline, isoprenaline, dopa	66
Rat brain	Noradrenaline, dopamine	69
Mouse brain, mouse kidney	Noradrenaline, dopamine	116
Cat brain	Noradrenaline	70
Rat brain	Noradrenaline	86
Rat brain	Noradrenaline	71
Beef cortex	Noradrenaline, isoprenaline, adrenaline, dopamine	78
Rat skeletal muscle	Noradrenaline, isoprenaline	77
Rat adrenals	Phenylephrine, naphazoline	117
Rat brain cervical ganglia	Noradrenaline, isoprenaline	76
Rat brain	Noradrenaline	67
Rat brain cortex	Noradrenaline, dopamine	73
Rat brain cortex	Noradrenaline, isoprenaline	74
Rat brain cortex	Noradrenaline, isoprenaline	75
Insect nerve cord	Methoxamine, noradrenaline	K. Moore (unpublished)

amines stimulate the (Na$^+$,K$^+$)ATPase activities of hypothalamic, cortical, cerebellar, caudate nucleus, and superior cervical ganglia homogenates of rat or rabbit. Cheng et al. (77) have demonstrated that catecholamines stimulate (Na$^+$,K$^+$) ATPase activity of a sarcolemmal fraction from rat skeletal muscle. A summary of the various reports of catecholamine-induced activation of (Na$^+$,K$^+$)ATPase is presented in Table 1.

The suggested mechanisms by which catecholamines activate (Na$^+$,K$^+$)ATPase have been many. Hexum (78) proposed that the activation could be a result of removal of divalent metal inhibition of the enzyme through the formation of a catecholamine–metal complex, since it is known that (Na$^+$,K$^+$)ATPase activity is inhibited *in vitro* by such biologically important divalent metal ions as Ca^{2+}, Cu^{2+},Zn^{2+}, and Fe^{2+} (79,80). Godfraind et al. (71) also proposed that catecholamine stimulation of (Na$^+$,K$^+$)ATPase might be due to suppression of the inhibition of (Na$^+$,K$^+$)ATPase exerted by calcium. Although Lee and Phillis (73) showed that Ca^{2+} was required for (Na$^+$,K$^+$)ATPase to respond to catecholamines, as removal of Ca^{2+} by ethyleneglycol-bis(β-aminoethylether)N,N'-tetraacetic acid (EGTA) abolished catecholamine stimulation of the enzyme, the functional role of Ca^{2+} in biological membranes is still uncertain. Noradrenaline stimulation of adenylate cyclase can be similarly reduced by EGTA (81). Calcium and other divalent metal ions, such as Cu^{2+}, Zn^{2+}, and Fe^{2+}, can be removed by edetic acid (EDTA) (a chelating agent). If catecholamine stimulation of (Na⁺,K⁺)ATPase was merely the result of removal of divalent metal ions, one would expect to observe the abolition of enzyme stimulation by catecholamines in the presence of high concentrations of EDTA. However, this does not appear to happen. Wu and Phillis (75) have demonstrated that although 10^{-4} M EDTA stimulates (Na⁺,K⁺)ATPase maximally, the addition of 10^{-5} M noradrenaline to an (Na⁺,K⁺)ATPase preparation containing 10^{-4} or 10^{-3} M EDTA can further enhance (Na⁺,K⁺)ATPase activity. The enzyme activation by noradrenaline is, however, substantially reduced under these conditions. Therefore, removal of divalent metal ion inhibition of the enzyme can only partially account for catecholamine stimulation of (Na⁺,K⁺) ATPase. It has been proposed that catecholamines simply reverse the inhibition of (Na⁺,K⁺)ATPase resulting from vanadium contamination of the substrate ATP (82). Reversal of vanadate inhibition of the enzyme by noradrenaline and other catecholamines is reportedly accomplished through complexation and reduction of the vanadate (83,84). We (75) have studied the interactions of vandate, noradrenaline, and (Na⁺,K⁺)ATPase in rat brain homogenates using vanadate-contaminated ATP, vanadate-free ATP, and synthetic ATP. The results of these experiments indicate that vanadium contamination of the substrate ATP cannot account for most of the stimulation of (Na⁺,K⁺)ATPase activity by noradrenaline (85).

The other proposed mechanism for catecholamine-induced stimulation of (Na⁺,K⁺)ATPase is that a catecholamine receptor is involved. Noradrenaline-elicited stimulation of the enzyme can be blocked by α- and β-adrenolytic agents (70,74,75,86), and dopamine stimulation of the enzyme is antagonized by chlorpromazine, an antipsychotic agent, which has been shown to block the dopamine receptor. 5-Hydroxytryptamine (5-HT) stimulation of (Na⁺,K⁺)ATPase from rat cerebral cortical homogenates can be effectively antagonized by metergoline (87), a known potent and specific 5-HT receptor antagonist (88–92). The evidence from studies with antagonists supports the suggestion that biogenic amine-mediated stimulation of (Na⁺,K⁺)ATPase is a receptor-linked reaction.

6   RECEPTOR-MEDIATED NORADRENALINE ACTIVATION OF (Na⁺,K⁺)ATPase

In an attempt to link activation of (Na⁺,K⁺)ATPase to a β-adrenergic receptor, Cheng et al. (77) were able to show that rat skeletal muscle membrane (Na⁺, K⁺)ATPase activity can be stimulated by various catecholic agents, but that this enhancement of (Na⁺,K⁺)ATPase activity in sarcolemmal membrane was not specifically a β-adrenergic response, because it was less specific in its structural requirements than the typical β-adrenergic receptor. However, agonists did require a certain molecular configuration reminiscent of those reported for catecholamine binding to various membrane preparations (93–97). Although the functional effect of this muscle catechol binding site has not been determined, it appears to be closely related to (Na⁺,K⁺)ATPase (77).

Wu and Phillis (74,75) have shown that (Na⁺,K⁺)ATPase in rat cerebral cortical homogenates or synaptosomal preparations is stimulated by noradrenaline, the threshold for this effect being less than 10^{-7} M. Adrenaline is slightly less potent than noradrenaline. The threshold concentration for the β-agonist, isoprenaline, is about 10^{-6} M, and the α-agonist, methoxamine, is 2.5×10^{-7} M, but phenylephrine has no stimulant action. The responses to noradrenaline could be antagonized by both the α-antagonist phentolamine and the β-antagonist propranolol (Figure 3). Although these results clearly demonstrate that (Na^+,K^+)ATPase activity evoked by noradrenaline can be blocked by adrenoceptor blockers, it is not sufficient to conclude from these studies that the stimulation of the enzyme is mediated by a conventional α- and/or β-adrenoceptor, as the results seem to suggest that the dissociation constant for either antagonist to block noradrenaline evoked (Na^+, K^+)ATPase activity is in the range 10^{-6} M, which is much greater than that of either antagonist in many known α- or β-receptor systems. Thus it is possible that the receptors involved may be different from those mediating other responses.

Roufogalis and Belleau (98) showed that phenoxybenzamine and dibenamine, two irreversible inhibitors of the adrenergic α-receptor, which were approximately equiactive on the inhibition of K^+-induced contractile response of smooth muscle (99,100), are equipotent as inhibitors of (Na^+,K^+)ATPase. However, the reactivity

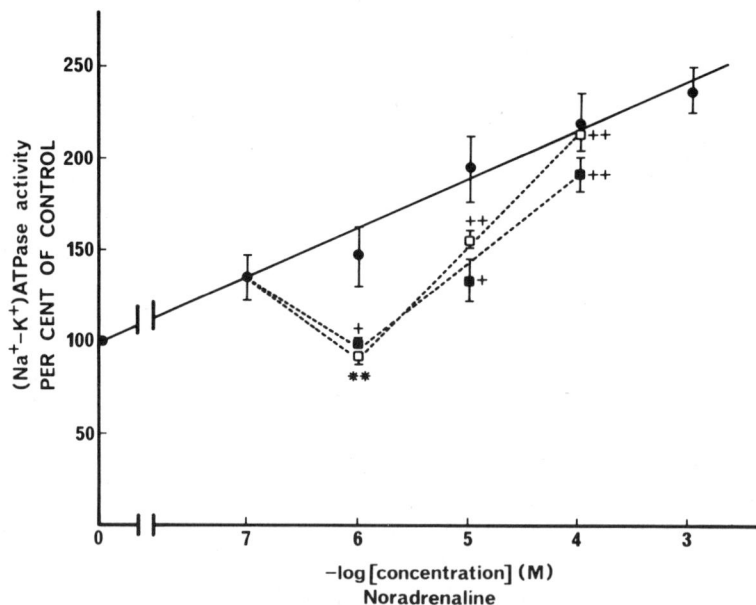

Figure 3   Effect of propranolol (10^{-5} M) and phentolamine (10^{-5} M) on the noradrenaline stimulated (Na^+,K^+)ATPase from rat cortical homogenate. Propranolol (10^{-5} M) or phentolamine (10^{-5} M) was preincubated for 10 min in the presence of noradrenaline (10^{-6}, 10^{-5}, 10^{-4} M). Details are as described (75). Data were calculated using Student's t-test. Each point represents the mean ± SE of six different experiments, each of which consisted of three identical samples. - ● -, Noradrenaline alone; . . . ■ . . . , noradrenaline and propranolol (10^{-5} M); . . . □ . . . , noradrenaline and phentolamine (10^{-5} M). Levels of significance were calculated by Student's t-test. ⁺⁺, not significant; ⁺, $0.05 > P > 0.01$; *⁎, $0.01 > P > 0.001$. From Ref. 75.

of (Na$^+$,K$^+$)ATPase toward dibenamine and related substances leaves open the question as to whether the membrane receptor itself might be an ATPase-like structure controlling calcium movement (101).

The antagonism by both α- and β-adrenergic blockers and the ineffectiveness of an α-adrenergic agonist, phenylephrine, to elicit a significant activation of (Na$^+$,K$^+$)ATPase activity prompted us to suggest that catecholamine stimulation of (Na$^+$,K$^+$)ATPase activity is mediated by a catecholic receptor(s), possibly resembling the α- and/or β-adrenergic receptors (75). As a careful evaluation of this "catecholic receptor" might be expected to yield a better understanding as to whether the receptor associated with (Na$^+$,K$^+$)ATPase is α- or β- or a mixture of both α- and β-receptors, agents with structural features analogous to noradrenaline were tested for their ability to activate rat cerebral cortical (Na$^+$,K$^+$)ATPase. Table 2 shows that phenolic compounds (compounds containing only one functional hydroxyl group on the phenyl ring) were inactive in stimulation of the enzyme. Catechol and pyrogallol were both active in stimulation of the enzyme. Catechol was more effective than pyrogallol, as indicated by the EC$_{20}$ values of 1.8×10^{-7} M and 1.3×10^{-6} M for catechol and pyrogallol, respectively. Addition of an ethanolamine side chain to catechol enhanced its ability to stimulate (Na$^+$,K$^+$)ATPase. This is apparent from the EC$_{20}$ value of noradrenaline, 1.0×10^{-7} M as compared to that of catechol, 1.8×10^{-7}M. The phenolic amines octopamine and normetanephrine, although they contain an ethanolamine side chain, were devoid of stimulatory effect on the enzyme. Modification of the side chain with a methylated amino group decreased the

TABLE 2 EC$_{20}$ and EC$_{50}$ Values of Various Agents in Stimulation of Rat Cerebral Cortex (Na$^+$,K$^+$)ATPase^a

Agent	EC$_{20}$	EC$_{50}$
Phenol	0	0
Catechol	1.8×10^{-7} M	2.2×10^{-5} M
Pyrogallol	1.3×10^{-6} M	1.6×10^{-5} M
Octopamine	0	0
l-Noradrenaline	1.0×10^{-7} M	7.9×10^{-6} M
l-Adrenaline	7.0×10^{-7} M	1.0×10^{-5} M
Protocatechuic aldehyde	5.6×10^{-7} M	3.1×10^{-6} M
3,4-Dihydroxymandelic acid	1.0×10^{-6} M	3.1×10^{-5} M
Metanephrine	0	0
Normetanephrine	0	0

^{a}Maximal stimulation of (Na$^+$,K$^+$)ATPase was obtained by catechol (10^{-3} M), $267 \pm 48\%$; pyrogallol (10^{-4} M), $220 \pm 30\%$; L-noradrenaline (10^{-3} M), $240 \pm 43\%$; L-adrenaline (10^{-4} M), $187 \pm 26\%$; protocatechuic aldehyde (10^{-4} M), $205 \pm 32\%$; 3,4,-dihydroxymandelic acid (10^{-3} M), $232 \pm 34\%$ of the control enzyme activity. Results are mean $\pm$ SE of five experiments.

activity of the compound to stimulate (Na^+,K^+)ATPase. Replacement of the ethanolamine side chain by an aldehyde or carboxylic side chain reduced the ability of
the agent to activate the enzyme system. Thus it appears that the catechol group is
a critical requirement for stimulation of (Na^+,K^+)ATPase (Table 2). Of the various
catechol analogs tested for their ability to stimulate (Na^+,K^+)ATPase, only 1,2-di-
hydroxybenzene and its derivatives were active, whereas the 1,3-dihydroxybenzene
series and 1,4-dihydroxybenzene and its derivatives were inactive. L-stereo-isomers
of the adrenergic agonists are more active than the D-isomers in eliciting pharmacological responses. Table 3 shows that (Na^+,K^+)ATPase can be activated by the D-
and L-isomers of noradrenaline and isoprenaline. However, L-noradrenaline is more
potent than D-noradrenaline; EC_{20} [the concentration of agonist which causes 20%
of maximal stimulation of (Na^+,K^+)ATPase activity] is 1×10^{-7} M and $1.2 \times
10^{-6}$ M for L-noradrenaline and D-noradrenaline, respectively, indicating that the L-
isomer is approximately 10 times more effective than the D-isomer. The difference
in the potency of enzyme activation by the D- and L-isomers of isoprenaline was less
apparent, as the EC_{20} values for L-isoprenaline and D-isoprenaline were 1×10^{-6} M
and 3×10^{-6} M, respectively (102). Thus activation of (Na^+,K^+)ATPase by catecholamines seems to show some characteristics of activation of adrenergic receptors.

Our findings on the relative potencies of the L- and D-isomers of isoprenaline are
less definitive. According to Harden et al. (104), L-isoprenaline is approximately
three orders of magnitude more potent than the D-isomer in inhibiting the binding of
iodohydroxybenzylpindolol in homogenates of rat cortex. The effects of D- and L-
isomers of various adrenergic agonists on (Na^+,K^+)ATPase activity, cyclic AMP
accumulation, and ligand binding are summarized in Table 3. Although the EC_{50}
value for L-isoprenaline stimulation of (Na^+,K^+)ATPase in rat cortex is very similar
to that for activation of adenylate cyclase from turkey erythrocytes (96), the lack of
stereo specificity of isoprenaline isomers in their stimulation of (Na^+,K^+)ATPase in
rat cortex homogenates raises some questions regarding the involvement of a specific
β-adrenergic receptor in noradrenaline stimulation of rat brain (Na^+,K^+)ATP-
ase. The possibility that there are two reaction mechanisms for noradrenaline stimulation of (Na^+,K^+)ATPase must be considered. A non-stereo-specific activation
mechanism may be involved in the activation of (Na^+,K^+)ATPase induced by catechol and D-noradrenaline, since activation of the enzyme system by these compounds is not blocked by α- or β-adrenergic blockers (Tables 4 to 6) and could, in
fact, be due to a non-receptor-mediated mechanism such as the chelation of inhibitory divalent cations proposed by Hexum (78). Activation of enzyme by methoxamine, a noncatecholic α-adrenergic agonist, and the blockade of methoxamine-in-
duced (Na^+,K^+)ATPase stimulation by 10^{-6} M phentolamine (unpublished
observations) further suggest the possible involvement of an α-receptor-mediated
(Na^+,K^+)ATPase activation by L-noradrenaline.

Kinetic analysis of catecholamine sensitive (Na^+,K^+)ATPase activity in mouse
brain synaptosomes (116) has revealed that noradrenaline increased (Na^+,K^+)ATP-
ase activity in the presence of Na^+ and K^+ in the reaction medium. In the absence
of Na^+ ions, the K^+-activated ATPase activity was stimulated by noradrenaline, but

TABLE 3 Comparison of EC$_{50}$ Values of Some Adrenergic Agonists on (Na$^+$,K$^+$)ATPase Activation, Cyclic AMP Accumulation, and Competition of Ligand Binding to the Receptor of Brain Tissue

	EC$_{50}$ (IC$_{50}$) for:		Competition for Receptor Binding	Type of Ligand[a]	References
Agent	(Na$^+$,K$^+$)ATPase Activation	Cyclic AMP Accumulation			
L-Noradrenaline	7.9×10^{-6} M	10×10^{-6} M	6.8×10^{-6} M	DHE	81, 103
D-Noradrenaline	5.0×10^{-5} M	245×10^{-6} M	66×10^{-6} M	DHE	103, 119
L-Isoprenaline	2.2×10^{-5} M	1×10^{-7} M	5.6×10^{-6} M	Clonidine	81, 118
			6.3×10^{-7} M	IHYP	104
D-Isoprenaline	2.5×10^{-5} M	1×10^{-6} M	38×10^{-6} M	Clonidine	118, 119
			1.0×10^{-4} M	IHYP	104

[a]DHE, dihydroergocryptine; IHYP, iodohydroxybenzylpindolol.

TABLE 4 Effect of Phentolamine and Propranolol on d-Noradrenaline Stimulation of (Na$^+$, K$^+$)ATPase in Rat Brain Cortical Homogenates[a]

Concentration (M)	Enzyme Activity (% of control)		
	D-Noradrenaline	Phentolamine (10^{-5} M) and D-Noradrenaline	Propranolol (10^{-5} M) and D-Noradrenaline
10^{-6}	104 ± 3.9 (8)	103 ± 2.6 (6)[b]	106 ± 8 (4)[b]
10^{-5}	132 ± 5.6 (8)	122 ± 3.7 (6)[b]	137 ± 7.1 (5)[b]
10^{-4}	169 ± 10 (6)	152 ± 7.5 (4)[b]	166 ± 9.8 (3)[b]

[a]Results are expressed as mean ± SE. The number of experiments is shown in parentheses. Student's t-test was applied to compare the results from samples containing agonist and antagonist to that of containing equal concentrations of agonist.
[b]Not significant.

TABLE 5 Effect of Propranolol on l- and d-isoprenaline Stimulation of Rat Cerebral Cortical (Na$^+$, K$^+$)ATPase[a]

Concentration (M)	Enzyme Activity (% of control)			
	L-Isoprenaline	*Propranolol (10^{-5} M) and L-Isoprenaline*	*D-Isoprenaline*	*Propranolol (10^{-5} M) and D-Isoprenaline*
10^{-6}	118 ± 4 (4)	103 ± 3.4 (3)[b]	110 ± 5.5 (4)	96 ± 7 (3)[b]
10^{-5}	134 ± 11 (4)	115 ± 9 (3)[c]	140 ± 4.0 (3)	106 ± 10 (3)[b]
10^{-4}	190 ± 10 (4)	167 ± 21 (3)[b]	163 ± 13 (4)	141 ± 11 (3)[b]

[a]Results are expressed as mean ± SE. The number of experiments is shown in parentheses. Student's t-test was applied to compare the results from the samples containing agonist and antagonist to that of containing equal concentration of agonist.
[b]$p < 0.001$.
[c]$0.001 < p < 0.01$.

TABLE 6   Effect of Phentolamine and Propranolol on Catechol Stimulation of Rat Cerebral Cortical (Na$^+$,K$^+$)ATPase[a]

	Enzyme Activity (% of control)		
		Phentolamine (10^{-5} M)	*Propranolol (10^{-5} M)*
Concentration (M)	*Catechol*	*Catechol*	*Catechol*
0	100 (8)	121 ± 11 (4)[b]	120 ± 10 (4)[b]
10^{-6}	130 ± 5.7 (4)	136 ± 10 (4)[c]	132 ± 23 (4)[c]
10^{-5}	145 ± 18 (8)	152 ± 10 (4)[c]	154 ± 18 (4)[c]

[a]Results are expressed as mean ± SE. The number of experiments is shown in parentheses. Student's *t*-test was applied to compare the results from samples containing both agonist and/or antagonist to that of containing same concentration of agonists alone.
[b]$p < 0.001$.
[c]Not significant.

in the absence of K$^+$ ions, Na$^+$ ATPase was not activated by noradrenaline, indicating that ATPase activity was more sensitive to catecholamine in the presence of K$^+$ than Na$^+$.

The possible involvement of adenosine 3′,5′-cyclic monophosphate (cyclic AMP) in the modulation of (Na$^+$,K$^+$)ATPase has been examined in liver plasma membrane (105,106) and (Na$^+$,K$^+$)ATPase of neural tissues (76). The consistent finding has been that in neural tissues, the (Na$^+$,K$^+$)ATPase activity was not modified either by cyclic AMP added to the incubation procedure (67,69,76), or after increasing endogenous cyclic AMP levels by pretreating membrane preparations with guanylyl-5′-imidodiphosphate [Gpp(NH)p], which has been shown to activate the cyclic AMP-generating system of membrane preparations (107).

Studies on the developing rat brain have revealed that (Na$^+$,K$^+$)ATPase activity increases during neonatal maturation together with dendrite formation and an increase in electrical activity (108–111), suggesting that the increase in (Na$^+$,K$^+$) ATPase activity is associated with a development of the cell membrane in which (Na$^+$,K$^+$)ATPase was located. The correlation between development of electroencephalogram (EEG) activity and (Na$^+$,K$^+$)ATPase activity in the brain has been studied (112). ATPase activity was detectable on the 21st day of gestation, at which time the EEG was also first detectable, and apparently reached adult levels by the twelfth day postpartum. Similar results were obtained in the cerebral cortex of kitten, where an increase in (Na$^+$,K$^+$)ATPase activity was noted between 1 and 6 weeks, which was also the period of rapid rise in spontaneous electrical activity in cortical neurons (113).

An increase in electrical activity of the immature cortex due to electrical stimulation also resulted in the induction of (Na$^+$,K$^+$)ATPase activity above its resting levels, suggesting an active participation of (Na$^+$,K$^+$)ATPase in the electrical events in the nervous system (113). A study of the response of rat brain (Na$^+$,K$^+$) ATPase to noradrenaline stimulation has shown that brain cortical (Na$^+$,K$^+$)ATPase during the prenatal and first 7 days of postnatal life is insensitive to noradrenaline activation, even at a noradrenaline concentration of 10^{-3} M. However, the degree of the sensitivity of the enzyme to noradrenaline stimulation gradually increases and eventually reaches the full response (adult-type response) at the 28th day of postnatal growth. This is interesting, especially in view of the ontogeny of adrenergic receptors in rat cerebral cortex (104). The concentration of β-adrenergic receptors is very low during the first week after birth. Between 7 and 14 days, there is a rapid increase in the density of receptors. Adult levels are reached by the end of the second week. Catecholamine stimulated cyclic AMP accumulation is barely detectable during the first week after birth and reaches adult levels between 7 and 14 days, suggesting that it is the development of β-adrenergic receptors which permits the expression of catecholamine-sensitive adenylate cyclase activity. A similar pattern was found in the development of rat cerebral cortical (Na^+,K^+)ATPase levels and its sensitivity toward catecholamine stimulation. Of course, as there is no firm support for the coupling of β-adrenergic receptor to (Na^+,K^+)ATPase, or evidence for an association of adenylate cyclase with (Na^+,K^+)ATPase, these data can only be viewed as an incentive to further investigation of possible relationships between the adrenoceptors and (Na^+,K^+)ATPase in the central nervous system.

7 CONCLUSION

In this chapter, we have provided evidence that in both the peripheral and central nervous systems, the hyperpolarizing depressant action of the biogenic amines involves an ouabain-sensitive membrane ATPase, (Na^+,K^+)ATPase. Although the nature of the linkage between amine receptor and the enzyme is currently uncertain, it appears that enzyme can be stimulated by catecholamines. Characterization of this amine-induced activation of the enzyme indicates that membrane fragments are necessary for the mediation of amine effect, as the enzyme will no longer respond once it is disassociated from the membrane.

Many proposals have been put forward to explain the mechanism of catecholamine stimulation of (Na^+,K^+)ATPase. Josephson and Cantley (82) proposed that catecholamines simply reverse the vanadium inhibition of the enzyme by a complexation and reduction of vanadate; Hexum (78) and Schaefer et al. (114) proposed that catecholamines reverse divalent cation inhibition of the enzyme, and Schaefer et al. (115) suggested that inhibition of lipid peroxidation is a possible explanation of catecholamine activation of (Na^+,K^+)ATPase. However, we are in favor of an alternative explanation, which is that catecholamine stimulation of (Na^+,K^+)ATPase is a receptor-linked reaction. This explanation is supported by the finding that vanadium contamination of substrate ATP cannot account for a large part of the stimulation of (Na^+,K^+)ATPase activity by noradrenaline (75) and that metal-ion chelating agents only partially reduce enzyme activation by noradrenaline, even in the presence of 10^{-3} M of EDTA. The reduction, but not abolition, of the enzyme activation by noradrenaline in the presence of chelating agents indicates that the mechanism of activation cannot be explained solely by removal of metal ions. Our observations demonstrate that stimulation of brain (Na$^+$,K$^+$)ATPase by catecholamines has many of the characteristics of a typical pharmacological response (i.e., stereo specificity, structure–activity relationships, low concentrations of catecholamines required to produce enzyme activation, and a parallelism between the onset of enzyme activation and the appearance of adrenoceptors). Activation of (Na$^+$,K$^+$) ATPase and the pharmacological action of catecholamines on neurons therefore appear to be related events.

NOTES ADDED IN PROOF

Török and Vizi (120) have demonstrated that the Ca^{2+}-dependent hyperpolarization of guinea-pig taenia coli smooth muscle cells is due to the combined effects of stimulation of a membrane sodium pump and an increase in membrane potassium permeability. Studies on disaggregated smooth muscle cells of *Bufo marinus* indicate that the relaxant effect of β-adrenergic stimulation is due to sodium pump stimulation (121). In this instance it is suggested that the loss of intracellular sodium leads to enhanced Na$^+$/Ca^{2+} exchange with a fall in intracellular Ca^{2+}. The fall in intracellular Ca^{2+} would decrease muscle contractility.

The hypothesis that catecholamine evoked hyperpolarization of central neurons is due to stimulation of a membrane sodium pump is supported by the observation that fluxes of Na$^+$ and K$^+$ across the plasma membrane of the brain cells were affected by noradrenaline. Noradrenaline (10^{-6}M) increased ^{22}Na efflux from cortical slices by 79% and ^{45}K influx by 83%. The effects of noradrenaline were antagonized by ouabain and by the β-adrenergic blocker, propranolol (122).

REFERENCES

1   R. B. Dean, *Biol. Symp.*, **3**, 331 (1941).

2   H. J. Schatzmann, *Helv. Physiol. Pharmacol. Acta*, **11**, 346 (1953)

3   J. C. Skou, *Physiol. Rev.*, **45**, 596 (1965).

4   H. J. Schatzmann, *Biochim. Biophys. Acta*, **94**, 89 (1965).

5   P. V. Sulakhe and P. J. St. Louis, *Gen. Pharmacol.*, **7**, 313 (1976).

6   A. L. Hodgkin and P. Horowicz, *J. Physiol.*, **153**, 370 (1960).

7   A. S. Frumento, *Science*, **147**, 1442 (1965).

8   R. C. Thomas, *Physiol. Rev.*, **52**, 563 (1972).

9   E. T. Hays, P. Dwyer, P. Horowicz, and J. G. Swift, *Am. J. Physiol.*, **227**, 1340 (1974).

10   B. H. Bressler, J. W. Phillis, and W. Kozachuk, *Eur. J. Pharmacol.*, **33**, 201 (1975).

11   K. Koketsu and Y. Ohta, *Life Sci.*, **19**, 1009 (1976).

12   N. Tashiro, *Br. J. Pharmacol.*, **48**, 121 (1973).

13   T. Clausen and J. A. Flatman, *J. Physiol.*, **270**, 383 (1977).

14   J. W. Phillis, in R. Huxtable and A. Barbeau Eds., *Taurine*, Raven Press, New York, 1976.

15   K. Kuba and K. Koketsu, *Prog. Neurobiol.*, **11**, 77 (1978).

16   E. S. Vizi, *Neuroscience*, **3**, 367 (1978).

17 E. Bulbring and H. Kuriyama, *J. Physiol., 166,* 59 (1963).

18 D. D. Christ and S. Nishi, *J. Physiol., 213,* 107 (1971).

19 J. W. Phillis, *The Pharmacology of Synapses,* Pergamon Press, Oxford, 1970.

20 K. Krnjević, *Physiol Rev., 54,* 418 (1974).

21 A. K. Tebēcis, *Transmitters and Identified Neurons in the Mammalian Central Nervous System,* Scientechnica, Bristol, 1974.

22 R. C. A. Frederickson, L. M. Jordan, and J. W. Phillis, *Brain Res., 35,* 556 (1971).

23 P. B. Bradley and J. H. Wolstencroft, *Br. Med. Bull., 21,* 15 (1965).

24 G. C. Salmoiraghi, *Pharmacol. Rev., 18,* 717 (1966).

25 D. R. Curtis and J. M. Crawford, *Annu. Rev. Pharmacol., 9,* 209 (1969).

26 F. E. Bloom and N. J. Giarman, *Annu. Rev. Pharmacol., 8,* 229 (1968).

27 J. W. Phillis, A. K. Tebēcis, and D. H. York, *Eur. J. Pharmacol., 7,* 471 (1968).

28 I. Engberg, J. A. Flatman, and K. Kadzielawa, *Acta Physiol. Scand., 91,* 2A (1974).

29 J. W. Phillis, *Can. J. Neurol. Sci., 4,* 151 (1977).

30 G. R. Siggins, B. J. Hoffer, A. P. Oliver, and F. E. Bloom, *Science, 171,* 192 (1971).

31 M. Segal and F. E. Bloom, *Brain Res., 72,* 99 (1974).

32 E. S. Vizi, *J. Physiol., 226,* 95 (1972).

33 J. W. Phillis and J. R. Kirkpatrick, *Gen. Pharmacol., 9,* 239 (1978).

34 E. Bulbring and T. Tomita, *Proc. R. Soc. Lond. Ser. B., 172,* 121 (1969).

35 E. Bulbring and T. Tomita, *Proc. R. Soc. Lond. Ser. B, 172,* 103 (1969).

36 E. Bulbring and H. Kuriyama, *J. Physiol., 166,* 59 (1963).

37 Y. Hotta and R. Tsukiu, *Nature (Lond.), 217,* 867 (1968).

38 A. Tomiyama, I. Takayanagi, M. Salki, and K. Takagi, *Jap. J. Pharmacol., 23,* 889 (1973).

39 J. P. Edstrom and J. W. Phillis, *Gen. Pharmacol., 12,* 57 (1981).

40 S. Nishi and K. Koketsu, *J. Neurophysiol., 31,* 717 (1968).

41 F. F. Weight and A. Padjen, *Brain Res., 55,* 219 (1973).

42 K. Koketsu, T. Shoji, and S. Nishi, *Life Sci., 13,* 453 (1973).

43 S. Nishi and K. Koketsu, *Life Sci., 6,* 2049 (1967).

44 M. Nakamura and K. Koketsu, *Life Sci., 11,* 1165 (1972).

45 J. W. Phillis, *Life Sci., 15,* 213 (1974).

46 G. G. Yarbrough, N. Lake, and J. W. Phillis, *Brain Res., 67,* 77 (1974).

47 J. W. Phillis and A. K. Tebēcis, *Life Sci., 6,* 1621 (1967).

48 N. Lake, G. G. Yarbrough, and J. W. Phillis, *J. Pharm. Pharmacol., 25,* 582 (1973).

49 J. W. Phillis, *Life Sci., 14,* 1189 (1974).

50 H. Bradford and H. McIlwain, *J. Neurochem., 13,* 1163 (1966).

51 J. C. Skou, *Biochim. Biophys. Acta, 23,* 394 (1957).

52 A. Schwartz, G. E. Lindenmayer, and J. C. Allen, *Pharmacol. Rev., 27,* 3 (1975).

53 P. J. Garrahan and I. M. Glynn, *J. Physiol., 192,* 237 (1967).

54 A. K. Sen and R. L. Post, *J. Biol. Chem., 239,* 345 (1964).

55 J. Kyte, *J. Biol. Chem., 246,* 4157 (1971).

56 J. Kyte, *Biochem. Biophys. Res. Commun., 43,* 1259 (1971).

57 O. Hansen, J. Jensen, J. G. Nørby, and P. Ottolenghi, *Nature (Lond.), 280,* 410 (1979).

58 R. L. Post, C. R. Merritt, C. R. Kinsolring, and C. D. Albright, *J. Biol. Chem., 235,* 1796 (1960).

59 L. J. Mullins and F. J. Brinley, *J. Gen. Physiol., 53,* 704 (1969).

60 R. A. Sjodin and L. A. Beaugé, *J. Gen. Physiol., 52,* 389 (1968).

61 L. J. Opit and J. S. Charnock, *Nature (Lond.)*, **208**, 471 (1965).

62 R. W. Albers, S. Fahn, and G. J. Koval, *Proc. Natl. Acad. Sci. U.S.A.*, **50**, 474 (1963).

63 J. C. Skou, *Prog. Biophys. Chem.*, **14**, 131 (1964).

64 R. Tanaka, in J. H. Biel and L. G. Abood, Eds., *Biogenic Amines and Physiological Membranes in Drug Therapy*, Medical Research Series, Vol. 5, Part 1, New York, 1971.

65 P. A. Herd, B. A. Horwitz, and E. R. Smith, *Experientia*, **26**, 825 (1970).

66 A. Schaefer, G. Unyi, and A. K. Pfeifer, *Biochem. Pharmacol.*, **21**, 2289 (1972).

67 G. Rodriguez de Lores Arnaiz and M. Mistrorigo de Pacheco, *Neurochem. Res.*, **3**, 733 (1978).

68 R. G. Coffey, J. W. Hadden, E. M. Hadden, and E. Middleton, *Fed. Proc.*, **30**, 497 (1971).

69 K. Yoshimura, *J. Biochem. (Tokyo)*, **74**, 389 (1973).

70 P. Iwangoff, A. Enz, and A. Chappeus, *Experientia*, **30**, 688 (1974).

71 T. Godfraind, M. C. Koch, and N. Verbeke, *Biochem. Pharmacol.*, **23**, 3505 (1974).

72 J. C. Logan and D. J. O'Donovan, *J. Neurochem.*, **27**, 185 (1976).

73 S. L. Lee and J. W. Phillis, *Can. J. Physiol. Pharmacol.*, **55**, 961 (1977).

74 P. H. Wu and J. W. Phillis, *Gen. Pharmacol.*, **9**, 421 (1978).

75 P. H. Wu and J. W. Phillis, *Gen. Pharmacol.*, **10**, 189 (1979).

76 P. V. Sulakhe, S. H. Jan, and S. J. Sulakhe, *Gen. Pharmacol.*, **8**, 37 (1977).

77 L. L. Cheng, E. M. Rogus, and K. Zierler, *Biochim. Biophys. Acta*, **464**, 338 (1977).

78 T. D. Hexum, *Biochem. Pharmacol.*, **26**, 1221 (1977).

79 J. Donaldson, T. St. Pierre, J. Minnich, and A. Barbeau, *Can. J. Biochem.*, **49**, 1217 (1971).

80 T. Tobin, T. Akera, S. I. Baskin, and T. M. Brady, *Mol. Pharmacol.*, **9**, 336 (1973).

81 U. Schwabe and J. W. Daly, *J. Pharmacol. Exp. Ther.*, **202**, 134 (1977).

82 L. Josephson and L. C. Cantley, Jr., *Biochemistry*, **16**, 4572 (1977).

83 L. C. Cantley, Jr., J. H. Ferguson, and K. Kustin, *J. Am. Chem. Soc.*, **100**, 5210 (1978).

84 L. C. Cantley, Jr., M. D. Resh, and G. Guidotti, *Nature (Lond.)*, **272**, 552 (1978).

85 J. W. Phillis and P. H. Wu, *J. Pharm. Pharmacol.*, **31**, 556 (1979).

86 J. C. Gilbert, M. B. Wyllie, and D. V. Davison, *Nature (Lond.)*, **255**, 237 (1975).

87 P. H. Wu and J. W. Phillis, *J. Pharm. Pharmacol.*, **31**, 782 (1979).

88 C. Beretta, R. Ferrini, and A. H. Glässer, *Nature (Lond.)*, **207**, 421 (1965).

89 C. Beretta, A. H. Glässer, M. B. Nobili, and R. Silvestri, *J. Pharm. Pharmacol.*, **17**, 423 (1967).

90 K. Fuxe, L. Agnati, and B. Everitt, *Neurosci. Lett.*, **1**, 283 (1975).

91 B. V. Clineschmidt and V. J. Lotti, *Br. J. Pharmacol.*, **50**, 311 (1974).

92 B. S. R. Sastry and J. W. Phillis, *Can. J. Physiol. Pharmacol.*, **55**, 130 (1977).

93 V. Tomasi, S. Koretz, T. K. Ray, J. Dunnick, and G. V. Marinetti, *Biochim. Biophys. Acta*, **211**, 31 (1970).

94 R. J. Lefkowitz and E. Haber, *Proc. Natl. Acad. Sci. U.S.A.*, **68**, 1773 (1971).

95 M. Schramm, H. Feinstein, E. Naim, M. Lang, and M. Lasser. *Proc. Natl. Acad. Sci. U.S.A.*, **69**, 523 (1972).

96 J. P. Bilezikian, and G. D. Aurbach, *J. Biol. Chem.*, **248**, 5575 (1973).

97 P. Cuatrecasas, G. P. E. Tell, S. Vincenzo, I. Parikh, and K. J. Chang, *Nature (Lond.)*, **247**, 92 (1974).

98 B. D. Roufogalis and B. Belleau, *Life Sci.*, **8**, 911 (1969).

99 S. Shibata and O. Carrier, Jr., *Can. J. Physiol. Pharmacol.*, **45**, 587 (1967).

100 S. Shibata, O. Carrier, Jr., and J. Frankenheim, *J. Pharmacol. Exp. Ther.*, **160**, 106 (1968).

101 B. Belleau, *Ann. N.Y. Acad. Sci.*, **139**, 580 (1967).

102 P. H. Wu and J. W. Phillis, *Int. J. Biochem.*, **12,** 353 (1980).

103 J. N. Davis, W. J. Strittmatter, E. Hoyler, and R. J. Lefkowitz, *Brain Res.*, **132,** 327 (1977).

104 T. K. Harden, B. B. Wolfe, J. R. Sporn, J. P. Perkins, and P. B. Molinoff, *Brain Res.*, **125,** 99 (1977).

105 E. Tria, P. Luly, V. Tomasi, A. Trevisani, and O. Barnabei, *Biochim. Biophys. Acta,* **343,** 297 (1974).

106 K. Kanike, B. J. R. Pitts, and A. Schwartz, *Biochim. Biophys. Acta,* **483,** 294 (1977).

107 N. Narayanan and P. V. Sulakhe, *Mol. Pharmacol.*, **13,** 1033 (1977).

108 L. J. Côté, *Life Sci.*, **3,** 899 (1964).

109 F. E. Samson, Jr., H. Dick, and W. M. Balfour, *Life Sci.*, **3,** 511 (1964).

110 F. E. Samson, Jr., and D. J. Quinn, *J. Neurochem.*, **14,** 421 (1967).

111 F. E. Samson, Jr., *Prog. Brain Res.*, **16,** 216 (1965).

112 A. A. Abdel-Latif, J. Brody, and H. Ramahi, *J. Neurochem.*, **14,** 1133 (1967).

113 P. R. Huttenlocher and M. D. Rawson, *Exp. Neurol.*, **22,** 118 (1968).

114 A. Schaefer, P. M. Komlós, and A. Seregi, *Biochem. Pharmacol.*, **28,** 2307 (1979).

115 A. Schaefer, M. Komlós, and A. Seregi, *Biochem. Pharmacol.*, **24,** 1781 (1975).

116 D. Desaiah and I. K. Ho, *Eur. J. Pharmacol.*, **40,** 255 (1976).

117 Y. Gutman and P. Boonyaviroj, *J. Neural Transm.*, **40,** 245 (1977).

118 D. C. U'Prichard, D. A. Greenberg, and S. H. Snyder, *Mol. Pharmacol.*, **13,** 454 (1977).

119 S. E. Robinson, P. L. Mobley, H. E. Smith, and F. Sulser, *Naunyn-Schiedebergs Arch. Pharmacol.*, **303,** 175 (1978).

120 T. L. Török and E. S. Vizi, *Acta Physiol. Acad. Sci. Hung.*, **55,** 233 (1980).

121 C. R. Scheid, T. W. Honeyman and F. S. Fay, *Nature*, **277,** 32 (1979).

122 P. H. Wu and J. W. Phillis, *Eur. J. Pharmacol.* In press.

CHAPTER TEN

MODULATION OF ADRENERGIC REACTIVITY AND ADRENOCEPTORS BY THYROID HORMONES

George Kunos

Department of Pharmacology and Therapeutics, McGill University, Montreal, Quebec, Canada

George Kunos is recipient of a Chercheur-Boursier award of the Conseil de la Recherche en Santé du Québec.

It has been known for over a century that thyroid hormones can modify the effects of sympathetic stimulation or catecholamines in various organs, or even mimic these actions to some extent. In an extensive review of the field, Harrison (1) noted in 1964 that "current confusion in adrenal medullary–thyroid relationships is great." Much new knowledge has accumulated since that time (see reviews 2–4), but the mechanisms of this relationship are still not clear. An obvious source of difficulty is the complexity of the actions of both thyroid hormones and catecholamines. A vast array of tissue components and physiologic functions appear to be influenced by thyroid hormones. This multiplicity of effects is probably related to the fact that thyroid hormones have specific receptors in the cell nucleus (5), which can mediate a generalized increase in messenger RNA production (6). The resulting change in protein synthesis could include various cell components that are involved in the actions of catecholamines. In addition, there are high-affinity binding sites for thyroid hormones in nonnuclear cell components (7), such as the plasma membrane (8,9). Although the functional relevance of these sites is not known, they may be involved in modifying membrane receptors for other hormones through short-term effects not involving protein synthesis. Thyroid hormones can also influence catecholamine-sensitive physiological processes indirectly by affecting homeostatic regulatory mechanisms.

Catecholamines also affect a large number of biological functions, and there are several levels of possible interactions between the two groups of hormones. Responses to catecholamines may be modified by changes in their synthesis, release, or disposition, as well as changes in the affinity or number of their receptors. Coupling of receptors to effector systems and biochemical mechanisms intermediate between receptor activation and the final tissue response are other possible sites of interaction. As an additional and intriguing possibility, it has been suggested that triiodothyronine (T_3) or some of its metabolites may act as "false" adrenergic transmitters and thus modulate sympathetic neurotransmitter effects; this suggestion was based on the demonstration of selective, synaptosomal accumulation of T_3 in rat brain (10,11).

As this chapter concentrates on possible interactions at the level of the adrenergic receptor, pre- and postreceptor effects of thyroid hormones will be discussed only in terms of whether or not they are sufficient to account for reported changes in physiological responses to catecholamines. Special attention is paid to the results of numerous studies that show reciprocal changes in α- and β-adrenoceptor reactivity in many but not all tissues. Examples of this and possible underlying mechanisms will be discussed in the sections dealing with individual organ systems. The recent development of radioligands with high specific activity has allowed direct identification and quantitative analysis of receptor-like binding sites in tissue fragments or, occasionally, in intact cells. An attempt will be made to correlate the findings of such studies with changes in physiological responses as detected by classical methods.

1   CARDIOVASCULAR SYSTEM

1.1   Heart

The influence of thyroid state on the adrenergic reactivity of the heart has been extensively studied not only because myocardial preparations provide convenient assay systems, but also because the similarities between the cardiac manifestations of hyperthyroidism and the effects of sympathetic stimulation have intrigued many investigators. A large number of clinical and experimental observations suggest that the sensitivity of the heart to catecholamines is increased in hyperthyroidism and decreased in hypothyroidism (1–4), although not all workers have been able to show such changes (12–15). Several factors can account for these contradictory observations. First, complete dose–response curves necessary to quantitate changes in drug sensitivity have not been determined in many studies. Also, there is inconsistency in the way sensitivity to drugs is defined. For example, the baseline myocardial contractility and heart rate are reduced in hypothyroidism and usually increased in hyperthyroidism, whereas the maximal developed tension or rate in response to catecholamines remains the same, so that the absolute change in the force or rate of contraction is greatest at low and smallest at high thyroid hormone levels (14). However, when sensitivity to an agonist is expressed as its half-maximally effective concentration (EC_{50}), significantly lower sensitivity to noradrenaline could be observed in hypothyroid than in control cats, even though low concentrations of the agonist produced similar absolute changes in force (15). Similarly, the dose of isoproterenol that produces a fixed increase in heart rate is an unreliable indicator of cardiac sensitivity to catecholamines in thyroid dysfunction. A reported lack of change in this index in human thyroid disease (15a) does not rule out increased catecholamine sensitivity in conditions where the basal heart rate is increased, such as hyperthyroidism, or following thyroxine treatment.

Second, there may be important species or strain differences as to whether cardiac sensitivity to catecholamines in the euthyroid state is closer to conditions in the hypothyroid or in the hyperthyroid state, and this could account for some of the conflicting conclusions. For example, the inotropic and chronotropic sensitivity to isoproterenol was reduced and sensitivity to phenylephrine was increased in hypothyroid rats, whereas opposite changes in the hyperthyroid state were much smaller for force responses or were completely absent for rate responses (16). On the other hand, in rabbit left atria the inotropic potency of agonists was not altered after thyroidectomy, whereas thyroxine (T_4) treatment significantly increased sensitivity to isoproterenol and decreased sensitivity to phenylephrine (17). The response pattern of normal rabbit atria to adrenoceptor agonists and antagonists was similar to conditions observed in hypothyroid rat atria. This illustrates that qualitatively similar changes in different species may be more apparent at different levels of thyroid state relative to the euthyroid condition. Interestingly, a similar species-dependent difference between rats and rabbits have been observed in the effects of thyroid state on myocardial myosin adenosine triphosphatase (ATPase) activity (18), which could suggest some similarity between myosin ATPase and cardiac adrenoceptors.

The third, and probably most important, source of conflicting results could be the fact that in most early studies, naturally occurring catecholamines were tested that can activate both α- and β-adrenoceptors. More recently, studies with selective agonists and antagonists for α- and β-adrenoceptors have indicated that changes in thyroid state produce opposite changes in α- and β-receptor-mediated mechanical responses of the heart. Since such reciprocal effects can also be induced by factors other than altered thyroid state (see Ref. 19), they may represent a general feature of the regulation of adrenoceptor reactivity. It is therefore appropriate to summarize what changes in classical pharmacological parameters are compatible with a change in the ratio of the two receptors. In the heart, where α- and β-receptors can mediate the same end response, there should be a change in the relative potencies of various agonists with different relative activities on α- and β-adrenoceptors. Unlike changes in potency, changes in the efficacy of agonists are generally thought to reflect postreceptor events. However, if the number of receptors is limiting the response (i.e., there is no significant receptor "reserve"), changes in the total number of receptors could conceivably lead to a change in the maximal response to an agonist. Antagonists have zero efficacy and their blocking potency is influenced only by their affinity to the receptor. Therefore, blocking of the effect of a selective agonist by a selective antagonist of the same receptor should not be influenced by changes in the absolute or relative number of receptors. However, when blocking of the effect of a mixed α/β-agonist is tested in such a two-receptor system, the slope of the Schild plot (log of the concentration of an antagonist plotted against log of the dose ratio − 1 for the agonist) is expected to be less than unity over part of the antagonist concentration range (20), but it will approach unity if the second type of receptor, resistant to the antagonist, is eliminated either by an antagonist of its own (21) or by induced changes in receptor ratios (22). This means that when the ratio of the two receptors is altered, block of a mixed agonist by a selective antagonist will also change, and the difference in blocking will be greatest at high antagonist concentrations, where the residual response to the agonist is determined by the unblocked receptors of the other type. In other words, differential blocking of the effect of a mixed agonist by a selective antagonist can indirectly indicate a change in receptor numbers (22).

The inotropic and chronotropic potency of phenylephrine and methoxamine increased and the potency of isoproterenol decreased in atria from thyroidectomized (16,23,24) or propylthiouracil-fed rats (25–27). This shift in response pattern of the hypothyroid myocardium from β to α was also supported by corresponding reciprocal changes in the blocking effectiveness of α- and β-receptor antagonists against the mixed stimulants noradrenaline and phenylephrine in the same studies. Block of responses to the relatively pure β-stimulant isoproterenol by propranolol

(28) and to the α-stimulant methoxamine by phentolamine (27) were not altered by thyroid state, and even with the mixed agonist phenylephrine there was no thyroid-dependent change in the blocking potency (pA_2) of α-receptor blocking drugs, when this was determined in the presence of a β-receptor antagonist (29). As discussed above, these observations indirectly indicate a change in the relative number or availability of α- and β-receptors without significant changes in their affinities. Changes in the opposite direction [i.e., increased β and decreased α effects in the hyperthyroid rat (16), guinea pig, and rabbit myocardium (28)] have also been reported.

Thyroidectomy is associated with atrophy of the adrenal cortex (30,31), reduced levels of calcitonin, and owing to partial removal of the parathyroids, decreased parathormone levels. However, these changes do not appear to contribute to the altered receptor response pattern of the hypothyroid heart. Adrenalectomy did not influence adrenoceptor responses of the rat (26) or dog heart (32); hypophysectomy produced changes similar to those after thyroidectomy in rats, and the altered cardiac response pattern was reversed only by T_4 and not by cortisone treatment (24).

Changes in thyroid state can affect various "prereceptor" processes, but these could not account for the observed reciprocal changes in agonist potencies in the heart. There is no clear-cut picture of the effects of thyroid state on the catechol-amine-metabolizing enzymes monoamine oxidase (MAO) and catechol-O-methy-transferase (COMT); the reported changes show considerable variations with species, tissue, age, and sex. In the adult rat heart, MAO activity is unaltered (33) or slightly reduced by T_4 treatment (34), and increased after hypophysectomy (35). Phenylephrine is metabolized by MAO, but the small changes in enzyme activity are opposite to those that could account for the altered potency of phenylephrine under these conditions. The increased potency of methoxamine in the hypothyroid myocardium (24,27) also cannot be explained by altered disposition of the drug, as methoxamine is not substrate to MAO, COMT, or neuronal uptake (36). The activity of COMT, the enzyme that metabolizes isoproterenol, is not influenced in the rat heart either by T_4 treatment (37) or by hypophysectomy (35). It was suggested (37) that decreased tissue uptake of noradrenaline (38) or its increased delivery to the myocardium may account for its increased effectiveness in hyperthyroidism. However, hyperthyroidism can also potentiate the mechanical (16,28) and metabolic effects of isoproterenol (39), an amine not readily taken up by sympathetic nerve endings, and inotropic responses to adrenaline were similarly potentiated by cocaine in ventricular strips of euthyroid and T_4-treated rabbits (40). Altered drug delivery through increased coronary circulation can also be ruled out as the cause of the potentiation of metabolic responses to catecholamines, as this could be observed in hearts perfused at constant flow (41). Changes in extraneuronal uptake can influence the apparent potency of various catecholamines, including isoproterenol. T_4 treatment of guinea pigs was shown to reduce extraneuronal catecholamine uptake in some tissues, but not significantly in the myocardium (42). Even so, inhibition of extraneuronal uptake by estradiol in rat atria did not influence the decrease in the inotropic potency of isoproterenol after hypophysectomy (24), which eliminated altered extraneuronal uptake as the possible mechanism of this change.

Several studies show that the turnover rate of noradrenaline in the myocardium is increased in hypothyroidism (43–45), and an increased "tonic" release of the neurotransmitter may indirectly mediate the effect of the hypothyroid state on end-organ sensitivity. Downregulation by homologous hormone (46) may account for the reduced β- but not for the increased α-receptor sensitivity of the hypothyroid myocardium. Depletion of noradrenaline stores by reserpine in hypothyroid rats failed to reverse the altered myocardial response pattern (24), which also argues against an indirect effect involving endogenous noradrenaline. Also, there is no sign of increased myocardial stimulation in the hypothyroid rat in vivo; this may be related to an observed shift from O-methylated to deaminated metabolites of noradrenaline in the hearts of hypophysectomized rats, which suggests that the increased noradrenaline turnover may be intraneuronal, resulting in the release of inactive metabolites (43). A direct rather than an indirect effect of thyroid hormones is also suggested by findings that in vitro incubation of isolated, cultured myocardial cells with T_3 can increase β-receptor sensitivity (47–49). Some effect of sympathetic innervation is indicated, however, by the finding that denervation of hypothyroid rats by 6-hydroxydopamine shortly after thyroidectomy prevented the shift from β- to α-receptors of the isolated atria (24). The failure of reserpine pretreatment to simulate this effect suggests that it must involve neuronal factors other than normal stores of the neurotransmitter (24).

Altered postreceptor mechanisms may also contribute to apparent changes in receptor reactivity, but there is no unequivocal evidence for such a change in the heart in different thyroid states. Although the role of adenosine $3',5'$-cyclic monophosphate (cyclic AMP) as a second messenger for the β-adrenergic increase in cardiac contractility is not universally accepted (50), adenylate cyclase is activated through β-receptors, and the subsequent cyclic AMP-dependent enzyme cascade does provide an assay system for testing post-β-receptor events in the heart. In the hypothyroid myocardium, basal adenylate cyclase activity was reported not to change (15), to increase (51–53), or to decrease (54,54a), whereas in the hyperthyroid state neither basal nor fluoride-stimulated activity was altered (52,53,55,56). The noradrenaline-stimulated enzyme activity was found only slightly depressed in the hypothyroid (15) or unaltered in the hyperthyroid myocardium (57,58). Other workers found, however, that the potency of catecholamines in activating adenylate cyclase was increased in the hyperthyroid and decreased in the hypothyroid rat heart (52,54a,56a,59). In atria of thyroidectomized rats, the potency of isoproterenol for raising cyclic AMP levels was reduced (54a,60), and this effect was reversed by T_4 treatment of hypothyroid rats (60). Adrenaline-induced cyclic AMP accumulation was also shown to be increased in isolated rat heart cells cultured in the presence of T_3 (49). These changes cannot be explained by altered cyclic AMP metabolism, since no change in phosphodiesterase activity could be detected in the myocardium of animals in different thyroid states (15,54a,56), although a small increase in the activity of the particulate enzyme has been recently reported to occur in the hyperthyroid rat heart (56a). A selective reduction in isoproterenol and guanine nucleotide stimulation of adenylate cyclase in hypothyroid rat heart has recently been reported (60a). As basal and fluoride-stimulated activities as well as the number of guanine nucleotide-binding sites on the enzyme were unaltered, the changes were attributed to changes at either the β-receptor or the coupler site of adenylate cyclase (60a).

Potentiation of the cardiac glycogenolytic effect of catecholamines in thyroid hormone-treated animals is well established (39,41,61). Since some authors found the effects of catecholamines on cardiac adenylate cyclase unchanged, the possibility arose that the sensitizing effect of thyroid hormones may be localized at one of the components of the cyclic AMP-dependent cascade system. This possibility was, in fact, supported by a finding that cardiac phosphorylase activation by dibutyryl cyclic AMP (dbcAMP) was increased in hyperthyroid rats (62). However, the results of other studies failed to localize the site of this potentiation; hyperthyroidism did not influence the activity of cardiac phosphorylase b kinase (63), did not increase the activation of protein kinase by cyclic AMP or noradrenaline (64) and even decreased its activation by isoproterenol (56a). The increased response to dbcAMP was probably due to its increased uptake into the hyperthyroid myocardium or its increased conversion into the active monobutytyl analog (64). Therefore, it appears that the major site for the potentiation of cardiac β-adrenoceptor responses by thyroid hormones must reside at a step before cyclic AMP formation, as indicated by studies demonstrating altered potencies of catecholamines in increasing cyclic AMP.

Since calcium can also activate cardiac phosphorylase (65), increased sensitivity of phosphorylase to calcium or increased release of calcium from intracellular stores may also account for the supersensitivity to catecholamines. The former possibility was eliminated by the finding that calcium-activation of cardiac phosphorylase was not affected by T_4 treatment of rats (66). The sensitivity of the contractile system to calcium is also unaffected by thyroid state, as indicated by the unchanged inotropic potency of calcium (28,54a,66,67). However, an increased release of calcium from intracellular stores has not been tested, and this mechanism may be possible in view of findings of an increased transport and exchange of calcium by the sarcoplasmic reticulum in hyperthyroid rabbits (68) and dogs (69).

Recent studies indicate that β-adrenoceptor stimulants can increase membrane fluidity by activation of one of the enzymatic steps in the formation of phosphatidylcholine from phosphatidylethanolamine in the cell membrane (70). As this event appears to precede adenylate cyclase activation (70), demonstration of its presence in myocardial tissue and a study of its sensitivity to thyroid manipulation may help to further localize the site of thyroid–catecholamine interaction.

In contrast to β-adrenoceptors, activation of α-adrenoceptors has not yet been unequivocally associated with a biochemical mechanism present in all tissues. Evaluation of the mechanical effects of α-adrenoceptor stimulation in the heart is also complicated by the fact that the response is often complex, displaying both positive and negative inotropic components (71,72). The negative component has been attributed to an α-receptor-mediated inhibition of cardiac adenylate cyclase (73), but potentiation or even the presence of such an effect in the hypothyroid myocardium has not yet been reported. The α-receptor-mediated positive inotropic effect appears to develop independently of adenylate cyclase activation. In the hypothyroid rat heart, phenylephrine, in the presence of a β-blocking drug, was found to increase contractile force without a concomitant increase in cyclic AMP levels (29,72,74). In electrically driven left atria from hypothyroid rats, phenylephrine, in the absence of β-blockade, did increase cyclic AMP levels; the potency of phenylephrine was higher than in euthyroid controls and the effect on cyclic AMP was blocked by high concentrations of an α- but not by a β_1-receptor antagonist (60). However, this effect appeared to be a consequence rather than the cause of increased contractile activity; when phenylephrine was tested in quiescent atria, there was no increase in cyclic AMP in the hypothyroid preparations, whereas an increase, blocked by propranolol, was still evident in control atria (60). These results are also compatible with a lack of direct involvement of cyclic AMP in the increased α-receptor activity in the hypothyroid myocardium. Phenylephrine can increase myocardial guanosine 3'5'-cyclic monophosphate (cyclic GMP) levels (60,75). However, correlation of this effect with α-receptor activation is not clear: it was found to be blocked not only by α- but also by β-antagonists (60,75), although selective block by an α-receptor antagonist has also been reported (76). Moreover, the phenylephrine-induced rise in cyclic GMP level was reduced rather than increased in the hypothyroid rat myocardium (60).

The greater sensitivity of the effect of phenylephrine than that of isoproterenol to inhibition by a calcium antagonist suggested that α-receptor stimulation may increase the myocardial force of contraction by enhancing transmembrane calcium movements (77). An alternative explanation of these findings is that calcium antagonists may have α-receptor blocking properties in the heart, as was recently found in rat liver cells (78). Another process that appears to be under α-adrenergic control in the myocardium is glucose uptake (76). Recent studies in several tissues have indicated that α-adrenoceptor stimulation will lead to increased turnover of phosphatidylinositol in the cell membrane (79). Studies of the effects of altered thyroid states on these processes in the myocardium could lead to new information on the mechanism of thyroid modulation of α-adrenergic reactivity of the heart.

From the discussion above it is clear that although thyroid state can influence a variety of enzymes and physiological functions in the heart, such nonreceptor mechanisms cannot satisfactorily and fully explain the altered adrenergic reactivity. There has been sufficient indirect evidence to indicate changes in the receptor system itself. Findings that the potency of the pure α-agonist, methoxamine, increased, and the potency of isoproterenol decreased in the hypothyroid myocardium indicate that both α- and β-receptors are affected and the altered response pattern is not due to changes in one receptor simply masking or unmasking the unchanged activity of the other. The absence of changes in the pA_2 values for either α- or β-receptor antagonists indicate that the affinity of the receptors is not significantly altered. The remaining possibility is a change in the number or availability of α- and β-receptors. However, changes in the coupling of receptors to effector systems is an alternative explanation of all of the foregoing observations, inasmuch as selective "coupling" processes for different receptors can be postulated. Since the nature of coupling is unknown, a change in this process can only be postulated by exclusion, by demonstration of no change in the receptor recognition site.

The recent development of radioligands with high specific activity made it possible to measure directly the affinity and concentration of binding sites that show specificity similar to α- and β-adrenoceptors (see Chapters 5 and 8). There have been several reports of altered ligand binding to myocardial tissue of rats in different thyroid states. Some of the observed changes correlated with the altered *in vivo* reactivity of receptors, but there were also discrepancies. We will briefly review the available information and point out several problems in interpreting the results of binding studies and in trying to extrapolate from them to possible changes in functional receptors in intact tissues.

The first problem relates to the way "specific" binding is determined and to the equation of such sites with receptors mediating a well-defined tissue function. Hyperthyroidism induced by thyroid hormone treatment of normal rats increased the density of binding sites for the β-antagonist [³H]dihydroalprenolol ([³H]DHA) (53,80,81), and decreased the density of binding sites for the α-receptor antagonist [³H]dihydroergocryptine ([³H]DHEC) (80,82). The affinities of binding sites were not changed or were only minimally affected. Hypothyroidism induced by propylthiouracil treatment or thyroidectomy decreased the density of [³H]DHA binding sites, but also decreased the density of [³H]DHEC binding sites (53,80,82). Although the foregoing findings in the hyperthyroid preparations correlate qualitatively with reciprocal changes in the reactivity of receptors in intact atria, the affinity of binding for both ligands was relatively low (K_D's between 5 and 19 nM), and in some cases the binding of [³H]DHEC was not even saturable (80). Recent observations indicate that [³H]DHEC preferentially labels postsynaptic α-receptors at concentrations below 3 nM, whereas a large part of the binding at higher concentrations is unrelated to such receptors (83) and it may represent presynaptic α-receptors (84,85). For [³H]DHA it was also shown that its binding at concentrations above 4–5 nM may be largely nonspecific when an unlabeled antagonist rather than an agonist is used as the suppressing ligand (86,87). When [³H]DHA and [³H]DHEC binding in cardiac tissue was determined at low ligand concentrations and at close concentration intervals, a saturable component with low density (30–50 fmol/mg protein) and high affinity ($K_D < 3$nM) could be detected for both ligands (24,88–90). Although the density of such high affinity [³H]DHA sites still increased and the density of [³H]DHEC sites decreased in hyperthyroid hearts, these changes were small and statistically not significant (88,91). The absence of major changes in receptor numbers or affinity is compatible with findings that the receptor response pattern of euthyroid rat atria is much closer to that in hyperthyroid than in hypothyroid preparations (16).

Another problem of interpretation of binding studies could arise from possible changes in the physicochemical properties of tissue components in altered thyroid states, which would result in different yields of receptor-rich cell components when the same subcellular fractionation method is used in different thyroid states. The absence of a change in the overall yield of crude "membrane" protein does not rule out this possibility, because the composition of the membrane fraction may have been altered by dilution with nonmembrane proteins. Differences in the way membranes are prepared may explain why hypothyroidism in one study was found to increase rather than decrease [³H]DHA binding-site density in rat hearts (92).

A third complicating factor is the significant atrophy of the hypothyroid myocardium, which may be associated with the loss of cells or tissue components rich in both α- and β-receptors. Such a change may be independent from a direct action of thyroid hormones on receptors, but it could accentuate a decrease in β or mask an increase in α-receptor density. Hypothyroidism was shown to nonspecifically reduce the concentration of membrane-bound tissue components (54,82), which may indicate a generalized decrease in membrane density. Thus a comparison of euthyroid and hypothyroid preparations could lead to erroneous conclusions on the effects of thyroid hormone on receptor densities, particularly when the yield of the membrane fraction is also different, as in a recent study (91). On the other hand, thyroid hormone treatment is more selective in its effect on adrenoceptors (82). This greater receptor specificity may be related to the fact that the altered receptor response pattern develops rapidly within a few days after thyroid hormone treatment and before the onset of cardiac hypertrophy, whereas the opposite changes observed in the hypothyroid myocardium take several weeks to appear (24), which allows more time for nonspecific changes to develop. This difference in specificity of changes is also apparent when mechanical responses of intact atria are measured; hypophysectomy was found to result in reciprocal changes in the inotropic potency of α- and β-receptor agonists, and also in decreased basal contractility and changes in agonist efficacy (24). Short-term treatment with T_4 completely reversed the changes in agonist potencies without significantly affecting the other parameters (Figure 1). It appears therefore that the most specific way to assess the direct effect

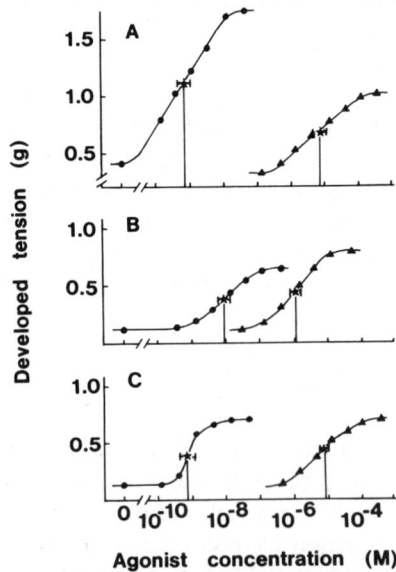

Figure 1 Inotropic responses of left atria to isoprenaline (●) and to methoxamine (▲) in normal *(A)*, hypophysectomized *(B)*, and hypophysectomized rats treated for 2 days with 200 µg/kg/day thyroxine *(C)*. Number of experiments: 4 *(A)*, 4 *(B)*, and 3 *(C)*. The asterisks on the dose–response curves and the horizontal bars indicate the mean $EC_{50} \pm 2 \times SE$. The vertical lines were drawn to illustrate mean EC_{50}'s on the abscissas. Reproduced by permission from Ref. 24.

of thyroid hormones on adrenoceptors is by comparing conditions in the hypothyroid tissue with those after a short-term treatment of hypothyroid animals with thyroid hormones. Banerjee and co-workers found that T_3 treatment of thyroidectomized rats (500 μg/kg/48 hr for 6 days) caused an increase in the density of high-affinity [^{3}H]DHA (88) and a matching decrease in the density of high affinity [^{3}H]DHEC binding sites (89) in the total ventricular homogenate of thyroidectomized rats. No changes in binding site affinities were noted. In the study mentioned above, Kunos et al. (24) found that when hypophysectomized rats were treated for only 2 days with 200 μg/kg/day of T_4, the increase in β- and decrease in α-receptor responses of atria were associated with an increase in the density of [^{3}H]DHA binding sites in ventricular homogenates from the same rats and a matching decrease in the density of prazosin suppressible [^{3}H]WB-4101 binding sites (Figure 2, Table 1). These findings confirm Banerjee's observations and extend them in three ways. First, the T_4 treatment schedule (lower dose, T_4 rather than T_3, shorter treatment) was such that it did not lead to significant cardiac hypertrophy (24); therefore, the observed changes must reflect true changes in receptor density (24). Second, both [^{3}H]WB-4101 and prazosin are selective antagonists for postsynaptic, α_1-type receptors (see Chapter 5), which eliminated the possibility that the labeled sites included presynaptic α-receptors not involved in the physiological functions meas-

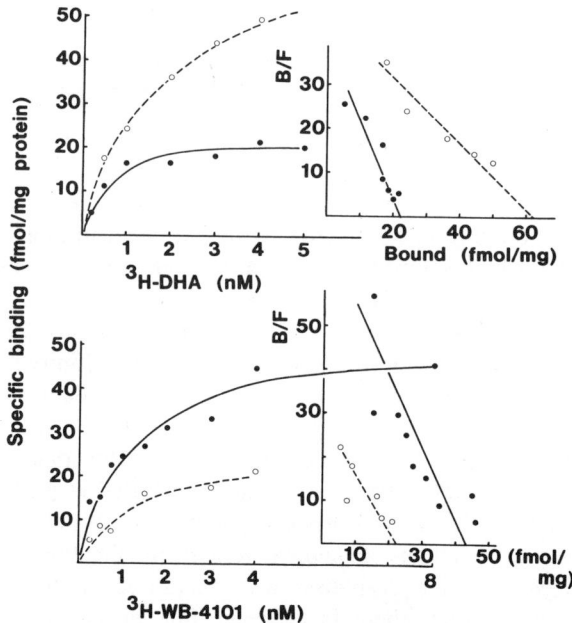

Figure 2   Effect of thyroxine treatment of a hypophysectomized rat on the specific binding of [^{3}H]WB-4101 and [^{3}H]dihydroalprenolol to cardiac membrane fragments. ●——●, hypophysectomized rat; O---O, hypophysectomized rat treated with 200 μg/kg/day T_4 for 2 days. Insets are Scatchard plots. Specific binding represent total binding minus binding in the presence of 1 μM (±)-propranolol for [^{3}H]DHA or 1 μM of prazosin for [3H]WB-4101, determined in triplicate. Reproduced by permission from Ref. 24.

TABLE 1   Effect of Thyroxine Treatment (200 μg/kg/day, 2 Days) on Cardiac Adrenoceptors in Hypophysectomized Rats[a]

	Hx[b]	$Hx + T_4$
Receptor Response (Intropic)		
Alpha (methoxamine, pD_2)	5.93 ± 0.15 (4)**	5.09 ± 0.12 (3)
Beta (isoprenaline, pD_2)	7.65 ± 0.10 (4)**	9.12 ± 0.14 (3)
Receptor Binding		
Binding-site density (fmol/mg protein)		
Alpha ([³H]WB-4101–prazosin)	38.7 ± 3.1 (4)**	18.7 ± 2.5 (4)
([³H]DHEC–phentolamine)	42.6 ± 3.1 (5)**	19.2 ± 3.5 (4)
Beta ([³H]DHA–propranolol)	27.5 ± 2.7 (6)*	45.5 ± 5.7 (5)
Total $\alpha_1 + \beta$	66.2	64.2
Binding-site affinity (K_d, nM)		
Alpha ([³H]WB-4101)	0.48 ± 0.08 (4)	0.60 ± 0.18 (4)
([³H]DHEC)	1.79 ± 0.41 (5)	1.03 ± 0.21 (4)
(Methoxamine–[³H]WB-4101)	6.5 μM (2)	8.9 μM (2)
Beta ([³H]DHA)	1.59 ± 0.27 (6)	1.89 ± 0.28 (5)
(*l*-Isoproterenol–[³H]DHA)	0.45 μM (2)	0.32 μM (2)

[a]For details, see Ref. 24. Number of experiments in parentheses.
[b]Asterisks indicate significant differences between adjacent values: *, $p < 0.05$; **, $p < 0.005$.

ured. Interestingly, when cardiac α-receptors were determined as phentolamine (10 μM) suppressible binding of [³H]DHEC at ligand concentrations *below* 3–4 nM, the density of these high-affinity ($K_D < 2$ nM) binding sites was almost the same as those labeled with [³H]WB-4101 (see Table 1). This confirms previous findings indicating that at low concentrations [³H]DHEC preferentially labels postsynaptic α-receptors (83–85). β-Adrenoceptors, identified as high-affinity [³H]DHA binding sites, are also predominantly postsynaptic in the heart, as their numbers increased rather than decreased after chemical sympathectomy (24). Thus the inverse changes in α- and β-receptors probably occurred in the same postsynaptic membrane. Third, receptor binding and receptor function changed in parallel when measured in preparations obtained from the same hearts and under identical pH, ionic, and temperature conditions, and both types of changes were maximal within 2 days of treatment (24). Although physiological responses were measured in atria and binding was tested in ventricular tissue, there is evidence that in altered thyroid states the inotropic response pattern of rat atrial and ventricular tissue is qualitatively similar (29). Thus the specific bindings measured probably represent α- and β-receptors mediating mechanical responses of the rat heart.

    The parallel time course of changes in α- and β-adrenoceptors in the rat heart is in contrast to results in a report on rabbit and guinea pig left atria, where thyroid hormone treatment of the euthyroid animals was found to increase the sensitivity to isoproterenol more slowly than it decreased the sensitivity to phenylephrine (28); no binding studies were carried out in the latter study and the reason for the different results is not clear. The authors of the latter study (28) argue that the opposite changes in α- and β-receptor responses after thyroxine treatment must be through different and independent mechanisms, because while the dose–response curve for isoproterenol (β) was shifted to the left in a parallel manner, the dose–response curve for phenylephrine in the presence of propranolol (α) was shifted not only to the right but also downward. However, it is extremely tenuous to draw conclusions on molecular mechanisms on the sole basis of dose–response curves in intact tissues. Even so, the response pattern described above is compatible with reciprocal changes in the numbers of α- and β-receptors, if one considers that there is evidence for a very large "receptor reserve" for cardiac β-receptors (see Chapter 7), whereas studies with phenoxybenzamine did not indicate significant spare α-receptors in the myocardium (16). In the absence of spare receptors, changes in receptor numbers could conceivably lead to changes in the maximal response to an agonist.

In spite of their relatively rapid development, the thyroid-dependent changes in cardiac α- and β-receptors could still be mediated by changes in the rate of their synthesis or degradation. However, the possibility that thyroid hormones may influence adrenoceptors independently of their effects on protein synthesis was indicated by findings that *in vitro* incubation of cultured myocardial cells with concentrations of T_3 that do not affect cell growth increased both the reactivity and the number of β-adrenoceptors (49). The density of [^{3}H]DHA binding sites was also increased in rat heart slices incubated with nanomolar concentrations of T_3, and the early increase in binding-site numbers that occurred within 60 min was not inhibited by cycloheximide or puromycin, which blocked protein synthesis in the same preparations (89). In the latter study, the effect of T_3 was proposed to involve an increase in the incorporation of soluble β-adrenoceptors into the cell membrane, since incubation of purified membranes with T_3 did not result in a change in binding (93). However, binding of [^{3}H]DHA was tested only at a single high concentration (15 nM), where a large fraction of propranolol suppressible binding may be nonspecific (see above). The functional role of these acute changes in binding sites is not yet clear. Studies with isolated cat hearts perfused with thyroid hormones *in vitro* for 45 min failed to detect any change in the adrenergic reactivity of the myocardium (94), although pretreatment of the cats with 6-hydroxydopamine in this study may have influenced the results by removing neurogenic components that may be involved in mediating the effects of thyroid hormones (24,95). *In vitro* incubation of guinea pig left atria with 1 μg/ml T_4 for 1 hr also did not influence the positive inotropic effect of phenylephrine, although *in vivo* treatment with a single dose of T_4 1 day before the experiment significantly attenuated the phenylephrine response (28).

It has been proposed that the pharmacological properties of myocardial adrenoceptors are determined by the metabolic state of the tissue (96), changes of which may result in an "interconversion" of α- and β-adrenoceptors. Although the demonstration of rapidly developing, matching, reciprocal changes in α- and β-receptor reactivity and parallel changes in binding-site densities do not definitely prove such a hypothesis, they are compatible with it. Alternatively, such changes may reflect independent changes in the two receptors or independent receptors functionally coupled by the same regulatory signal. Obviously, such changes do not exclude and may occur simultaneously with additional effects of thyroid hormones on receptor coupling and various pre- and postreceptor processes.

1.2 Blood Vessels

There is a large body of literature on the effects of thyroid state on pressor responses to injected catecholamines in humans and intact animals (1–4). Since blood pressure responses are too complex to allow conclusions on possible changes in adrenoceptor sensitivity in vascular smooth muscle, they are not discussed here. Studies of local circulatory responses to exogenous catecholamines seem to indicate that β-receptor mediated vasodilatation is increased in hyperthyroidism and decreased in hypothyroidism. In hyperthyroid dogs, infusion of noradrenaline and adrenaline into the femoral artery produced less vasoconstriction than in controls, but the constrictor response was restored after β-blockade with pronethalol (97). As no difference in responses to nonadrenergic vasoconstrictors was noted, the effects of hyperthyroidism could be attributed to increased β-receptor sensitivity masking the α-adrenergic effect of catecholamines (97). Reduced β-receptor responsiveness in the hypothyroid state is indicated by a decrease in the isoprenaline-induced rise in tail skin temperature in rats (98). It is not certain, however, whether this response is a direct vascular effect of isoproterenol or whether it is centrally mediated. Various other centrally mediated effects of isoproterenol were also found to be reduced in this study (98). Thyroidectomy was found to reduce the isoproterenol-induced relaxation of isolated aortic strips from genetically hypertensive rats. Nonadrenergic relaxation was unaffected and the change in the effect of isoproterenol was reversed by thyroxine treatment (99). However, T_4 treatment of rabbits also reduced rather than increased the isoproterenol sensitivity of isolated aortic strips (100).

There is no clear-cut relationship between thyroid state and α-adrenergic reactivity of vascular smooth muscle. Some studies show that thyroid hormone treatment nonspecifically decreases the maximal tension development of isolated vascular strips in response to various constrictor agents, including α-receptor agonists (28,100,101). Others report a small but, again nonspecific increase in such responses (102). Hypothyroidism was shown to selectively increase the efficacy but not the potency of noradrenaline in rabbit aortic strips (103), but no change (100) or a decreased response to catecholamines was reported in rat aortic strips (102). Decreased diastolic pressure elevation by α-receptor agonists in pithed hypothyroid rats also indicated reduced vasoconstrictor responsiveness (104,105).

In summary, the effects of thyroid state on vascular α- and β-receptor reactivity is equivocal. The same intervention is often reported to produce opposite changes in different or even in the same species, and most studies suggest that the changes that do occur are quite nonspecific.

2   METABOLIC EFFECTS OF CATECHOLAMINES

2.1   Adipose Tissue

The physiologically most important effect of catecholamines in adipose tissue is the activation of triglyceride lipase and the resulting release of free fatty acids, although activation of glycogen phosphorylase and inactivation of glycogen synthase can be also demonstrated. Increased lipolysis by catecholamines appears to be mediated by a β_1-type receptor, whereas in humans (106,107), rabbit (108), and hamster (109), but not in rat, dog, or guinea pig adipose tissue (110), the presence of inhibitory α-adrenoceptors can also be demonstrated. The effects of catecholamines on glycogen metabolism of rat adipocytes involve α- and β-receptors that mediate the same end response (111), a situation similar to that seen in the liver (see below).

The β-receptor-mediated stimulation of lipolysis involves the same cyclic AMP dependent cascade mechanism as found for other β-receptor-mediated tissue responses, which provides a basis for the search of the site of the modulatory action of thyroid hormones. Debons and Schwartz (112) were the first to show that thyroid hormone treatment potentiates and hypothyroidism suppresses the *in vitro* lipolytic action of catecholamines in adipose tissue. A blunting of the cyclic AMP accumulation by catecholamines in fat cells from hypothyroid rats (113) or human subjects (114) is also well documented. However, the mechanism of the apparently reduced β-receptor response is not yet clear. Treatment of rats with high doses of T_3 was found to increase and hypothyroidism to decrease basal adenylate cyclase activity in adipose tissue, and this was proposed to account for the modification of the catecholamine response (115). However, smaller doses of T_3 that still potentiate glycerol release and cyclic AMP accumulation by catecholamines failed to alter adenylate cyclase activity in fat cell ghosts (116), and fluoride-stimulated adenylate cyclase activity was similar in fat cell membranes from control and hypothyroid rats (117). The total ATP content of adipocytes was also unchanged by T_3 treatment (118). A thyroid-induced change distal to the formation of cyclic AMP is suggested by findings that the lipolytic response to dbcAMP is increased in hyperthyroid fat cells (116). However, increased cellular uptake of dbcAMP or its decreased conversion to the active monobutyryl analog may be alternative explanations of this observation. In fat cells of hypothyroid rats, no decrease in the lipolytic response to dbcAMP was observed, although responses to noradrenaline were reduced (119). Although the mechanism of a possible change in sensitivity to cyclic AMP is not clear, altered sensitivity of fat cell protein kinase is discounted by the finding that activation of protein kinase by cyclic AMP in fat cells from thyroidectomized rats was the same as in controls (120).

Considerable attention has been focused on the possibility that the altered lipolytic activity of catecholamines in different thyroid states is due to changes in the metabolism of the cyclic AMP generated by hormonal stimuli. This hypothesis was based on findings that the activity of the high-affinity particulate phosphodiesterase was higher in fat cells from hypothyroid than in those from control rats (121), and the increase was reversed by treatment of hypothyroid rats with T_3 (120). In agreement with this finding, the potentiation of the adrenaline-induced lipolysis by theophylline was greater in hypothyroid than in normal rat fat cells (122). In the latter study it was also found that EGTA increased the blunted lipolytic response to adrenaline in hypothyroid cells and a calcium ionophore suppressed the adrenaline response of normal fat cells, although this finding has not been confirmed in a recent work (122a). These findings indirectly indicate an inverse relationship between thyroid hormone levels and intracellular calcium concentration in fat cells. Thus both the increased phosphodiesterase and decreased adenylate cyclase and triglyceride lipase activities in hypothyroid fat cells may be secondary to increased calcium levels (122). Incubation of tissues in sodium-free buffer can increase intracellular calcium concentration by preventing Na^+–Ca^{2+} exchange. When human adipose tissue was incubated *in vitro* in a buffer where some of the Na^+ was replaced with Li^+ (123), noradrenaline-induced lipolysis was inhibited, and the inhibition was partially reversed by phentolamine. The lipolytic effect of dbcAMP was not affected by Li^+ and the inhibitory effect of the Li^+ buffer was blocked by lanthanum (123). These results also suggest the possibility that calcium may be the intracellular signal that mediates altered adrenoceptor reactivity of fat cells. However, in other studies the thyroid-dependent difference in the catecholamine-induced lipolysis and cyclic AMP accumulation persisted in the presence of high concentrations of theophylline (116), and there was no difference in the activity of the high-affinity phosphodiesterase in fat cell ghosts from normal and hypothyroid rats (113). These results argue against the role of phosphodiesterase and, indirectly, of calcium in the altered adrenergic response of fat cells.

As α- and β-receptors mediate opposite effects on lipolysis, the decreased response to catecholamines in the hypothyroid state may be the result of decreased activity of β-receptors, increased activity of α-receptors, or both. Increased α-receptor activity in adipose tissue of hypothyroid human subjects has been proposed to explain the finding that phentolamine increased the effect of adrenaline and noradrenaline on lipolysis and cyclic AMP accumulation significantly more in hypothyroid than in normal individuals (106). Whether the change in α-receptor response is mediated by a change in the affinity or number of binding sites or by a change in the pool of intracellular calcium that can be released by α-receptor stimulation is not known. In contrast to the effect of hypothyroidism, T_3 treatment of hamsters reduced the density of high and low affinity binding sites for ^{3}H-DHEC and also reduced the potentiating effect of phenotlamine on the adrenaline-induced rise in cAMP (123a). The data in this study also show opposite changes in adipocytes from hypothyroid hamsters, although these changes were small and were not statistically analyzed. These findings indicate that, as in the rat heart, thyroid hormone can "down" regulate α receptors. It has been suggested that the α receptors that mediate inhibition of adenylate cyclase in adipocytes are of the α_2 type, as opposed to α_1 receptors that mediate calcium-dependent processes (123b). The effect of thyroid state on this latter type of α-receptor in adipose tissue is less clear. It was reported that the effect of methoxamine on the incorporation of ^{32}P into phosphatidylinositol, an α_1 type response, is similar in euthyroid and hypothyroid rat fat cells (123c), suggesting the lack of thyroid regulation of α_1 receptors. However, results of the same study show that inactivation of glycogen synthase by isoproterenol was nearly abolished, whereas inactivation by adrenaline was unchanged in hypothyroid adipocytes. As this response is mediated by both β and α_1 receptors, the unchanged response to adrenaline can be explained only by a combination of a decreased β *and* an increased α_1 receptor component in the hypothyroid state (Table V in ref. 123c). Thyroid regulation of α_1 receptors in heart and liver (see sections 1.1 and 2.2) is also in conflict with the suggested lack of such regulation in fat tissue. Further studies are required to resolve this issue.

A decreased activity of β-receptors is suggested by the decreased potency of noradrenaline, not reversed by phentolamine, in raising cAMP levels in hypothyroid rat fat cells (124). Although maximal responses to the relatively pure β-stimulant isoproterenol were not altered by hypothyroidism in fat cells of rats or human subjects (107), complete dose–response curves, necessary to assess receptor sensitivity, have not been reported in this study. Ligand binding studies with the β-receptor antagonist [^{3}H]DHA did not give clear-cut results. In two recent studies, changes in thyroid state were reported not to affect either the density or the affinity of binding sites in rat fat cells (113,122), whereas a decreased density of binding sites in hypothyroid (125) and increased density in hyperthyroid rats (53) have also been reported. Binding of [^{3}H]DHA in fat cells appears to be to heterogeneous sites, as indicated by the upward concavity of Scatchard plots (113) and the absence of negative cooperativity that may account for it (126). It will be important to determine which of the binding components corresponds to functional receptors mediating lipolysis. Another source of discrepancies may be the finding that adipocytes from hypothyroid rats are considerably smaller than cells in control rats (125). Thus possible changes in binding-site densities will be different when values are expressed as per milligram of protein, cell number, or cell surface area. The yield of the receptor-rich subcellular cell fraction may also be different in different thyroid states.

In one study the absence of changes in either binding-site density or the catalytic activity of adenylate cyclase or phosphodiesterase led to the conclusion that, by exclusion, thyroid hormones must regulate the transduction of information between hormone receptors and adenylate cyclase in fat cells (113). Although this process is not yet clearly understood, its defect in the hypothyroid state does not seem to involve a decrease in the amount of guanyl nucleotides involved in coupling of receptors and adenylate cyclase, as the nucleotide analog Gpp(NH)p did not reduce the difference in the effect of noradrenaline on adenylate cyclase between control and hypothyroid cells (113). A defect in the nucleotide binding site of adenylate cyclase remains a possibility. In a more recent study Malbon reported (113a) that guanyl nucleotides reduce the affinity of agonists to [^{3}H]DHA binding sites in normal adipocytes, but do not influence affinity in cells from either hypothyroid or hyperthyroid rats. Since the K_D for isoproterenol was more than 10 times lower in hyperthyroid than in hypothyroid preparations, the above results indicate that guanyl nucleotides are not directly involved in the effects of thyroid hormone on adipocyte adrenoceptor responsiveness.

Two other mechanisms deserve further discussion. The calorigenic action of thyroid hormones has been associated with a thyroid-induced increase in (Na^+, K^+) ATPase activity in various thermogenically responsive tissues (127,128). Several studies indicate that hormone-induced lipolysis is sensitive to inhibition by ouabain (129,130), and activation of (Na^+, K^+)ATPase has been proposed to mediate β-adrenergic relaxation of smooth muscle (131). Therefore, it would be reasonable to postulate that the effect of thyroid hormones on catecholamine sensitive lipolysis is mediated by changes in (Na^+, K^+)ATPase. This possibilty is not supported, however, by findings that ouabain did not abolish the T_3-induced increment in the lipolytic effect of adrenaline (118,122). Another interesting possiblity is suggested by findings that the decreased lipolytic effect of adrenaline in hypothyroid rat adipocytes was restored to normal by incubation of the cells with adenosine deaminase, whereas a nonmetabolizable analog of adenosine reversed the effect of the enzyme and also suppressed the adrenaline response of cells from normal rats (132). This may suggest that the reduced lipolytic effect of adrenaline in hypothyroidism is mediated either by increased sensitivity to or increased production of adenosine. This inhibitory or modulatory effect of adenosine was found to be independent of extracellular calcium (122a).

Phosphorylase activation in adipose tissue by catecholamines is also affected by thyroid state; hypophysectomy was reported to severely blunt the response to adrenaline (133). However, whereas the parallel decrease in the lipolytic response was readily reversed by corticosteroid treatment, the decrease in phosphorylase activation was corrected only by throxine, and not by corticosteroids (133). This indicates that two cyclic AMP-mediated processes in the same cell may be separately modulated. The role of corticosteroids in the regulation of the lipolytic response to catecholamines is also illustrated by the finding that adrenalectomy in dogs was shown to reduce the β-receptor-mediated lipolysis by isoproterenol and increase the α-receptor mediated antilipolytic effect of adrenaline (134). These reciprocal changes are similar to those seen in hypothyroidism and show that different hormonal stimuli may result in similar modification of the adrenoceptor response. Further examples of this will be shown in the liver.

Thyroid hormones play an important role in the control of thermogenesis, and can enhance the so-called calorigenic effect of catecholamines. Several processes that involve utilization of ATP contribute to this calorigenic effect (135). One of the processes sensitized by thyroid hormones is the catecholamine stimulated uptake of free fatty acids by tissues, whereas in the hypothyroid state both the mobilization and the uptake of free fatty acids is suppressed (136). An important site of the calorigenic actions of catecholamines is the brown adipose tissue. Thyroid hormones control not only the morphogenesis of brown fat, but can also sensitize it to the lipolytic action of catecholamines (137). Hypothyroidism, on the other hand, was shown to suppress the effect of noradrenaline, whereas the lipolytic effect of dbcAMP was much less affected (138). This selectivity suggests that the site of the action of thyroid hormone is before the formation of the second messenger signal.

In summary, hypothyroidism reduces and thyroid hormones enhance the lipolytic effects of catecholamines. In some species, these changes reflect reciprocal changes in stimulatory β_1- and inhibitory α_2-receptor reactivity. The exact molecular mechanism of these changes is not clear, although several possibilities have been proposed and received partial support.

2.2   Liver

Catecholamines can stimulate or inhibit the same metabolic pathways in liver as in adipose tissue and thus increase the formation and release of metabolic fuels. In contrast to fat tissue, mobilization of glucose is quantitatively more important than lipolysis. A large number of earlier studies indicated that hyperthyroidism enhanced and hypothyroidism suppressed catecholamine-induced glucose mobilization (1). However, in intact animals or humans the hyperglycemic response to catecholamines is the result of complex homeostatic regulatory mechanisms, involving effects on not only liver, but pancreatic islets, adipose tissue, muscle, and pituitary as well (135). Therefore, results of such studies do not allow any conclusions on the nature of the liver adrenoceptor involved and they will not be discussed here.

There are relatively few reports on the influence of thyroid state on catecholamine responses of isolated liver preparations. Hagino and Nakashima (139,140) used a perfused rat liver preparation to study the effects of adrenergic agonists and antagonists on gluconeogenesis from lactate. In the euthyroid liver, adrenaline, isoproterenol, and phenylephrine, each tested in a single dose, were found to stimulate gluconeogenesis by activation of the enzyme pyruvate carboxylase, whereas in the hypothyroid liver, only adrenaline and phenylephrine were found to be effective (140). These findings were interpreted to suggest that, as was found earlier in the myocardium, β-receptors become less important and α-receptors more important in mediating responses of the hypothyroid liver. However, in isolated liver cells, where the metabolic effects of catecholamines are not complicated by possible vascular effects, it has been unequivocally demonstrated that in euthyroid rats catecholamine-induced gluconeogenesis is a predominantly α-receptor-mediated event, involving both increased mitochondrial pyruvate carboxylation and inhibition of pyruvate kinase (141–143). Results of the Japanese workers (139,140) are also difficult to evaluate because of the absence of full dose–response curves in their studies and because of their finding that responses of the normal liver to any of the agonists tested were not inhibited by either propranolol or phentolamine, although extremely high concentrations were used (0.4 mM) which are sufficient to nonspecifically suppress basal rates of gluconeogenesis (141,142). In a more recent study by the same group (144), *in vivo* injection of isoproterenol and adrenaline produced a greater rise in hepatic cyclic AMP levels in hypothyroid than in control rats, which is contradictory to their conclusions and is compatible with the results of more recent studies discussed below.

Two independent studies that appeared within a week of each other showed that catecholamine-induced glycogenolysis, a predominantly α-receptor-mediated event in hepatocytes of euthyroid rats, became a predominantly β-receptor-mediated response of hypothyroid rats (145,146), a change opposite in direction to that suggested by the Japanese workers cited above. The major change found was a potentiation of the isoproterenol-induced activation of phosphorylase in both studies (Figure 3) and a shift in the block of the mixed α/β-agonist adrenaline by α-receptor antagonists in the euthyroid to propranolol only in the hypothyroid cells (146). Hyperthyroidism did not alter the receptor response pattern and treatment of thyroidectomized rats with thyroid hormones but not with cortisone reversed the observed changes (146,147). The potentiation of the β-receptor response in the hypothyroid state was evident from measurements of both phosphorylase activation and cyclic AMP accumulation (145,147), and the selectivity of these changes was indicated by the lack of similar changes in the effects of glucagon (145,148).

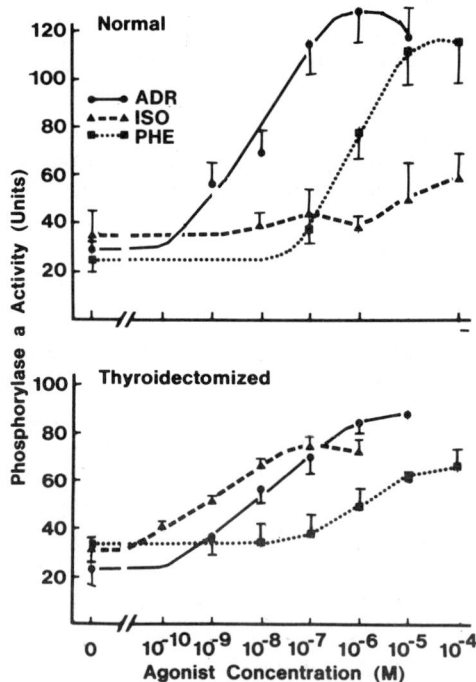

Figure 3 Effects of adrenergic agonists on glycogen phosphorylase activity in isolated hepatocytes from euthyroid and hypothyroid rats. Units of enzyme activity are nanomoles of [^{14}C]glucose incorporated into glycogen per minute per milligram of protein. Each point represents the mean of 3 to 13 experiments. Vertical bars indicate SE. Note the reversal of the relative agonist potencies from (-)-adrenaline > (-)-phenylephrine > (±)-isoprenaline in normal rats (α-type response) to (±)-isoprenaline > (-)-adrenaline > (-)-phenylephrine in hypothyroid rats (β-type response). Reproduced by permission from Ref. 146.

Whether or not a reciprocal decrease in the α-receptor response also occurs in the hypothyroid liver could not be established with certainty from the effect of phenylephrine on phosphorylase, since this involves not only α- but β-receptors as well, particularly in the hypothyroid state. Changes in α-receptor reactivity can be better evaluated by a response that is mediated exclusively by α-receptors. Adrenergic agonists can cause the release of calcium from liver cells, an effect mediated only by α- and not by β-adrenoceptors (149–151), and vasopressin can also release calcium through its own receptors (150,151). By measuring calcium release from isolated rat hepatocytes, it was found that previous thyroidectomy did not alter the total releasable pool of calcium, indicated by the amount of calcium released by the ionophore A23187 (148). The maximal amount of calcium released by phenylephrine or vasopressin showed the same slight reduction in the hypothyroid liver. However, hypothyroidism reduced the potency of phenylephrine for inducing calcium release tenfold, whereas the potency of vasopressin remained unchanged (148). These observations clearly indicate a selective decrease in α-receptor reactivity, in addition to the selective increase in β-receptor reactivity discussed above. The shift in receptor response from α to β in the hypothyroid rat liver is opposite in direction to the shift from β toward α-receptor responses in the heart, adipose tissue, and in some endocrine glands (see below). Although the reasons for this puzzling tissue-specific difference are not clear at present, its existence illustrates that the effects of hormonal influences on adrenoceptors require careful analysis and that results obtained in a given tissue cannot be generalized.

The biochemical mechanisms underlying the altered adrenoceptor response of the hypothyroid liver are not yet clear. Although small decreases in the activities of hepatic COMT and MAO have been noted (152), these are unlikely to account for the opposite changes in the potencies of isoproterenol and phenylephrine in intact cells and cannot explain the corresponding change in the effects of adrenoceptor antagonists against mixed α/β-agonists. Moreover, in a recent study thyroidectomy did not influence hepatic MAO activity in rats (152a). The effects of β-receptor stimulation in the liver are mediated through the cyclic AMP system, whereas metabolic responses mediated by α-receptors proceed along cyclic AMP-independent pathways (141–143,153) that involve the release of mitochondrial calcium (154). Basal and fluoride-stimulated adenylate cyclase activity in the liver are not significantly altered by thyroid hormones (155), and a small increase in phosphodiesterase activity in the hypothyroid rat liver (156) is opposite to what may account for the enhanced β-receptor responsiveness. Another study could not demonstrate any difference in phosphodiesterase activity between normal and hypothyroid rat liver (156a). Total glycogen content is also unchanged (157), and the basal activity of glucagon phosphorylase is only slightly reduced by hypothyroidism (157). Calcium-efflux studies in liver cells suggested a decrease in a slow-turnover calcium pool in hepatocytes from hypothyroid rats (158). However, the functional role of this calcium pool is not clear, and calcium releasable by the ionophore A23187 was shown not to significantly decrease after thyroidectomy (148). The finding that reciprocal and selective changes of α- and β-receptor responses were also apparent when cyclic AMP accumulation or calcium release was measured strongly suggests

TABLE 2   Effect of Thyroid State on α_1 and β-Adrenoceptor Binding Sites in Rat Liver[a]

	Control	p	Hypothyroid	p	Hypothyroid + T_3
α_1-Receptors ([^{3}H]prazosin - 2 μM phentolamine)					
binding site density (fmol/mg)	580 ± 42 (5)	<0.01	348 ± 58 (8)	<0.05	498 ± 23 (5)
K_d prazosin (pM)	66 ± 8 (5)	n.s.	53 ± 11 (8)	n.s.	51 ± 14 (5)
K_d *l*-adrenaline (μM)	0.5 (3)		0.8 (3)		1.1 (2)
β-Receptors					
([^{125}I]hydroxybenzylpindolol - 0.3					
μM propranolol)					
binding site density (fmol/mg)	24 ± 3 (8)	< 0.05	60 ± 13 (5)		—
K_d hydroxybenzylpindolol (nM)	0.2 ± 0.05 (8)	n.s.	0.2 ± 0.03 (5)		—
K_d *l*-isoproterenol (nM)	15	n.s.	15		—

[a]Data for α_1 receptors are unpublished results from the author's laboratory, those for β-receptors were taken from ref. 156a. In both studies hepatocyte membranes prepared by a modified version of Neville's method (see ref. 156a) were used to determine ligand binding. Data are means ± SE, number of experiments are shown in parentheses. n.s. stands for not significant. Hypothyroidism was induced by thyroidectomy (A) or propylthiouracil (B). Treatment with T_3, 0.2 mg/kg/day, was for 4 days.

that the site of these changes is either at the respective receptors or at their specific coupling to systems generating the second messenger signals. Recent ligand binding studies have provided evidence for the first possibility. As shown by the data in Table 2, the density of β-receptor binding sites increased and the density of α_1-receptors decreased in the hypothyroid rat liver, without any change in binding site affinities. Also, the decrease in α_1 receptors was reversed by *in vivo* treatment of hypothyroid rats with T_3. These data clearly indicate reciprocal regulation of hepatic α_1 and β-receptors by thyroid hormones.

It is interesting to note that adrenalectomy of rats results in an increase in the β-receptor-mediated cyclic AMP accumulation and phosphorylase activation and a selective decrease in α-receptor-mediated calcium release and phosphorylase activation (159), changes that are very similar to the changes observed in thyroidectomized rats. A search for tissue functions affected by either corticosteroids or thyroid hormones may be helpful in determining the mechanisms involved in these changes. The increased β-receptor response after adrenalectomy is associated with an increased density of binding sites for the β-receptor antagonist [125I]hydroxybenzylpindolol (160), which was completely reversed by a cortisone treatment schedule (160) identical to that found to be ineffective in reversing the increased β-adrenergic response after thyroidectomy (147). On the other hand, the density of α-receptor binding sites labeled with [3H]DHEC (161) or [3H]noradrenaline (159) did not change after adrenalectomy. This was interpreted to indicate that the selective decrease in the α-receptor response may be due to a change in receptor coupling to the calcium pool (159). However, neither of the above two ligands differentiate between α_1 and α_2 receptors, and a decrease in α_1 receptors could have been offset by an increase in α_2 receptors. α_2-Receptors are not involved in liver phosphorylase activation (161a). Affinity labeling of α-adrenoceptor binding sites with [3H]phenoxybenzamine in intact liver cells has been recently reported (162), and the use of this ligand would allow one to avoid the possible effects on receptors of tissue disruption and purification, when studying the effects of hormonal manipulations.

In summary, hypothyroidism in rats increases β- and decreases α_1-receptor-mediated responses in isolated hepatocytes, and both changes are rapidly reversed by *in vivo* treatment with thyroid hormones. These changes are opposite in direction to the effects of hypothyroidism on the relative dominance of α- and β-receptor responses of heart, adipose tissue, and some endocrine glands, but similar to changes in liver adrenoceptor responses of adrenalectomized rats. These changes appear to be selective to adrenoceptors, and in the hypothyroid rat liver they are associated with corresponding inverse changes in the densities of α_1 and β adrenoceptor binding sites.

3 ENDOCRINE SYSTEM

In addition to their cardiovascular and metabolic effects, catecholamines can modulate the synthesis and release of various endocrine hormones (4,163). In most cases α- and β-receptors in endocrine glands mediate opposite effects on hormone secretion, and in some systems thyroid state appears to influence α- and β-receptor reactivity in a reciprocal manner.

3.1 Thyroid Gland

Thyroid follicles are sympathetically innervated (see Ref. 4) and catecholamines or sympathetic nerve stimulation can affect both the synthesis and release of thyroid hormones (164,165), although the physiological importance of these effects is not clear. There has been considerable controversy about the nature of the adrenergic receptors involved in these effects. In thyroid slices or thyroidal membrane preparations, catecholamines can stimulate adenylate cyclase through activation of β-adrenoceptors (166,167). This effect is similar but not identical to the action of thyroid-stimulating hormone (TSH) on the thyroid.

On the other hand, i.v. administration of adrenaline or noradrenaline or stimulation of the cervical sympathetic nerves of T_4-treated mice were shown to increase the uptake and release of ^{131}I, effects selectively inhibited by α-receptor antagonists. Isoproterenol produced similar effects in the same T_4-treated mice, but these were blocked by propranolol (168). However, when the effect of isoproterenol was tested in isolated calf thyrocytes incubated without thyroid hormone, the same effects of isoproterenol were blocked only by α-, and not by β-antagonists (see Ref. 169). These findings led to the proposal (169) that the pharmacological properties of adrenoceptors mediating hormone secretion in the thyroid gland may depend on the level of thyroid hormones, in a way similar to the proposed "interconversion" myocardial α- and β-adrenoceptors (16). This possibility was supported by subsequent findings in another laboratory. When isolated porcine thyrocytes were cultured in the presence of TSH or dbcAMP, catecholamines stimulated the efflux of ^{131}I through β-adrenoceptors coupled to adenylate cyclase (170). However, when cells were cultured in the absence of such stimulators, catecholamines were unable to stimulate iodine efflux, but rather enhanced the uptake of ^{131}I via an α-type receptor (170).

3.2 Insulin Secretion in the Pancreas

Activation of β-receptors on pancreatic β-cells can enhance, and activation of α-receptors can inhibit the secretion of insulin (163). It has been reported that isoproterenol-induced hyperinsulinaemia in rats is potentiated in the hyperthyroid state but is absent in the hypothyroid state. On the other hand, adrenaline, a mixed α/β-agonist potentiated the glucose-induced hyperinsulinaemia in hyperthyroid rats, did not influence it in controls, and suppressed it in hypothyroid animals (171). These results indicate that the relative dominance of adrenoceptors controlling insulin release is dependent on the thyroid state: β-receptors dominate in hyperthyroid and α-receptors in hypothyroid rats (171).

4   OTHER TISSUES

4.1   Exocrine Glands

In vitro incubation of rat parotid gland with 1 μM T_4 was shown to rapidly ($<$60 min) increase the adrenaline-induced β-amylase secretion, a response associated with β-receptor-mediated adenylate cyclase activation (172). An analogous finding in submaxillary glands is the increase in the number of [^{3}H]DHA binding sites upon T_3 treatment of thyroidectomized rats (173). Hypothyroidism, on the other hand, was shown to reduce the glycogenolytic effect of adrenaline in rat submaxillary glands (174). These findings suggest that thyroid hormones control β-adrenoceptors in salivary glands in a way similar to that seen in heart. α-Adrenoceptors in salivary glands are involved in water and K^+ secretion (175). The effect of the thyroid state on these responses have not yet been tested.

4.2   Skeletal Muscle

Sharma and Banerjee (176) reported that in rats, thyroidectomy reduced the density but not the affinity of [^{3}H]DHA binding sites in skeletal muscle membranes, and the change was reversed by *in vivo* treatment with T_3. Hyperthyroidism induced by T_3 treatment of normal rats did not influence binding (176); thus the pattern of changes was similar to that observed by the same group in the heart (88). Since hypothyroidism was found to increase serum catecholamine levels whereas hyperthyroidism did not significantly alter it as compared to age-matched controls (177), the changes in β-receptor binding sites in the sympathetically noninnervated skeletal muscle were suggested to be mediated by altered circulating catecholamine levels (176).

4.3   Blood Cells

Changes similar to those found in skeletal muscle were noted in another noninnervated cell, the turkey erythrocyte (1y8). β-Adrenoceptors were labeled with [^{125}I]hydroxybenzylpindolol. Both binding-site numbers and catecholamine-stimulated cyclicAMP levels were decreased in cells from hypothyroid turkeys. Hyperthyroidism did not alter binding, but increased catecholamine-induced cyclicAMP accumulation suggesting that, in the latter situation, thyroid hormones must have affected the coupling between binding sites and adenylate cyclase (178). No difference in either [^{3}H]DHA binding or isoproterenol-induced adenylate cyclase activation was found in lymphocytes from normal and hyperthyroid human subjects (179). The effects of hypothyroidism were not tested in the latter study.

4.4   Bone Marrow

Catecholamines are known to stimulate erythropoiesis through a β_2-adrenergic mechanism. In cultured bone marrow cells from euthyroid dogs, this effect was manifested as an increase in erythroid colony growth upon incubation with isopro-terenol but not with noradrenaline or phenylephrine (180). In cells from hypothyroid dogs, the effect of isoproterenol was lost, whereas noradrenaline and phenylephrine became effective agonists and their effects could be inhibited by phentolamine (1 nM) but not by propranolol (1 μM), indicating a loss of β- and emergence of α-adrenergic receptors mediating the same end response. Incubation of hypothyroid cells with 100 nM of l-T_4 for as short as 30 min completely reversed the reaction pattern from α to β (180).

4.5   Central Nervous System

Thyroxine treatment of rats was found to increase the locomotor stimulation by noradrenaline (181). In mice, T_3 treatment increased locomotor stimulation by clonidine but not by dopamine, and the effect of clonidine persisted after reserpine pretreatment (182). These observations suggest hyperresponsiveness of some central α-adrenoceptors in the hyperthyroid state. This possibility is supported by a finding of increased spike frequency in response to iontophoretically applied noradrenaline but not acetylcholine or glycine in brain-stem neurones of T_3-treated rats, whereas decreased responses were found in hypothyroid animals (183).

4.6   Catecholamine Effects on DNA Synthesis

β-Adrenoceptor agonists can increase the synthesis of DNA in a number of tissues, an effect thought to contribute to tissue hypertrophy after chronic administration of catecholamines. DNA synthesis in renal tissue, measured by the incorporation into DNA of radioactive thymidine, was reduced by hypothyroidism and the re-duction reversed by T_3 treatment of the hypothyroid mice (184). These changes in β-receptor reactivity are similar to those observed in heart, adipose tissue, and a number of other systems.

5   ALTERED ADRENERGIC REACTIVITY IN CONDITIONS ASSOCIATED WITH CHANGES IN THYROID STATE

Changes in adrenoceptor response pattern similar to those observed in altered thyroid states have been noted in a number of physiological conditions. Some of these conditions are known to be associated with changes in thyroid activity, and thus the observed changes in adrenoceptor responses may be indirectly related to thyroid hormones.

5.1   Fasting and Obesity

Fasting in obese human subjects was reported to change the β-receptor-mediated increase in lipolysis by noradrenaline into an α-receptor-mediated inhibition of the lipolytic response, which was reversed after refeeding (185). The altered response pattern during fasting was similar to the response of adipose tissue from hypothyroid human subjects (106). In fasting the peripheral deiodination of T_4 into the metabolically active T_3 is diminished, and in overfeeding this conversion is increased (186). This could mean that the shift from β- to α-adrenoceptor reactivity in fasting may be mediated by a relative hypothyroidism. Fasting is known to markedly suppress the activity of the sympathetic nervous system (186), whereas hypothyroidism is known to increase it (43–45,177). This makes it unlikely that the similar effects of the two conditions on the adrenoceptor response pattern of adipose tissue are mediated by changes in sympathetic neurotransmitter input on receptors.

In genetically obese (ob/ob) mice the reduced sensitivity of adipose tissue to the lipolytic action of catecholamines can be reversed by thyroid hormone treatment of the obese animals (187–189). This suggested that the reduced thyroid function in these animals (187) may be responsible for the altered lipolytic response.

5.2   Cold Adaptation

Chronic exposure of animals to cold elicits a series of adaptative responses aimed to increase heat production and decrease heat loss. The main source of heat in cold-acclimated animals is nonshivering thermogenic processes. Catecholamines can increase nonshivering thermogenesis by mobilizing glucose and free fatty acids for heat-generating metabolic processes. There appears to be a significant increase in the sensitivity to the metabolic effects of catecholamines in cold-acclimated animals (190), and cold acclimation is also associated with a TSH surge and "functional" hyperthyroidism (191). Although some *in vitro* studies did not show altered adrenergic reactivity in isolated tissues from cold-acclimated rats (192), other studies using both isolated cardiac and vascular preparations and intact rats show an increased sensitivity of β and decreased sensitivity of α-receptor-mediated responses in the cardiovascular system (193,194). It has been proposed that these changes may be mediated indirectly, by secondary changes in the thyroid status of the animal (193).

5.3   Aging

An age-dependent desensitization of vascular β-adrenoceptors is well established (195). In a recent study both old age and hypothyroidism were found to reduce relaxation of aortic strips by β-receptor agonists in rats, and in both cases normal responses were restored after T_4 treatment of old or hypothyroid animals (196). This suggests that the age-dependent decrease in β-receptor responsiveness is mediated by a relative hypothyroid state. The same conclusion is suggested by the results of a preliminary study of ligand binding in hearts from young and old rats (197). In old animals the density of cardiac β-receptor binding sites was lower and the density of α-receptor binding sites was higher than in hearts from young animals (197). The age-dependent reciprocal change in the two receptors was similar to that found by the same group in the hearts of hypothyroid rats (88,89). An age-dependent change from β- to α-receptor response pattern also occurs in rat liver (198).

5.4 Hibernation and Temperature Effects

Studies from several laboratories have demonstrated that lowering the temperature of isolated, spontaneously beating amphibian hearts below a critical range 17–20°C can increase the reactivity of α- and decrease the reactivity of β-receptors (19). Although some studies failed to detect such changes in electrically driven preparations (see Ref. 19), more recently a similar temperature dependent "conversion" of β to α receptor response has been demonstrated to occur on *in vitro* cooling of dog kidney (198a). It was also shown that the *in vitro* effects of low temperature, on frog heart, which are similar to the effect of hypothyroidism on mammalian heart adrenoceptors, were present only in winter and not in summer (96,199), and that treatment of winter frogs with anterior pituitary extracts abolished the sensitivity of the hearts to α-adrenoceptor blockade and made them react as summer frog hearts (200). Since hibernation is associated with reduced thyroid activity (201), these observations suggest that the relative hypothyroid state in hibernation may have a permissive effect on *in vitro* temperature alterations of the adrenoceptor response pattern of the heart. Although the mechanism of such an effect is not yet clear, it may be related to differences in the thermal response of membrane lipids. Both hibernation (202) and hypothyroidism (203) were reported to increase the relative unsaturation of membrane fatty acids, which allows membranes to retain their fluidity when exposed to low temperature.

5.5 Partial Hepatectomy

Hornbrook has recently shown that partial hepatectomy in adult rats changes the adrenaline-induced glycogenolytic response from an α-receptor-mediated effect in the normal rat liver to a mixed α/β effect in the residual tissue remaining after hepatectomy (198). The mechanism of this rapid change (<24 hr) has not yet been established, but an interesting possibility may be proposed. Both hepatectomy and starvation are associated with increased serum glucagon levels (204,205), and similar increases of glucagon were shown to suppress some actions of thyroid hormones by decreasing nuclear receptors for T_3 or influencing some post-T_3 receptor event (206). Hepatectomy may therefore result in functional hypothyroidism by indirectly interfering with normal thyroid hormone action in target tissues.

6 POSSIBLE MECHANISMS OF THYROID MODULATION OF ADRENOCEPTORS

The results discussed in this chapter clearly demonstrate that thyroid state can influence the adrenergic reactivity of various tissues. Because of the multiplicity of actions of thyroid hormones, it is not surprising that the mechanisms of these interactions are not clear and that they can occur at various steps involved in the effects of catecholamines. Although there is evidence that some pre- and postreceptor events are influenced by thyroid hormones in some tissues, these changes are insufficient to explain the altered adrenergic reactivity. Changes in the number and affinity of adrenoceptors as well as possible changes in the coupling of receptors to effector systems have been suggested by a number of recent studies. At present, the molecular mechanism of such changes is still a matter of speculation. In most cases the effects of hypothyroidism on adrenoceptors develop slowly over several weeks. This may be related to the long biological half-life of thyroid hormones or of their effects in tissues (207). Reversal of these changes by thyroid hormone treatment or the effects of hyperthyroidism develop much faster, but still require 6–12 hr. This could suggest that thyroid hormones act by influencing the synthesis or degradation of receptor protein at the transcriptional or translational level. Thyroid hormones were shown to affect the numbers of or the responses mediated by a number of other hormone receptors, including muscarinic cholinergic (208), glucagon (209), estrogen (210), glucocorticoid (211), vasopressin (212), or prolactin receptors (213). In view of the rather nonselective increase in messenger RNA production by thyroid hormones (6), such multiple effects are also compatible with a change in protein synthetic rate. On the other hand, the results of several studies discussed in this chapter indicate that, in addition to possible changes in receptor synthesis, incubation of tissues with thyroid hormones *in vitro* can produce rapid changes in receptors which are apparently independent of protein synthesis. The mechanism of such rapid changes is not clear. Incorporation of soluble receptors into the cell membrane (93) or changes in membrane fluidity (214,215) that could unmask cryptic receptors (216) are possibilities. Thyroid hormones may mediate such effects through interaction with specific high-affinity receptors identified recently in the cell membrane (8,9).

The existence of such acute effects of thyroid hormones on adrenoceptors and the numerous examples of reciprocal changes in α- and β-receptor responses in various tissues justifies a brief discussion of another possibility, the proposed interconversion of α- and β-adrenoceptors (19,96). This was first suggested on the basis of rapid, temperature-induced, reciprocal changes in α- and β-receptor responses of isolated, spontaneously beating frog hearts (96,199). For a more detailed discussion of temperature-induced changes in the balance between cardiac α- and β-receptor responses, the reader is referred to recent reviews (19,22,217,218).

A narrow definition of "interconversion" would imply that the same molecular entity changes its conformation. However, in a broader sense it could include the possibility of an interaction between different subunits of a macromolecule or a close functional coupling between distinct membrane entities. Although the numerous examples of rapid changes in adrenoceptors upon *in vitro* incubation with thyroid hormones are compatible with an interconversion mechanism, they do not prove it. Definitive evidence for this mechanism should include (*a*) reciprocal, rapid changes in α- and β-receptor responses mediated by matching, inverse changes in the numbers of functional receptors; (*b*) demonstration of a functional coupling between the two receptors; and (*c*) clarification of the molecular basis of such coupled, inverse regulation.

The results of studies in rat heart (24) and liver (see section 2.2) meet the first criterion, although the minimum time required for these changes to develop is not yet clear. In the rat heart a functional coupling of α- and β-receptors was indicated by the observation that the presence of a low concentration of a β-blocking drug during incubation of hypothyroid rat left atria with phenoxybenzamine significantly potentiated the irreversible block of inotropic responses to phenylephrine (16). The finding that prolonged washing of the tissue, sufficient to reverse the small inhibition produced by the β-antagonist alone, did not reverse the potentiation of the block by phenoxybenzamine was interpreted to indicate that block of β-receptors shifted the balance in favor of α-receptors, which were then trapped by the irreversible antagonist (16). A similar interaction was observed in spontaneously beating frog hearts tested at low temperature (217). A functional coupling of β- and α_2-receptors is also indicated in a recent study using rat brain membranes (219; also Chapter 5). Incubation of membranes with a desensitizing concentration of isoproterenol rapidly reduced the number of β-receptors and increased the number of α_2-receptors. These opposite changes had a parallel time course and were both prevented by preincubation of the membranes with a β-receptor antagonist (219). As α_2-receptors are postsynaptic in the brain (see Chapter 5), this observation could indicate an allosteric coupling of α_2- and β-receptors in the same postsynaptic membrane area. The fact that these changes occurred in isolated membrane fragments eliminates the possibility that their mechanism involves internalization of receptors, demonstrated to occur during desensitization of β-receptors (220). The underlying mechanism may be similar to those responsible for reciprocal changes in α- and β-receptors in peripheral tissues, induced by temperature, hormonal factors, or disease states (221).

As for the third criterion, no definitive evidence has been presented yet on the molecular relationship between α- and β-receptors. Wood et al. (222) provided evidence that in solubilized rat liver membranes removal of [^{3}H]DHA binding sites (β) by affinity chromatography did not alter the amount of [^{3}H]WB-4101 binding sites (α) recovered. These results suggest that hepatic α_1- and β-receptors either reside on distinct molecules or are different, separable subunits of the same macromolecule. They do not prove or disprove the functional coupling of the two receptors in the intact membranes, just as separation of β-receptors and adenylate cyclase did not contradict their close functional coupling. Furthermore, the very low recovery (3–10%) of solubilized WB-4101 binding sites in the study cited above makes it uncertain whether these sites are representative of the large majority of α-receptors lost during solubilization. In another study using rat liver membranes, Guellaen and Hanoune (223) reported that an SH- reagent decreased the density of [^{3}H]DHEC α binding sites and dithiothreitol decreased the density of [^{3}H]DHA labeled β-receptors, but neither agent affected the other type of receptor. These results suggest that a sulfhydryl-disulfide transition is not the mechanism of a possible conformational transition or functional coupling of α- and β-receptors. They do not address the question of molecular identity or nonidentity of the two receptors. However, interpretation of these findings is tempered by the fact that very high concentrations of phentolamine (0.05 M) were required to produce a very modest suppression of DHEC binding (20–25%), which raises questions about the identity of 'specific' binding sites with functional α-receptors. Further evidence is required to clarify the molecular basis for the apparent interconversion of α- and β-receptors under various conditions, including altered thyroid states.

Thyroid hormones play an important role in tissue and cell differentiation and their effects on the relative dominance of α- and β-adrenoceptors may be part of the cell differentiation process. In fact, there is evidence that in several conditions associated with dedifferentiation of cells the adrenoceptor response pattern is altered in a way similar to that seen in hypothyroidism. The shift from α- to β-type responses in rat liver following partial hepatectomy (198) has been discussed above. Treatment of rats with a chemical carcinogen, which also leads to dedifferentiation of liver cells, was shown to produce a similar shift from α- to β-receptor responses of hepatocytes (224). A lower state of cell differentiation may be the common denominator between the changes noted above and the shifts from α- to β- type responses of liver cells from juvenile (225) or weanling rats (198). The opposite direction of thyroid-dependent changes in adrenoceptors of heart and liver is also apparent for ontogenic changes in α/β-receptor balance. α-Receptor binding sites were shown to be present in fetal but not in adult sheep myocardium, whereas the density of β-receptor binding sites decreased from fetal to the adult state (226). This change may reflect the differentiation of the myocardium from vascular smooth muscle during the embryological development of the heart; motor responses of vascular smooth muscle remain to be mediated by α-adrenoceptors.

As changes in thyroid state thoroughly alter the metabolic activity of tissues, the associated changes in adrenoceptor reactivity may be a mechanism of adaptation of tissue function to altered metabolic requirements. Further studies are necessary to identify the molecular mechanism(s) underlying such changes.

ACKNOWLEDGMENT

Work from the author's laboratory was supported by grants from the Medical Research Council of Canada and the Canadian Heart Foundation.

REFERENCES

1   T. S. Harrison, *Physiol. Rev.*, **44**, 161 (1964).

2   S. S. Waldstein, *Annu. Rev. Med.*, **17**, 123 (1966).

3   S. W. Spaulding and R. H. Noth, *Med. Clin. N. Am.*, **59**, 1123 (1975).

4   L. Landsberg, *Clin. Endocrinol. Metab.*, **6**, 697 (1977).

5   J. H. Oppenheimer, W. H. Dillmann, H. L. Schwartz, and H. C. Towle, *Fed. Proc.*, **38**, 2154 (1979).

6   H. C. Towle, W. H. Dillmann, and J. H. Oppenheimer, *J. Biol. Chem.*, **254**, 2250 (1979).

7   J. R. Tata, *Nature (Lond.)*, **257**, 18 (1975).

8   N. B. Pliam and I. D. Goldfine, *Biochem. Biophys Res. Commun.*, **79**, 166 (1977).

9   J. Gharbi and J. Torresani, *Biochem. Biophys. Res. Commun.*, **88**, 170 (1979).

10 M. B. Dratman, F. L. Crutchfield, J. Axelrod, R. W. Colburn, and N. Thoa, *Proc. Natl. Acad. Sci. U.S.A.*, **73**, 941 (1976).

11 M. B. Dratman and F. L. Crutchfield, *Am. J. Physiol.*, **235**, E638 (1978).

12 J. B. Van der Schoot and N. C. Moran, *J. Pharmacol. Exp. Ther.*, **149**, 336 (1965).

13 H. S. Margolius and T. E. Gaffney, *J. Pharmacol. Exp. Ther.*, **149**, 329 (1965).

14 R. A. Buccino, J. F. Spann, Jr., P. E. Pool, E. H. Sonnenblick, and E. Braunwald, *J. Clin. Invest.*, **46**, 1669 (1967).

15 G. S. Levey, C. L. Skelton, and S. E. Epstein, *J. Clin. Invest.*, **48**, 2244 (1969).

15a D. G. McDevitt, J. A. Riddel, D. R. Hadden, and D. A. D. Montgomery, *Br. J. Clin. Pharmacol.*, **6**, 297 (1978).

16 G. Kunos, *Br. J. Pharmacol.*, **59**, 177 (1977).

17 G. Kunos, C. Brass, W. H. Kan, and L. Mucci, *Fed. Proc.*, **37**, 684A (1978).

18 Y. Yazaki and M. S. Raben, *Circ. Res.*, **36**, 208 (1975).

19 G. Kunos, *Annu. Rev. Pharmacol. Toxicol.*, **18**, 291 (1978).

20 R. F. Furchgott, in C. Owman and L. Edvinsson, Eds., *Neurogenic Control of Brain Circulation*, Pergamon Press, Oxford, 1977, p. 155.

21 H. J. Schümann, M. Endoh, and J. Wagner, *Naunyn-Schmiedebergs Arch. Pharmacol.*, **284**, 133 (1974).

22 G. Kunos, *Trends Pharmacol. Sci.*, **1**, 282 (1980).

23 G. Kunos, I. Vermes-Kunos, and M. Nickerson, *Nature (Lond.)*, **250**, 779 (1974).

24 G. Kunos, L. Mucci, and S. O'Regan, *Br. J. Pharmacol.*, **71**, 371 (1980).

25 M. Nakashima, K. Maeda, A. Sekiya, and Y. Hagino, *Jap. J. Pharmacol.*, **21**, 819 (1971).

26 M. Nakashima and Y. Hagino, *Jap. J. Pharmacol.*, **22**, 227 (1972).

27 M. Nakashima, H. Tsuru, and T. Shigei, *Jap. J. Pharmacol.*, **23**, 307 (1973).

28 H. Hashimoto and M. Nakashima, *Eur. J. Pharmacol.*, **50**, 337 (1978).

29 J. Wagner and O.-E, Brodde, *Naunyn-Schmiedebergs Arch. Pharmacol.*, **302**, 239 (1978).

30 C. Fortier, F. Labrie, G. Pelletier, J.-P. Raynaud, P. Ducommun, A. Delgado, R. Labrie and M. A. Ho-Kim, in G. E. W. Wolstenholme and J. Knight, Eds., *Multicellular Organisms*, Churchill, London, 1970, p. 178.

31 H. Johansson and L. E. Jönsson, *Acta Chir. Scand.*, **137**, 59 (1971).

32 H. S. Margolius, P. Reid, S. Mohammed, and T. E. Gaffney, *J. Pharmacol. Exp. Ther.*, **155**, 415 (1967).

33 J. H. Tong and A. d'Iorio, *Endocrinology*, **98**, 761 (1976).

34 A. Ho-Van-Hap, L. M. Babineau, and L. Berlinguet, *Can. J. Biochem.*, **45**, 355 (1967).

35 L. Landsberg, J. de Champlain, and J. Axelrod, *J. Pharmacol. Exp. Ther.*, **165**, 102 (1969).

36 S. Kalsner and M. Nickerson, *Br. J. Pharmacol.*, **35**, 428 (1969).

37 R. J. Wurtman, I. J. Kopin, and J. Axelrod, *Endocrinology*, **73**, 63 (1963).

38 H. J. Dengler, H. E. Spiegel, and E. O. Titus, *Nature (Lond.)*, **191**, 816 (1961).

39 J. H. McNeill and T. M. Brody, *J. Pharmacol. Exp. Ther.*, **161**, 40 (1968).

40 A. H. Anton and J. S. Gravenstein, *Eur. J. Pharmacol.*, **10**, 311 (1970).

41 K. R. Hornbrook and A. Cabral, *Biochem. Pharmacol.*, **21**, 897 (1972).

42 D. J. Jacobowitz and R. Brus, *Eur. J. Pharmacol.*, **15**, 274 (1971).

43 L. Landsberg and J. Axelrod, *Circ. Res.*, **22**, 559 (1968).

44 T. Tu and C. W. Nash, *Can. J. Physiol. Pharmacol.*, **53**, 74 (1975).

45 J. L. Tedesco, K. V. Flattery, and E. A. Sellers, *Can. J. Physiol. Pharmacol.*, **55**, 515 (1977).

46 M. Raff, *Nature (Lond.)*, **259**, 265 (1976).

47 K. Wildenthal, *J. Clin. Invest.*, **51**, 2702 (1972).

48 K. Wildenthal, *J. Pharmacol. Exp. Ther.*, **190**, 272 (1974).

49 J. S. Tsai and A. Chen, *Nature (Lond.)*, **275**, 138 (1978).

50 E. H. Hu and J. C. Venter, *Mol. Pharmacol.*, **14**, 237 (1978).

51 J. Brockhuysen and M. Ghislain, *Biochem. Pharmacol.*, **21**, 1493 (1972).

52 A. Wollenberger and L. Will-Shahab, *Recent Adv. Stud. Card. Struct. Metab.*, **9**, 193 (1976).

53 T. P. Ciaraldi and G. V. Marinetti, *Biochim. Biophys. Acta*, **541**, 334 (1978).

54 R. M. Smith, W. S. Osborne-White, and R. A. King, *Biochem. Biophys. Res. Commun.*, **80**, 715 (1978).

54a O.-E. Brodde, H.-J. Schümann, and J. Wagner, *Mol. Pharmacol.*, **17**, 180 (1980).

55 R. Brus and M. E. Hess, *Endocrinology*, **93**, 982 (1973).

56 B. E. Sobel, P. J. Dempsey, and T. Cooper, *Proc. Soc. Exp. Biol. Med.*, **132**, 6 (1969).

56a J. Tse, R. W. Wrenn, and J. Kuo, *Endocrinology*, **107**, 6 (1980).

57 J. H. McNeill, L. D. Muschek, and T. M. Brody, *Can. J. Physiol. Pharmacol.*, **47**, 913 (1969).

58 B. A. Young and J. H. McNeill, *Can. J. Physiol. Pharmacol.*, **52**, 375 (1974).

59 O. E. Brodde, H. J. Schümann, and J. Wagner, *Naunyn Schmiedebergs Arch. Pharmacol.*, **302**, Suppl. R25 (1978).

60 G. Kunos, L. Mucci, and V. Jaeger, *Life Sci.*, **19**, 1597 (1976).

60a R. V. Sharma, O. Habhab, and R. C. Bhalla, *Biochem. Pharmacol.*, **28**, 2858 (1979).

61 K. R. Hornbrook, P. V. Quinn, J. H. Siegel, and T. M. Brody, *Biochem. Pharmacol.*, **14**, 925 (1965).

62 J. H. McNeill, *Res. Commun. Chem. Pathol. Pharmacol.*, **16**, 735 (1977).

63 A. Frazer, M. E. Hess, and J. Shanfeld, *J. Pharmacol. Exp. Ther.*, **170**, 10 (1969).

64 S. Katz, D. Hamilton, T. Tenner, and J. H. McNeill, *Res. Commun. Chem. Pathol. Pharmacol.*, **18**, 777 (1977).

65 A. J. D. Friesen, G. Allen, and J. R. E. Valadares, *Science*, **155**, 1108 (1967).

66 E. J. Hartley and J. H. McNeill, *Can. J. Physiol. Pharmacol.*, **54**, 590 (1976).

67 M. Murayama and M. J. Goodkind, *Circ. Res.*, **23**, 743 (1968).

68 J. Suko, *J. Physiol. (Lond.)*, **228**, 563 (1973).

69 W. G. Nayler, N. C. R. Merrilees, S. Chipperfield, and J. B. Kurtz, *Cardiovasc. Res.*, **5**, 469 (1971).

70 F. Hirata, W. J. Strittmatter, and J. Axelrod, *Proc. Natl. Acad. Sci. U.S.A.*, **76**, 368 (1979).

71 D. G. Wenzel and J. L. Su, *Arch. Int. Pharmacodyn.*, **160**, 379 (1966).

72 J. B. Osnes and I. Oye, *Adv. Cyclic Nucleotide Res.*, **5**, 415 (1975).

73 A. M. Watanabe, D. R. Hathaway, H. R. Besch, Jr., B. B. Farmer, and R. A. Harris, *Circ. Res.*, **40**, 596 (1977).

74 J. B. Osnes, *Acta Pharmacol. Toxicol.*, **39**, 232 (1976).

75 M. S. Amer and J. E. Byrne, *Nature (Lond.)*, **256**, 421 (1975).

76 S. L. Keely, J. D. Corbin, and T. Lincoln, *Mol. Pharmacol.*, **13**, 965 (1977).

77 M. Endoh, J. Wagner, and J. H. Schümann, *Naunyn Schmiedebergs Arch. Pharmacol.*, **287**, 61 (1975).

78 P. F. Blackmore, M. F. El-Refai, and J. H. Exton, *Mol. Pharmacol.*, **15**, 598 (1979).

79 L. M. Jones and R. H. Michell, *Biochem. Soc. Trans.*, **6**, 673 (1978).

80 T. Ciaraldi and G. V. Marinetti, *Biochem. Biophys. Res. Commun.*, **74**, 984 (1977).

81 L. T. Williams, R. J. Lefkowitz, A. M. Watanabe, D. R. Hathaway, and H. R. Besch, Jr., *J. Biol. Chem.*, **252**, 2787 (1977).

82 M. M. McConnaughey, L. R. Jones, A. M. Watanabe, H. R. Besch, Jr., L. T. Williams, and R. J. Lefkowitz, *Fed. Proc.*, **37**, 914A (1978).

83 G. Kunos, B. Hoffman, Y. N. Kwok, W. H. Kan, and L. Mucci, *Nature (Lond.)*, **278**, 254 (1979).

84 P. Guicheney, R. P. Garay, C. Levy-Marchal, and P. Meyer, *Proc. Natl. Acad. Sci. U.S.A.*, **75**, 6285 (1978).

85 D. F. Story, M. S. Briley, and S. Z. Langer, *Eur. J. Pharmacol.*, **57**, 423 (1979).

86 S. R. Nahorski and A. Richardson, *Br. J. Pharmacol.*, **66**, 469P (1979).

87 R. Winek and R. Bhalla, *Biochem. Biophys. Res. Commun.*, **91**, 200 (1979).

88 S. P. Banerjee and L. S. Kung, *Eur. J. Pharmacol.*, **43**, 207 (1977).

89 V. K. Sharma and S. P. Banerjee, *J. Biol. Chem.*, **253**, 5277 (1978).

90 R. S. Williams and R. J. Lefkowitz, *Circ. Res.*, **43**, 721 (1978).

91 R. S. Williams and R. J. Lefkowitz, *J. Cardiovasc. Pharmacol.*, **1**, 181 (1979).

92 L. Will-Shahab, S. Bartel, A. Wollenberger, and I. Kuttner, *Proc. 7th Int. Congr. Pharmacol., Paris, 1978,* p. 967.

93 S. Kempson, G. V. Marinetti, and A. Shaw, *Biochim. Biophys. Acta,* **540**, 320 (1978).

94 P. Wahlberg, E. Carlsson, and U. Brandt, *Acta Endocrinol.*, **85**, 220 (1977).

95 S. P. Banerjee, V. K. Sharma, and L. S. Kung-Cheung, *Proc. 7th Int. Congr. Pharmacol., Paris, 1978,* p. 370.

96 G. Kunos and M. Szentivanyi, *Nature (Lond.)*, **217**, 1077 (1968).

97 T. Zsoter, H. Tom, and C. Chappel, *J. Lab. Clin. Med.*, **64**, 433 (1964).

98 M. J. Fregly, E. L. Nelson, Jr., G. E. Resch, F. P. Field, and L. O. Lutherer, *Am. J. Physiol.*, **229**, 916 (1975).

99 F. Rioux and B. A. Berkowitz, *Circ. Res.*, **40**, 306 (1977).

100 P. F. Coville and J. M. Telford, *Br. J. Pharmacol.*, **39**, 49 (1970).

101 J. Wagner, D. Reinhardt, and W. Huppertz, *Arcb. Int. Pharmacodyn.*, **218**, 40 (1975).

102 F. P. Field, R. A. Janis, and D. J. Triggle, *Can. J. Physiol. Pharmacol.*, **51**, 344 (1973).

103 U. Rosenquist and L. O. Boréus, *Life Sci.*, **11**, (I) 595 (1972).

104 G. A. Bray, *Endocrinology,* **79**, 554 (1966).

105 H. J. Schümann, E. Borowski, and G. Gross, *Eur. J. Pharmacol.*, **56**, 145 (1979).

106 U. Rosenquist, S. Efendit, B. Jereb, and J. Ostman, *Acta Med. Scand.*, **189**, 381 (1971).

107 J. P. D. Reckless, C. H. Gilbert, and D. J. Galton, *J. Endocrinol.*, **68**, 419 (1976).

108 M. Lafontan and R. Agid, *Comp. Biochem. Physiol.*, **55**, 85 (1976).

109 R. Recquery, R. Malagrida, and Y. Giudicelli, *FEBS Lett.*, **98**, 241 (1979).

110 T. W. Burns, P. E. Langley, and G. A. Robison, *Adv. Cyclic Nucleotide Res.*, **1**, 63 (1972).

111 J. C. Lawrence, Jr., and J. Larner, *Mol. Pharmacol.*, **13**, 1060 (1977).

112 A. F. Debons and I. L. Schwartz, *J. Lipid Res.*, **2**, 86 (1961).

113 C. C. Malbon, F. J. Moreno, R. J. Cabelli, and J. N. Fain, *J. Biol. Chem.*, **253**, 671 (1978).

113a C. C. Malbon, *Mol. Pharmacol.*, **18**, 193 (1980).

114 V. Grill and U. Rosenquist, *Acta Med. Scand.*, **194**, 129 (1973).

115 G. Krishna, S. Hynie, and B. B. Brodie, *Proc. Natl. Acad. Sci. U.S.A.*, **59**, 884 (1968).

116 A. Caldwell and J. N. Fain, *Endocrinology,* **89**, 1195 (1971).

117 C. Correze, M. H. Laudat, P. Laudat, and J. Nunez, *Mol. Cell. Endocrinol.*, **1**, 309 (1974).

118 J. N. Fain and J. W. Rosenthal, *Endocrinology,* **89**, 1205 (1971).

119 M. B. Maude, R. E. Anderson, K. J. Armstrong, and J. E. Stouffer, *Arch. Biochem. Biophys.*, **161**, 628 (1974).

120 R. G. Van Inwegen, G. A. Robison, W. J. Thompson, K. J. Armstrong, and J. E. Stouffer, *J. Biol. Chem.*, **250**, 2452 (1975).

121 K. J. Armstrong, J. E. Stouffer, R. G. Van Inwegen, W. J. Thompson, and G. A. Robison, *J. Biol. Chem.*, **249**, 4226 (1974).

122 A. Goswami and I. N. Rosenberg, *Endocrinology*, **103**, 2223 (1978).

122a J. J. Ohisalo, *FEBS Letters*, **116**, 91 (1980).

123 U. Rosenquist, *Acta Med. Scand.*, **196**, 69 (1974).

123a Y. Giudicelli, D. Lacasa, and B. Agli, *Biochem. Biophys. Res. Commun.*, **94**, 1113 (1980).

123b J. N. Fain and J. A. Garcia-Sainz, *Life Sci.*, **26**, 1183 (1980).

123c J. A. Garcia-Sainz and J. N. Fain, *Mol. Pharmacol.*, **18**, 72 (1980).

124 V. Grill and U. Rosenquist, *Acta Endocrinol.*, **78**, 39 (1975).

125 Y. Giudicelli, *Biochem. J.*, **176**, 1007 (1978).

126 C. C. Malbon and R. J. Cabelli, *Biochim. Biophys. Acta*, **544**, 93 (1978).

127 C.-S. Lo and I. S. Edelman, *J. Biol. Chem.*, **251**, 7834 (1976).

128 K. D. Philipson and I. S. Edelman, *Am. J. Physiol.*, **232**, C196 (1977).

129 R. J. Ho, B. Jeanrenauld, and A. E. Renold. *Experientia*, **22**, 86 (1966).

130 J. Kypson, L. Triner, and G. G. Nahas, *J. Pharmacol. Exp. Ther.*, **159**, 8 (1968).

131 C. R. Scheid, T. W. Honeyman, and F. S. Fay, *Nature (Lond.)*, **277**, 32 (1979).

132 J. J. Ohisalo and J. E. Stouffer, *Biochem. J.*, **178**, 249 (1979).

133 D. E. W. Hellman, H. J. Eisen, and H. M. Goodman, *Horm. Metab. Res.*, **3**, 331 (1971).

134 M. Berlan and L. Dang-Tran, *C. R. Soc. Biol.*, **171**, 970 (1977).

135 J. Himms-Hagen, in H. Blaschko and E. Muscholl, Eds., *Handbook of Experimental Pharmacology*, Vol. 23, Springer-Verlag, New York, 1972, p. 363.

136 K. Moriya, H. Maekubo, K. Honma, and S. Itoh, *Jap. J. Physiol.*, **25**, 733 (1975).

137 J. LeBlanc, A. Villemaire, and A. Vallières, *Am. J. Physiol.*, **218**, 1742 (1970).

138 P. Hemon, D. Ricquier, and G. Mory, in L. Jansky and X. J. Musacchia, Eds., *Regulation of Depressed Metabolism and Thermogenesis*, Charles C. Thomas, Springfield, Ill., 1976, p. 174.

139 Y. Hagino and M. Nakashima, *Jap. J. Pharmacol.*, **23**, 543 (1973).

140 Y. Hagino and M. Nakashima, *Jap. J. Pharmacol.*, **24**, 373 (1974).

141 B. E. Kemp and M. G. Clark, *J. Biol. Chem.*, **253**, 5147 (1978).

142 J. C. Garrison and M. K. Borland, *J. Biol. Chem.*, **254**, 1129 (1979).

143 T. M. Chan and J. H. Exton, *J. Biol. Chem.*, **253**, 6393 (1978).

144 Y. Hagino and T. Shigei, *Jap. J. Pharmacol.*, **26**, 535 (1976).

145 C. C. Malbon, S.-Y, Li, and J. N. Fain, *J. Biol. Chem.*, **253**, 8820 (1978).

146 H. G. Preiksaitis and G. Kunos, *Life Sci.*, **24**, 35 (1979).

147 H. G. Preiksaitis and G. Kunos, *Pharmacologist*, **20**, 245A (1978).

148 H. G. Preiksaitis, C. Peterfy, and G. Kunos, *Fed. Proc.*, **39**, 399A (1980).

149 D. G. Haylett, *Br. J. Pharmacol.*, **57**, 158 (1976).

150 J.-L. J. Chen, D. F. Babcock, and H. A. Lardy, *Proc. Natl. Acad. Sci. U.S.A.*, **75**, 2234 (1978).

151 P. F. Blackmore, F. T. Brumley, J. L. Marks, and J. H. Exton, *J. Biol. Chem.*, **253**, 4851 (1978).

152 R. P. Zimon, E. V. Flock, G. M. Tyce, S. G. Sheps, and C. A. Owen, Jr., *Endocrinology*, **80**, 808 (1967).

152a T. L. Sourkes, K. Missala, G. H. Bastomsky, and T. Y. Fang, *Canad. J. Physiol. Pharmacol.*, **55**, 789 (1977).

153 A. D. Cherrington, F. D. Assimacopoulos, S. C. Harper, J. D. Corbin, C. R. Park, and J. H. Exton, *J. Biol. Chem.*, **251**, 5209 (1976).

154 P. F. Blackmore, J.-P. Dehaye, and J. H. Exton, *J. Biol. Chem.*, **254,** 6945 (1979).

155 J. K. Jones, F. Ismail-Beigi, and I. S. Edelman, *J. Clin. Invest.*, **51,** 2498 (1972).

156 K. A. Gumaa, J. S. Hothersall, A. L. Greenbaum, and P. McLean, *FEBS Lett.*, **80,** 45 (1977).

156a C. C. Malbon, *J. Biol. Chem.*, **255,** 8692 (1980).

157 T. Takahashi and M. Suzuki, *Endocrinol. Jap.*, **22,** 187 (1975).

158 U. Rosenquist, *Mol. Cell. Endocrinol.*, **12,** 111 (1978).

159 T. M. Chan, P. F. Blackmore, K. E. Steiner, and J. H. Exton, *J. Biol. Chem.*, **254,** 2428 (1979).

160 B. B. Wolfe, T. K. Harden, and P. B. Molinoff, *Proc. Natl. Acad. Sci. U.S.A.*, **73,** 1343 (1976).

161 G. Guellaen, M. Yates-Aggerbeck, G. Vauquelin, D. Strosberg, and J. Hanoune, *J. Biol. Chem.*, **253,** 1114 (1978).

161a B. B. Hoffman, T. Michel, D. Mullikin Kilpatrick, R. J. Lefkowitz, M. E. M. Tolbert, H. Gilman, and J. N. Fain, *Proc. Natl. Acad. Sci. U.S.A.*, **77,** 4569 (1980).

162 W. H. Kan, C. Farsang, H. G. Preiksaitis, and G. Kunos, *Biochem. Biophys. Res. Commun.*, **91,** 303 (1979).

163 J. B. Young and L. Landsberg, *Clin. Endocrinol. Metab.*, **6,** 657 (1977).

164 M. L. Maayan and S. H. Ingbar, *Science,* **162,** 124 (1968).

165 A. Melander, E. Nilson, and F. Sundler, *Endocrinology,* **90,** 194 (1972).

166 N. J. Marshall, S. Von Barke, and P. G. Malan, *Endocrinology,* **96,** 1520 (1975).

167 S. W. Spaulding and G. N. Burrow, *Nature (Lond.),* **254,** 347 (1975).

168 L. E. Ericson, A. Melander, C. Owman, and F. Sundler, *Endocrinology,* **87,** 915 (1970).

169 A. Melander, E. Ranklev, F. Sundler, and U. Westgren, *Endocrinology,* **97,** 332 (1975).

170 D. Dumas and M. Guibout, *FEBS Lett.,* **88,** 287 (1978).

171 F. Okajima and M. Ui, *Am. J. Physiol.,* **234,** E106 (1978).

172 T. E. Nelson and J. E. Stouffer, *Biochem. Biophys. Res. Commun.,* **48,** 480 (1972).

173 S. E. Pointon and S. Banerjee, *Biochim. Biophys. Acta.,* **583,** 129 (1979).

174 A. E. Dominguez, O. L. Catanzaro, B. E. Fernandez, and N. A. Vidal, *Experientia,* **29,** 1291 (1973).

175 Z. Selinger, S. Eimerl, and M. Schramm, *Proc. Natl. Acad. Sci. U.S.A.,* **71,** 128 (1974).

176 V. K. Sharma and S. P. Banerjee, *Biochim. Biophys. Acta,* **539,** 538 (1978).

177 P. Coulombe, J. H. Dussault, and P. Walker, *J. Clin. Endocrinol. Metab.,* **44,** 1185 (1977).

178 J. P. Bilezikian, J. N. Leob, and D. E. Gammon, *J. Clin. Invest.,* **63,** 184 (1979).

179 R. S. Williams, C. E. Guthrow, and R. J. Lefkowitz, *J. Clin. Endocrinol. Metab.,* **48,** 503 (1979).

180 W. J. Popovic, J. E. Brown, and J. W. Adamson, *J. Clin. Invest.,* **64,** 56 (1979).

181 W. Emlen, D. S. Segal, and A. J. Mandell, *Science,* **175,** 79 (1972).

182 U. Strömbom, T. H. Svensson, D. M. Jackson, and G. Engstrom, *J. Neural Transm.,* **41,** 88 (1977).

183 J. A. Gonzales-Vegas and D. Fuenmayor, *Experientia,* **34,** 1527 (1978).

184 O. L. Catanzaro and A. Marzi, *Experientia,* **30,** 1334 (1974).

185 P. Arner and J. Östman, *Acta Med. Scand.,* **200,** 273 (1976).

186 L. Landsberg and J. B. Young, *N. Engl. J. Med.,* **298,** 1295 (1978).

187 W. Otto, T. G. Taylor, and D. A. York, *J. Endocrinol.,* **71,** 143 (1976).

188 N. Bégin-Heick and H. M. C. Heick, *Can. J. Physiol. Pharmacol.,* **55,** 1320 (1977).

189 S. W. Thenen and R. H. Carr, *Proc. Soc. Exp. Biol. Med.,* **159,** 116 (1978).

190 J. Leblanc, J. Vallières, and C. Vachon, *Am. J. Physiol.,* **222,** 1043 (1977).

191 L. D. Carlson, *Pharmacol. Rev.,* **18,** 291 (1966).

192 J. Himms-Hagen and I. M. Mazurkiewicz-Kwilecki, *Can. J. Physiol. Pharmacol.*, **48,** 657 (1970).

193 M. J. Fregly, F. P. Field, M. J. Katovich, and C. C. Barney, *Fed. Proc.*, **38,** 2162 (1979).

194 M. N. E. Harri, L. Melender, and R. Tirri, *Experientia,* **30,** 1041 (1974).

195 J. H. Fleisch and C. S. Hooker, *Circ. Res.*, **38,** 243 (1976).

196 R. J. Parker, B. A. Berkowitz, C.-H. Lee, and W. D. Denckla, *Mech. Ageing Dev.*, **8,** (1978).

197 V. K. Sharma, P. R. Sundaresan, and S. P. Banerjee, *Pharmacologist,* **21,** 238A (1979).

198 K. Hornbrook, *Pharmacologist,* **20,** 166A (1978).

198a E. J. Corwin, and R. L. Malvin, *Physiologist,* **23,** 180A (1980).

199 G. Kunos and M. Nickerson, *J. Physiol. (Lond.),* **256,** 23 (1976).

200 M. Nickerson and G. M. Nomaguchi, *Am. J. Physiol.*, **163,** 484 (1950).

201 A. J. Hulbert and J. W. Hudson, *Am. J. Physiol.*, **230,** 1211 (1976).

202 E. Lerner, A. L. Shug, C. Elson, and E. Shrago, *J. Biol. Chem.*, **247,** 1513 (1972).

203 J. F. Patton and W. S. Platner, *Am. J. Physiol.*, **218,** 1417 (1970).

204 C. G. D. Mørley, S. Koku, A. G. Rubenstein, and J. L. Boyer, *Biochem. Biophys. Res. Commun.*, **67,** 653 (1975).

205 N. Grey, J. E. McGuigan, and D. M. Kipnis, *Endocrinology,* **86,** 1383 (1970).

206 W. H. Dillmann and J. H. Oppenheimer, *Endocrinology,* **105,** 74 (1979).

207 H. Tamai, H. Suemastu, N. Kurokawa, M. Esaki, T. Ikemi, F. Matsuzuka, K. Kuma, and S. Nagataki, *J. Clin. Endocrinol. Metab.*, **48,** 54 (1979).

208 V. K. Sharma and S. P. Banerjee, *J. Biol. Chem.*, **252,** 7444 (1977).

209 S. N. Madsen and O. Sonne, *Nature (Lond.),* **262,** 793 (1976).

210 J. A. Cidlowski and T. G. Muldoon, *Endocrinology,* **97,** 59 (1975).

211 A.-M. Leseney, J.-J. Befort, N. Befort, M. Benmiloud, and N. Defer, *FEBS Lett.*, **99,** 239 (1979).

212 T. M. Harkcom, J. K. Kim, P. J. Palumbo, Y. S. F. Hui, and T. P. Dousa, *Endocrinology,* **102,** 1475 (1978).

213 A. Bhattacharya and B. K. Vonderhaar, *Biochem. Biophys. Res. Commun.*, **88,** 1405 (1979).

214 A. J. Hulbert, M. L. Augee, and J. K. Raison, *Biochim. Biophys. Acta,* **455,** 597 (1976).

215 D. Mendoza, H. Moreno, E. M. Massa, R. D. Morero, and R. N. Farias, *FEBS Lett.*, **84,** 199 (1977).

216 W. J. Strittmatter, F. Hirata, and J. Axelrod, *Science,* **204,** 1205 (1979).

217 G. Kunos and H. G. Preiksaitis, in E. Szabadi, C. M. Bradshaw, and P. Bevan, Eds., *Recent Advances in the Pharmacology of Adrenoreceptors,* Elsevier/North-Holland, Amsterdam, 1978, p. 219.

218 M. Nickerson and G. Kunos, *Fed. Proc.*, **36,** 2580 (1977).

219 A. Maggi, D. C. U'Prichard, and S. J. Enna, *Science,* **207,** 645 (1980).

220 D.-M Chuang and E. Costa, *Proc. Natl. Acad. Sci. U.S.A.*, **76,** 3023 (1979).

221 A. Szentivanyi, *J. Allergy Clin. Immunol.*, **65,** 5 (1980).

222 C. L. Wood, M. G. Caron, and R. J. Lefkowitz, *Biochem. Biophys. Res. Commun.*, **88,** 1 (1979).

223 G. Guellaen and J. Hanoune, *Biochim. Biophys. Acta,* **587,** 618 (1979).

224 K. R. Hornbrook, *Fed. Proc.*, **38,** 540A (1979).

225 J. B. Blair, M. E. James, and J. L. Foster, *J. Biol. Chem.*, **254,** 7579 (1979).

226 J. B. Cheng, L. E. Cornett, A. Goldfien, and J. M. Roberts, *Br. J. Pharmacol.*, **70,** 515 (1980).

INDEX